Biological and Medical Physics, Biomedical Engineering

More information about this series at http://www.springer.com/series/3740

Biological and Medical Physics, Biomedical Engineering

The fields of biological and medical physics and biomedical engineering are broad, multidisciplinary and dynamic. They lie at the crossroads of frontier research in physics, biology, chemistry, and medicine. The Biological and Medical Physics, Biomedical Engineering Series is intended to be comprehensive, covering a broad range of topics important to the study of the physical, chemical and biological sciences. Its goal is to provide scientists and engineers with textbooks, monographs, and reference works to address the growing need for information.

Books in the series emphasize established and emergent areas of science including molecular, membrane, and mathematical biophysics; photosynthetic energy harvesting and conversion; information processing; physical principles of genetics; sensory communications; automata networks, neural networks, and cellular automata. Equally important will be coverage of applied aspects of biological and medical physics and biomedical engineering such as molecular electronic components and devices, biosensors, medicine, imaging, physical principles of renewable energy production, advanced prostheses, and environmental control and engineering.

L. Ridgway Scott · Ariel Fernández

A Mathematical Approach to Protein Biophysics

 Springer

L. Ridgway Scott
Departments of Computer Science and
    Mathematics, Institute for Biophysical
    Dynamics, Computation Institute
University of Chicago
Chicago, IL
USA

Ariel Fernández
Collegium Basilea, Institute for Advanced Study
Basel
Switzerland

and

IAM – Argentine Institute of Mathematics,
    INQUISUR – Chemistry Institute
CONICET – National Research Council
Buenos Aires
Argentina

ISSN 1618-7210   ISSN 2197-5647   (electronic)
Biological and Medical Physics, Biomedical Engineering
ISBN 978-3-319-88158-4   ISBN 978-3-319-66032-5   (eBook)
DOI 10.1007/978-3-319-66032-5

Mathematics Subject Classification (2010): 92C05

This Springer imprint is published by Springer Nature
The registered company is Springer International Publishing AG
The registered company address is: Gewerbestrasse 11, 6330 Cham, Switzerland

Contents

Chapter 1
Understanding Proteins as Digital Widgets

Conventional assertions of biology have been qualitative, not quantitative. There is not yet a periodic table to codify protein behavior. Here, we explore quantitative aspects of protein biophysics in what at first seems to be fuzzy territory. Most drugs are small molecules that interact with proteins to interfere with the deleterious behavior of mutant proteins. We give examples of descriptions of protein–drug interactions in the literature that appear murky and confused but become clarified by our more complete analysis.

Although some fundamental operations of nature appear continuous (analog) at first, we will see that there is a discrete (digital) interpretation that is essential in explaining function. When examined at the molecular scale, certain biological entities have a natural continuum description with seemingly infinite possible behaviors, and yet they assemble to function in a discrete and repeatable (deterministic) manner.

We will see that protein biophysics can be understood in terms that are used to describe digital computers. We will use this terminology to differentiate between analog systems (such as the connection between a computer and its monitor) and digital systems, which talk in zeros and ones. Fundamental in this dichotomy is the notion of a switch that can change behavior discretely from one state to another. Not surprisingly, biology utilizes many switches at a molecular level.

Digital circuits on computer chips are also based on continuous mechanisms, namely electrical currents and voltages in wires and electronic components. The analogy with our topic is hopefully apparent. A book by Mead and Conway [342] written at the end of the 1970s transformed computer architecture by emphasizing design rules that simplified the task of converting a fundamentally analog behavior into one that was digital and predictable. We seek to do something analogous here, but we are not in a position to define rules for nature to follow. Rather, we seek to understand how some of the predictable, discrete behaviors of proteins can be explained as if certain digital rules were being followed.

1.1 Target Audience

Mathematicians are increasingly involved in biochemistry and biophysics research, and the involvement is increasingly related to the development of models. This book is intended to provide a basis for further engagement of this type, at least in the area of protein biophysics. Rational drug design is a goal that can be more easily attained by such an improvement in quantitative reasoning. Indeed, that goal alone is sufficient motivation for developing a more

© Springer International Publishing AG 2017
L.R. Scott and A. Fernández, *A Mathematical Approach to Protein Biophysics*,
Biological and Medical Physics, Biomedical Engineering,
DOI 10.1007/978-3-319-66032-5_1

mathematical approach, as we do here, since drug design is becoming both more expensive and less effective. But to gain a seat at the drug-design table, mathematicians must become immersed in the biophysics to a degree that is not yet common. This book is intended to aid in that process.

This book attempts to delineate certain rules of molecular behavior that make atomic scale objects behave in a digital and deterministic way. But it departs from the usual presentation of protein biophysics by taking a mathematical approach to the greatest extent possible. Trends for biology to become quantitative have been predicted and observed widely, but to support this there is a need for an entry point to the field that is accessible to people with a mathematically rigorous orientation. Anecdotal descriptions of biology leave mathematically oriented scientists with unanswered technical questions. We hope to establish a more rigorous foundation for protein biophysics both to entice mathematicians and computer scientists to the field and to promote a higher standard for all scientists in the field.

This book has been used successfully in both undergraduate and graduate courses. In undergraduate courses, the text is followed chapter by chapter with selected exercises done weekly. This is augmented by having students read papers and perform a capstone research project. The approach for graduate students is similar, but it is further augmented by a requirement to present research papers that expand upon the topics in this book, and the capstone research projects are often more advanced.

The ideal group of students would be a balance of mathematicians, computer scientists, and biologists, working together in teams. But this book has also been used as a big data analytics course for undergraduate computer science majors in which data mining is heavily emphasized and no background in biology is assumed. One could alternatively emphasize more complex mathematical issues for a different audience and use this book as the basis for a course focused on mathematical modeling.

Little advanced mathematics, statistics, or computer science is used in this book. Nevertheless, many exercises challenge typical computer science and biology students because of their mathematical depth, and the code development requirements similarly can be a challenge for typical mathematics and biology students. Thus, this book supports an interdisciplinary approach to a specific topic as opposed to one in which the main focus is tool building.

This book is *not* intended to be a complete introduction to protein biophysics. Someone wanting to do research in the area would want to become familiar with standard texts such as [59, 102, 393]. Those books, written from a more conventional point of view, contain much important information that is complementary to the contents here.

1.2 Proteins as Digital Components

Biology can be viewed as an information system. As a simple example, we are biological entities communicating via this book. More to the point, many types of signaling in biological systems involve interactions between proteins and ligands.[1] A type of physical baton-passing is used to communicate information, e.g., in neurotransmission and cellular signaling. But there are too many examples of information processing in biology to stop here to enumerate them. What is of interest here is to understand how certain biological systems (involving proteins) function as digital information systems despite the fact that the underlying processes are analog in nature.

---

[1] A ligand is anything that binds to something. We provide a glossary of terms like this in Chapter 20 rather than defining them in the text.

There is tremendous biotic diversity, and corresponding genomic diversity, but the number of distinct protein structures appears to be quite small according to existing measures, such as the SCOP classification (Section 5.2.2). Moreover, the observable diversity of the macrobiological world is evident. Thus, a central question in quantitative biology is how such diversity gets transferred through proteins, which would appear to introduce a bottleneck according to the current understanding.

It should not be strange that continuous behaviors can result in quantum components. Quantum mechanics explains continuous wave behavior in terms of discrete (atomistic) behavior. Proteins do this at a much larger length scale, but the principle is similar. Proteins are often involved in signaling and function in a discrete (or digital, or quantized) way. In addition, proteins are discrete building blocks of larger systems, such as viruses and cells. How they bind together (e.g., in a virus capsid) is also deterministic (repeatable) and precise. But the underlying chemical/physical mechanisms used are fundamentally continuous (analog).

The benefits of finding simple rules to explain complicated chemical properties are profound. The octet rule (Section 2.1) for electron shell completion allowed rapid prediction of molecular formulæ and inference of chemical structure by simple counting [383]. Resonance theory (Section 14.1) describes general bonding patterns as a combination of simple bonds (e.g, single and double bonds) [382]. The discrete behavior of DNA elucidated by Crick, Franklin, Watson, Wilkins, and others [190, 512, 517] initiated the molecular biology revolution. Our objective here is to provide an introduction to some basic properties of protein–ligand interactions with the hope of stimulating further study of the discrete nature of molecular interactions in biology.

Much of biology is about describing significant differences between things of interest that may look similar at first glance, as well as identifying similarities among things which look superficially different. Here, the main focus will be on differences between proteins. For example, we would like to identify the difference between proteins that support animal life and those that are toxic. A guide to understanding such extreme sensitivity to detail can be inferred from basic chemistry. The small difference between methanol and ethanol is a well-known example. Similarly, propylene glycol is a methylated form of ethylene glycol. The latter is toxic to animals, but the former is not, and both are effective as anti-freeze. We will see that such changes via a single methylation alone are able to have profound effects on the behavior of anticancer drugs [141, 170, 176]. We will explain simple rules that provided guidance in making such modifications. The key issue is that methylation changes the dielectric effect of surrounding water locally.

1.3 Physical Forces

The only force of interest in biochemistry is the electrostatic force, and it varies continuously throughout proteins. So this begs the question: How does the molecular machinery of biology discretize this continuum variation? This question is the guiding motivation for this book.

Electrostatic gradients in proteins are among the largest known in nature. Moreover, we are primarily interested in proteins operating in an aqueous, and thus dielectric, environment. We will devote significant space to the dielectric effect in subsequent chapters, but the main point to know for now is that a dielectric medium shields (diminishes) the effect of electric charges. The dielectric properties of water are among the strongest in nature, and indeed, water can be viewed as hostile to proteins. This leads to an interesting contention that we address in more detail in Section 2.2.2, but for now we depict these as opposing arrows in Figure 1.1.

One way to envisage the dielectric effect and protein charges is to imagine a harbor in fog. The red and green buoy lights correspond to the positive and negative charges of a protein, and the fog is the dielectric medium that tends to shield (obscure) the charges. Fog can be dispersed locally by some atmospheric change, and the lights will be suddenly more visible. In proteins, hydrophobic groups tend to reduce the fog of the dielectric. Their intervention helps to make decisions quicker and more decisive. The delicate balance of interactions leads

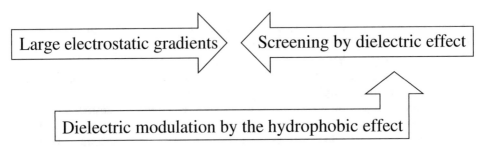

Fig. 1.1 Three competing effects that determine protein behavior. These conspire to weaken interactive forces, making biological relationships more tenuous and amenable to mutation.

to effective switches which are both sensitive and energetically efficient. Just how this balance is implemented via molecular machinery is a major focus of this book.

Not only is the dielectric coefficient of water remarkably large, but it is also capable of being strongly modulated in ways that are still being unveiled. In particular, hydrophobic effects modulate the dielectric properties of water [118]. Proteins are ideal molecular entities both to generate electrostatic fields and to modulate them via dielectrics because of their assembly of hydrophobic, hydrophilic, and amphiphilic sidechains. Moreover, the charge variation on proteins is so large that it is hard to make an analogy on larger scales, and the variation in hydrophobicity is equally extreme. Hydrophobic mediation of the dielectric properties of water appears to have a significant impact on protein function. Thus, we are faced with a series of counterbalancing and extreme forces, depicted in Figure 1.1, that must be comprehended in order to explain how digital behavior is implemented at the biomolecular level.

One take-home message is that the modulation of the dielectric properties of water by the hydrophobic parts of proteins is an essential aspect of molecular digitalization that needs to be considered carefully. Typical representations of proteins show only physical location, basic covalent bonds, and individual charges. However, the (non-covalent) hydrogen bond is recognized as the main determinant of structural biology. We are concerned therefore with its dielectric modulation brought about by hydrophobic parts of proteins, in consonance with the scheme in Figure 1.1.

1.4 Roadmap

We have intentionally alternated chapters between more mathematical aspects and more biological content. This way different students can excel periodically and not get completely lost during one phase of the course. In addition, we have chosen specific topics where details are sufficiently well worked out. This means that some subjects are developed in more or less detail than others, and nearby subjects may be omitted completely.

Chapter 2 introduces required notions from physics and chemistry with an emphasis on explaining what is meant by digital rules. That chapter then previews key concepts and results of this book. The next chapter presents the mathematical foundations of a class of chemical bonds. Much later, this topic will re-appear in more detail. Not surprisingly, two chapters are then needed to explain the basics of proteins, and an entire chapter is devoted to what is currently known about hydrogen bonds, the main determinant of protein structure. With these preliminaries in hand, this book addresses a key question: what determines how proteins interact with each other and with other molecules, such as drugs. Three chapters develop the concept of wrapping hydrogen bonds, although the third chapter can be skipped on a first reading. Many of the subsequent chapters are devoted to applications of these ideas to particular interactions. In addition, two different interactions with a quantum mechanical aspect are discussed. An additional chapter on electrostatic models completes this book, although it is followed by an important chapter on physical units.

1.5 Orientation

Although this book will present many positive results, it will also highlight many deficiencies in the field. We explain the lack of a model of hydrogen bonds, even though their importance in biology is unquestioned. Similarly, a nanoscale model of water is lacking, although the role of water in biology is pervasive and complex. Further developments at the level of basic physics and chemistry in areas like these will have a substantial benefit for the understanding of biological function.

This book is not a typical introduction to a well-developed field in which all the main results are already standard. Rather it is an invitation to join in the exploration of the detailed mechanisms of protein interactions. We expect this to require many hands. Our intention here is to help stimulate in particular the study of some more mathematical questions, many of which we leave open. To quote Mead and Conway [342], "And thus the period of exploration begins."

1.6 Exercises

Exercise 1.1. Pour cooking oil into a glass of water and stir it vigorously until the oil is well dispersed. Now wait and watch as the oil droplets coalesce. Do the individual droplets retain any sort of discrete form? Or does the hydrophobic force just create a blob in the end?

Exercise 1.2. Acquire a pair of polarized sunglasses and observe objects just below the surface of a body of water both with and without the sunglasses. Do these observations while facing the sun, when it is at a low angle with respect to the water surface. You should observe that the "glare" is greatly reduced by the polarizing lenses. Also make the same observations when the sun is overhead, and when looking in a direction away from the sun when it is at a low angle.

Exercise 1.3. Quantum mechanical computations suffer from the "curse of dimensionality" because each additional electron adds another three dimensions to the problem. Thus, a problem with k electrons requires the solution of a partial differential equation in $\mathbb{R}^{3k}$. If we require a Cartesian discretization with m grid intervals per dimension, then the resulting problem requires m^{3k} words of memory to store the discrete representation. Compare this with the number of atoms in the observable universe. Assuming we could somehow make a computer using all of these atoms with each atom providing storage for one of the m^{3k} words of memory required for the discrete representation, determine how large a value of k could be used. Try values of $m = 3$ and $m = 10$.

Exercise 1.4. Pour salt into a glass of water and watch what happens to the salt. Take a small amount out and put it under a microscope to see whether the picture stays the same.

Exercise 1.5. Pour salt into a glass of water and stir it until it dissolves. Now also add some oil to the water and stir it until small droplets form. Look at the surface of the oil droplets and see whether you can see salt crystals that have reformed due to the change in electrostatic environment there. This might best be done on a slide beneath a microscope objective. Do you see the reformation of salt crystals due to the removal of the dielectric material (water) by the hydrophobic material (oil)?

Chapter 2
Digital Rules for Proteins

Lise Meitner (1878–1968) was "the physicist first to recognize that experiments reported by two former colleagues in Berlin meant that atoms had been split, never got [the Nobel] prize, even though one of those colleagues, Otto Hahn, did in 1944." (October 20, 2003, New York Times.) Element 109, meitnerium, is named in her honor.

Digital rules are a hallmark of a mature science. Protein associations are the elementary events in biology when examined at the molecular level. This book attempts to develop digital rules for interactions involving proteins. We begin with a sketch of some of the main ideas that the book will cover. This is not an outline but rather is a narrative that introduces the main goals and challenges to be addressed and gives a glimpse of some of the major advances.

We describe some challenging features of modeling the interactions of proteins in biological systems as well as opportunities to be addressed in the future. This is meant to provide some orientation, but it is also meant to be a disclaimer. That is, we disclose what we see as limitations of standard approaches which have forced us to adopt new strategies. There may well be other approaches that will be even more successful in the future.

2.1 Digital Nature of Molecules

We begin by illustrating what we mean by digital, or discrete, behavior in analog, or continuous, systems. This gives us an opportunity to review some basic concepts from chemistry. The main entities of chemistry are molecules, and the building blocks of molecules are atoms. We begin by looking at digital rules for atoms and then move to molecules.

2.1.1 Digital Nature of Atoms

Atoms can be characterized by the number of electrons, protons, and neutrons of which they are composed. Some atoms of primary interest in protein biochemistry are listed in Table 2.1, and this table expresses digital rules at different scales: subatomic, atomic, and molecular.

© Springer International Publishing AG 2017
L.R. Scott and A. Fernández, *A Mathematical Approach to Protein Biophysics*,
Biological and Medical Physics, Biomedical Engineering,
DOI 10.1007/978-3-319-66032-5_2

There are other subatomic digital rules that we will generally take for granted here, like Pauli's electron spin exclusion principle and Hund's rule about electron orbital occupation.

Several rules are encoded in Table 2.1. The first rule is used to reduce the number of columns: the number of protons always equals the number of electrons (the net charge is zero). A second rule is that the typical number of neutrons in the dominant isotope is nearly the same as the number of protons. But the most important rule is the **octet rule**: the number of the electrons in the outer shell plus the number (listed in the "lacking" column) of electrons contributed by atoms covalently bonded to it is always eight (except for hydrogen), up to chlorine, and then, it is eighteen for the larger atoms (for even larger atoms, the magic number is thirty-two, but those atoms do not concern us). This simple rule facilitates the determination of molecular bond formation.

Atom	Symbol	+/−	neutrons	outer	lacking	ion	mass	radius
Hydrogen	H	1	0 (1)	1	1	+1	1.008	1.20
Carbon	C	6	6 (7)	4	4		12.01	1.70
Nitrogen	N	7	7 (8)	5	3		14.007	1.55
Oxygen	O	8	8 (9, 10)	6	2	−2	15.9994	1.52
Fluorine	F	9	10	7	1	−1	18.998	1.47
Sodium	Na	11	12	1	7	+1	22.9898	2.27
Magnesium	Mg	12	12 (13, 14)	2	6	+2	24.305	1.73
Phosphorus	P	15	16	5	3		30.974	1.80
Sulfur	S	16	16 (17, 18, 20)	6	2	−2	32.065	1.80
Chlorine	Cl	17	18 (20)	7	1	−1	35.4527	1.75
Potassium	K	19	20 (22)	1	17	+1	39.098	2.75
Calcium	Ca	20	20 (22–24)	2	16	+2	40.08	2.00
Iron	Fe	26	30 (28, 31, 32)	8	10	+3	55.845	1.10*
Copper	Cu	29	34 (36)	11	7	+1;2	63.55	1.40*
Zinc	Zn	30	34 (36–38, 40)	12	6	+2	64.4	1.39*
Selenium	Se	34	46 (40, 42–44)	16	2		78.96	1.90
Iodine	I	53	74	17	1	−1	126.90	1.98

Table 2.1 Subset of the periodic table. The column " + /−" denotes the number of protons and electrons in the atom. The column "outer" is the number of electrons in the outer shell. The column "lacking" is the number of electrons needed to complete the outer shell. The column "ion" gives the charge of the atomic ion (copper has two possible forms). The column "mass" gives the atomic mass in Daltons (see Chapter 18 for details), reflecting the naturally occurring isotopic distribution. The column "radius" lists the "mean" van der Waals radius [56], with the exceptions marked by *'s taken from various Web sites.

Another rule encoded in Table 2.1 is that the atomic mass is very close to the number of protons plus the number of neutrons (in the specified unit, the Dalton). The mass of the proton and neutron are approximately the same, and the rest mass of an electron is less than 0.0006 times this amount; see Table 2.2. The units of atomic mass are discussed in Section 18.2.1, but for now, the main point is that the mass of the proton and neutron are about one in the standard unit for atomic measurements, the Dalton. For reference, we list in Table 2.2 these masses in more familiar units.

As we see in Table 2.1, the number of neutrons can vary. We have listed what is known as the dominant isotope, followed (in parentheses) by the other possible stable isotopes (involving the listed numbers of neutrons). Neutrons add mass but not charge. Various isotopes are important in certain contexts; a hydrogen atom with an extra neutron is called deuterium. Atoms occur naturally in different isotopic forms, and the atomic mass reflects this natural variation. Otherwise, the atomic mass would be essentially the sum of the numbers of protons and neutrons, with a small correction for the electronic mass, as well as another correction that we will discuss shortly. For chlorine, about a quarter of the atoms have 20 neutrons, and thus, the atomic mass is about halfway between integer values. The given atomic masses are themselves only averages, and any particular set of atoms will vary in composition slightly; see the Periodic Table in [383] for more details.

We might expect that the atomic masses of a pure isotope would be given by

$$m \approx \mu(p, n) = p(m_p + m_e) + n m_n, \tag{2.1}$$

where m_p, m_e, and m_n are the mass of the proton, electron, and neutron, respectively, p is the number of protons (and electrons), and n is the number of neutrons. We list in Table 2.2 the masses of the proton, neutron, and electron in familiar units (10^{-27} grams), as well as the standard unit of atomic mass, the Dalton, in these units. However, we see that for Carbon-12 (for which the number of neutrons is six), the formula (2.1) would predict that

$$m \approx \mu(6, 6) = 6(m_p + m_e + m_n) = 20.091 \times 10^{-27} \text{ grams} = 12.0989 \text{ Daltons.} \tag{2.2}$$

However, it turns out that the definition of the Dalton is exactly one-twelfth of the mass of Carbon-12. Thus, the mass of the component parts is greater than the mass of the atom. The difference in mass corresponds to a difference in energy ($E = mc^2$), and the atom represents a lower energy configuration than the separated constituents. For reference, we give in Table 2.3 the ratios between the measured atomic mass and the prediction in (2.1) for a few atoms for which there is only one stable isotope, in addition to the ratio for Carbon-12. See Exercise 2.1 regarding similar computations for atoms with more complex isotopic combinations. Note that the mass of the neutron is 1.0087 Dalton, and the mass of the proton is 1.0073 Dalton.

The digital description of an atom is to be contrasted with the analog description of the Schrödinger equation [385]. This equation describes the electron (and proton) distribution, which is the key determinant of atomic interaction. It is a continuum equation predicting the electronic distribution at all points in space, and there is a separate three-dimensional space at the least for each electron and in the general case for the protons as well. Even if it were simple to solve this equation (which it is not), it would be difficult to determine simple facts from such a representation. We are forced to consider effects on this level in many cases, but operating at the atomic level has clear advantages.

proton	neutron	electron	Dalton
1672.622	1674.927	0.910938	1660.539

Table 2.2 Masses of atomic constituents, as well as the Dalton, listed in units of 10^{-27} grams. The Dalton is the standard unit of mass for atomic descriptions.

Hydrogen	Carbon-12	Fluorine	Sodium	Phosphorus	Iron	Iodine
1.000036	1.0083	1.0084	1.0087	1.0091	1.0095	1.0091

Table 2.3 Ratios of the atomic masses of different atoms to the mass predicted by formula (2.2). The isotopic fraction for hydrogen was taken to be 0.015% deuterium, and the isotopic fractions for iron were taken to be (28) 5.8%, (30) 91.72%, (31) 2.2%, and (32) 0.28%, where the numbers in parentheses indicate the number of neutrons.

There are other simple rules in chemistry that allow prediction of bond formation, such as the electronegativity scale (Section 8.2.1) and the resonance principle (Section 14.1). The electronegativity scale allows the determination of polarity of molecules (Section 8.2). The resonance principle states that observed states of molecular bonds are often a simple convex combination of two elementary states. For example, a benzene ring can be thought of as being made of alternating single and double bonds, whereas in reality, each bond is closely approximated by a convex combination of these two bonds. The resonance principle may be thought of as a Galerkin approximation to solutions of the Schrödinger equation.

The size data for atoms listed in the "radius" column in Table 2.1 provide another way to distinguish between different atoms in a simple way, although the lengths do not correspond to a tangible boundary. If we could look at atoms at this level, the nucleus would be a tiny dot, and the electrons would be a fuzzy cloud, extending beyond the stated radius with a certain (nonzero) probability (cf. Figure 1 in [56]). Rather, the length corresponds to the size of an "exclusion zone" to give an idea of the size of a region where other atoms would not (typically) be found. A similar notion of length will be discussed in Section 5.6 regarding the size of atomic groups that form proteins. The length variation may not seem extreme, but the corresponding volume variation is over an order of magnitude.

For simplicity, Table 2.4 gives the relevant volumes for boxes of various sizes, ranging from 1.2Å to 2.8Å on a side. Thus, in Table 2.1, we see that nitrogen and oxygen each have a volume twice that of hydrogen, and carbon has nearly three times the volume of hydrogen. Curiously, despite their similarity, potassium has a volume more than twice that of calcium. Moreover, the size relation in Table 2.1 seems to be going in the wrong direction. The size of atoms is a *decreasing* function of the number of electrons in the outer shell (for reference, the radius of Lithium is 1.82Å). When a shell becomes filled, the atom size jumps, but it then decreases as the new shell becomes populated. Of course, when the number of electrons in the outer shell is increasing, so is the number of positive charges in the nucleus, and this means the nucleus has an increasing ability to pull the electrons closer.

r	1.2	1.3	1.4	1.5	1.6	1.7	1.8	1.9	2.1	2.2	2.3	2.4	2.5	2.6	2.7	2.8
r^3	1.7	2.2	2.7	3.4	4.1	4.9	5.8	6.9	9.3	10.6	12.2	13.8	15.6	17.6	19.7	22.0

Table 2.4 Relation between volume r^3 and length r in the range of lengths (in Ångstroms) relevant for atoms.

2.1.2 Ionic Rules

One rule that bridges between atoms and molecules is related to ions. Certain ions are key to life in humans. There many important ionic effects that are specific to individual ions and are not easily explained by simple rules [291]. However, the charge of many ions can be deduced from a model that fits the data in Table 2.1 quite well for most, but not all, ions. The rule is that the atomic ion charge c is related to the number of electrons ℓ lacking in the outer shell and the total number of electrons O in a complete outer shell by

$$c = \begin{cases} O - \ell & \text{positive ions} \\ -\ell & \text{negative ions.} \end{cases} \tag{2.3}$$

Thus, the rule simply measures the distance to the nearest complete shell. However, these rules do not apply universally, and among the cases in Table 2.1, it works only when $|c| \leq 2$. Copper has two different ionic states; iron is an outlier. Nitrogen forms a complex ion N_3^-, and phosphorus forms an ionic complex with oxygen as well as with other atoms. Selenium forms different ionic complexes with oxygen.

2.1.3 Carbon/Hydrogen Rules

Another example of a simple rule at the molecular level relates to the common occurrence of hydrogens bonded to carbons. These occur so commonly in bio-organic chemistry that they are often only implied in graphical representations. In biomolecules, the hydrocarbon skeleton is typically represented without direct labeling of the atoms involved, with combinatorial rules that subsume the valence information implied in the electronic outer shell of hydrogen and carbon. For example, a benzene ring might be written as a simple hexagon, without any labeling of hydrogen and carbons. The chemical formula for benzene is C_6H_6. Implicit in the graphical representation is that a carbon is located at each vertex of the hexagon and that each carbon is bonded to a hydrogen. The rules apply to a wide variety of molecules, such as β-D-galactose (GAL) and N-acetyl-D-glucosamine (NAG) (Section 4.6.3); the three letter names for these molecules are their names in the Protein Data Bank. The three letter names for benzene in the Protein Data Bank are BNZ.

2.2 Digital Nature of Proteins

We have described simple rules for atoms and molecules, and their interactions, both because they will be used extensively in the following, but more importantly because they form a model of the type of rules, we will attempt to establish for proteins. We will provide an introduction to proteins starting in Chapter 4, but we now give a description of the type of rules for protein interactions that we will establish.

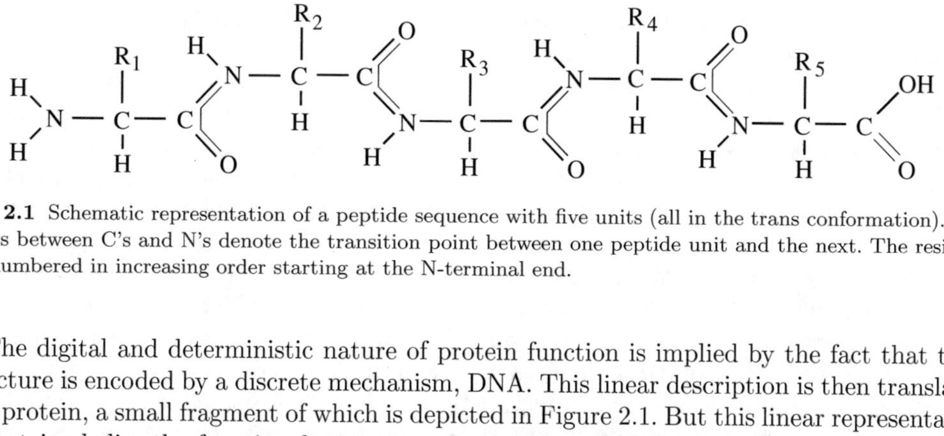

Fig. 2.1 Schematic representation of a peptide sequence with five units (all in the trans conformation). The bonds between C's and N's denote the transition point between one peptide unit and the next. The residues are numbered in increasing order starting at the N-terminal end.

The digital and deterministic nature of protein function is implied by the fact that their structure is encoded by a discrete mechanism, DNA. This linear description is then translated into protein, a small fragment of which is depicted in Figure 2.1. But this linear representation of proteins belies the functional geometry of a protein, which is three dimensional. It is the latter representation that appears at first to be anything but a discrete widget. However, we will see that it is so. There are posttranslational events (Section 4.3) which modify proteins and make their behavior more complex, but it is clear that nature works hard to make proteins in the same way every time.

What is striking about the fact that proteins act in digital ways is the significant role played by hydrophobic effects (Section 2.6) in most protein–ligand interactions. Such interactions account not only for the formation of protein complexes, but also for signaling and enzymatic processes. But the hydrophobic effect is essentially nonspecific. Thus, its role in a discrete system is intriguing.

We will see that it is possible to quantify the effect of hydrophobicity in discrete ways. The concept of *wrapping* (see Chapter 8) yields such a description, and we show that this can affect many important phenomena, including protein association (molecular recognition) (Chapter 7) and the flexibility of the peptide bond (Chapter 14). In these two examples, we will see behaviors that are essentially digital in nature that can be predicted based on quantitative measures. In particular, both of these effects yield what can be viewed as switches, things that can be turned on and off. There are other examples of switches in protein behavior, and even the fundamental hydrogen bond can be viewed in this way to a certain extent.

We will also study other features of protein systems that can be described by simple digital rules. We will consider different types of bonds that can be formed between proteins. These are all based on electronic interactions, and thus, we will study extensively different types of electronic interactions from a mathematical point of view, including van der Waals forces.

The effect of electronic interactions can be substantially modified by the dielectric effect. This is modulated by hydrophobicity, so we now discuss the main concepts related to hydrophobicity.

2.2.1 Hydrophobicity and Hydrophilicity

Any molecular material is termed hydrophobic if it does not interact favorably with water, typically because it cannot make appropriate hydrogen bonds with water. One major objective of the book is to clarify the effects of hydrophobicity in particular cases, including its role in protein–ligand interactions and other important phenomena. Hydrophobic moieties displace water and, as we will see, decrease the dielectric capacity of the environment. Hydrophobic molecular groups can reduce the dielectric effect locally, which in turn enhances certain nearby bonds. Particular atomic groups in protein sidechains will be identified as being hydrophobic. A major objective of the book is to clarify these concepts.

In one sense, hydrophilicity is the exact opposite of hydrophobicity. The latter means to repel water, and the former means to attract water. We will identify certain atom groups in protein sidechains as hydrophilic. However, one of the key points that we will emphasize is that hydrophilicity should not be thought of as a counterbalance to hydrophobicity. In particular, a hydrophilic group of atoms does not have a simple role of reversing the effect of hydrophobicity on the dielectric environment. Hydrophilic groups are always polar, which means that they represent an imbalance in charge. Thus, they contribute to modifying the electric environment. The dielectric effect tends to dampen the effect of electric charges, so hydrophobic groups can have the effect of removing the damper on the charges of hydrophilic groups. Thus, the effects of hydrophobicity and hydrophilicity are orthogonal in general and not opposite on some linear scale. In fact, hydrophobic groups can enhance the effect of a polar group, so they can correlate in just the opposite way from what the words seem to mean.

2.2.2 Solvation ≠ Salvation

The life of a protein in water is largely a struggle for the survival of its hydrogen bonds. The hydrogen bond (cf. Chapter 6) is the primary determinant of the structure of proteins. But water molecules are readily available to replace the structural hydrogen bonds with hydrogen bonds to themselves; indeed, this is a significant part of how proteins are broken down and recycled. We certainly cannot live without water [376], but proteins must struggle to live with it [146, 309].

Proteins are the fabric of life, playing diverse roles as building blocks, messengers, molecular machines, energy providers, antagonists, and more. Proteins are initiated as a sequence of amino acids, forming a linear structure. They coil into a three-dimensional structure largely by forming hydrogen bonds. Without these bonds, there would be no structure, and there would be no function. The linear (unfolded) structure of amino acid sequences is entropically more favorable than the bound state, but the hydrogen bonds make the three-dimensional structure energetically favorable.

Water, often called the matrix of life [191], is one of the best makers of hydrogen bonds in nature. Each water molecule can form hydrogen bonds with four other molecules and frequently does so. Surprisingly, the exact bonding structure of liquid water is still under discussion [3, 464, 490, 516], but it is clear that water molecules can form complex bond structures with other water molecules. For example, water ice can take the form of a perfect lattice with all possible hydrogen bonds satisfied.

But water is equally happy to bind to available sites on proteins instead of bonding with other water molecules. The ends of certain sidechains of amino acids look very much like water to a water molecule. But more importantly, the protein backbone hydrogen bonds can be replaced by hydrogen bonds with water, and this can disrupt the protein structure. This can easily lead to the break-up of a protein if water is allowed to attack enough of the protein's hydrogen bonds.

The primary strategy for protecting hydrogen bonds is to bury them in the core of a protein. But this goes only so far, and inevitably, there are hydrogen bonds formed near the surface of a protein. And our understanding of the role of proteins with extensive noncore regions is growing rapidly. The exposed hydrogen bonds are more potentially interactive with water. These are the ones that are most vulnerable to water attack.

Amino acids differ widely in the hydrophobic composition of their sidechains (Section 4.1.1). Simply counting carbonaceous groups (e.g., CH_n for $n = 0, 1, 2$ or 3) in the sidechains shows a striking range, from zero (glycine) to nine (tryptophan). Most of the carbonaceous groups are nonpolar and thus hydrophobic. Having the right amino acid sidechains surrounding, or **wrapping**, an exposed hydrogen bond can lead to the exclusion of water, and having the wrong ones can make the bond very vulnerable. The concept of wrapping an electrostatic bond by nonpolar groups is analogous to wrapping live electrical wires by nonconducting tape.

We refer to the under-protected hydrogen bonds, which are not sufficiently wrapped by carbonaceous groups, as **dehydrons** (Section 2.6.2) to simplify terminology. The name derives from the fact that these hydrogen bonds are stabilized and strengthened by being dehydrated.

2.2.3 Energetic Ambivalence

One could imagine a world in which all hydrogen bonds were fully protected. However, this would be a very rigid world. Biology appears to prefer to live at the edge of stability. Thus, it is not surprising that new modes of interactions would become more prevalent in biology than in other areas of physics. For example, it has been recently observed that exposed hydrogen bonds appear to be sites of protein–protein interactions [174]. Thus, what at first appears to be a weakness in proteins is in fact an opportunity.

One could define an **epidiorthotic force** as one that is associated with the repair of defects. The grain of sand in an oyster that leads to a pearl can be described as an epidiorthotic stimulant. Similarly, snowflakes and raindrops tend to form around small specs of dust. Such forces also have analogies in personal, social, and political interactions where forces based on detrimental circumstances cause a beneficial outcome. A couple who stay together because they do not want to be alone provides such an example. The defect of an under-protected hydrogen bond gives rise to just such an epidiorthotric force. The action of this force is indirect, so it takes some explaining.

An under-protected hydrogen bond would be much stronger if water were removed from its vicinity. The benefit can be understood first by saying that it is the result of removing a threat of attack (or the intermittent encounter of water forming hydrogen bonds with it). But there is an even more subtle (but mathematically quantifiable) effect due to the change in dielectric environment when water is removed, or even just structured, in the neighborhood. The dielectric constant of water is about eighty times that of the vacuum. Changing the dielectric environment near an under-protected hydrogen bond makes the bond substantially stronger.

If the removal of water from an under-protected hydrogen bond is energetically favorable, then this means there is a force associated with attracting something that would exclude water. Indeed, one can measure such a force, and it agrees with what would be predicted by calculating the change energy due to the change in dielectric (Section 9.1). You can think of this force as being somewhat like the way that adhesive tape works. Part of the force results from the removal of air between the tape and the surface, leaving atmospheric pressure holding it on. However, the analogy goes only so far in that there is an enhancement of electrical energy associated with the removal of water. For sticky tape, this would correspond to increasing the mass of the air molecules in the vicinity of the tape, by a factor of 80, without increasing their volume!

Thus, the epidiorthotric force associated with water removal from an under-protected hydrogen bond provides a mechanism to bind proteins together. This is a particular type of hydrophobic effect, because wrapping the bond with hydrophobic groups provides protection from water. It is intriguing that it arises from a defect which provides an opportunity to interact.

2.3 Pchemomics

The term "omics" refers to the use of biological databases to extract new knowledge by large-scale statistical surveys. The term "cheminformatics" is an accepted moniker for the interaction of informatics and chemistry, so there is some precedent for combining terms like pchem (a.k.a., physical chemistry) with a term like 'omics.' We do not suggest the adoption of the (unpronounceable) term pchemomics, but it serves to suggest the particular techniques being combined in a unique way. An example of pchemomics is the early study of the hydrogen bond [287]. Indeed, the original study of the structure of the peptide bond (see section 8-4 of [382]) used such an approach. But pchemomics involves a two-way interaction with data. In addition to providing a way to learn new properties in physical chemistry, it also involves using physical chemistry to look at standard data in new ways.

The Protein Data Bank (**PDB**) provides three-dimensional structures that yield continuing opportunities for proteomics discoveries. Using the perspective of physical chemistry in data mining in the PDB, some simple laws about protein families were determined by studying patterns of under-wrapped hydrogen bonds [162]. We examine just one such result in Section 2.3.4; many other results in physical chemistry can be likewise explored.

A simple view of the PDB gives a representation suitable only for Lagrangian mechanics (or perhaps just statics). If we keep in mind which atom groups are charged, we begin to see an electrostatic view of proteins, and standard protein viewers will highlight the differently charged groups. But the dielectric effect of the solvent is left to the imagination. And the crucial role of the modulation of the dielectric effect by hydrophobic groups is also missing. Adding such views of proteins involves a type of physical chemistry lens.

Given our understanding of what it means to be under-wrapped, it is not surprising that under-wrapped hydrogen bonds would appear more often in regions of proteins that are themselves not well structured. NORS (NO Recognizable Structure) regions [241] in proteins are large (at least seventy consecutive amino acids) sections which form neither α helices or β sheets. These appear more frequently among interactive proteins. Correspondingly, studies [179] have shown a strong correlation between the number of under-wrapped hydrogen bonds and interactivity.

A full understanding of wrapping and the related force associated with under-wrapping requires tools from physical chemistry. Interactions between physical chemistry and "omics" will offer further insights into biological systems. Indeed, precise modeling of water even by explicit solvent methods is still a challenge. Only recently have models begun to predict the temperature behavior of the density of liquid water [314]. This means that for very subtle issues, one must still be careful about even all-atom simulations. The mysteries of water continue to confront us. But its role in biology will always be central.

2.3.1 A New Tool?

Since we are seeking to answer new types of research questions, it may be comforting to know that there is a powerful tool that is being used. The combination of data mining and physical chemistry is not new, but its usefulness is far from exhausted. Moreover, it is not so common to see these utilized in conjunction with more conventional techniques of applied mathematics, as we do here. Thus, we take a moment to reflect on the foundations of the basic concepts that make up what we refer to as pchemomics.

Typical data mining in bioinformatics uses more discrete information, whereas the PDB uses continuous variables to encode chemical properties. The need for physical chemistry in biology has long been recognized. In the book [482], the following quote is featured:

> The exact and definite determination of life phenomena which are common to plants and animals is only one side of the physiological problem of today. The other side is the construction of a mental picture of the constitution of living matter from these general qualities. In the portion of our work, *we need the aid of physical chemistry.*

The emphasis at the end was added as an aid to the eye. These words were written by Jacques Loeb in "The biological problems of today: physiology" which appeared in the journal *Science* in volume 7, pages 154–156, in 1897. So our theme is not so new, but the domain of physical chemistry has advanced substantially in the last century, so there continues to be an important role for it to play in modern biology.

2.3.2 Data Mining Definition

It is useful to reflect on the nature of **data mining**, since this is a relatively new term. It is a term from the information age, so it is suitable to look for a definition on the Web. According to WHATIS.COM,

> Data mining is sorting through data to identify patterns and establish relationships.
> Data mining parameters include:

- Association — ┌looking┐ for patterns where one event is connected to another event

- Sequence or path analysis — ┌looking┐ for patterns where one event leads to another later event

- Classification — ┌looking┐ for new patterns (may result in a change in the way the data is organized but that's ok)

- Clustering — finding and ┌visually┐ documenting groups of facts not previously known

Our conclusion? Data mining involves **looking** at data (the ┌boxes┐ have been added for emphasis). If data mining is looking at data, then **what type of lens do we use?**

2.3.3 Data Mining Lens

There are many ways to look at the same biological data. In the field of data mining, this might be called using different *filters* on the data. However, it is not common to look at the same data with many different filters, so we prefer the different metaphor of a lens. It could be a telescope, a microscope, polarized sunglasses, or just a good pair of reading glasses.

All proteins have chemical representations, e.g., the protein

$$C_{400}H_{620}N_{100}O_{120}P_1S_1.$$

In the early research on proteins [482], discovering such formulæ was a major step. But a much bigger step came with the realization that proteins are composed of sequences of amino acids. This allowed proteins to be described by alphabetic sequences, and the descriptions come in different forms: DNA, RNA, amino acid sequences. One can think of these from a linguistic perspective, and indeed, this has been a productive approach [254].

The function of DNA is largely to store sequence information, but proteins operate as three-dimensional widgets. Not all proteins have a stable three-dimensional representation, but most biologically relevant proteins function via three-dimensional structures. Indeed, random proteins would be expected not to form stable three-dimensional structures [120]. The PDB is a curated database of such structures that provides a starting point to study protein function from a physical chemistry perspective.

But structure alone does not explain how proteins function. Physical chemistry can both simplify our picture of a protein and also allow function to be more easily interpreted. In particular, we will emphasize the role of the modulation of the dielectric environment by hydrophobic effects. We describe a simple way this can be done to illustrate the effect on individual electronic entities, such as bonds. But there is need for better lenses to look at such complex effects.

One way to describe what we are doing in modern terms is "big data analytics." The definition of what is "big" is subjective, but the exercises in the book require handling thousands of PDB files. Following such small steps in a research direction can lead to much larger datasets, but we do not discuss here the issues related to high-performance computing necessary to do computing at that scale [444]. The "analytics" aspect requires the implementation of many mathematical ideas. Most of these require at most calculus, but the need to understand the three-dimensional geometry of proteins forces a review of concepts that are accessible in the precalculus curriculum. In using this book as a text, such mathematical concepts can be emphasized to a greater or lesser extent, as desired.

2.3.4 Hydrogen Bonds are Orientation-Dependent

The hydrogen bond provides a good starting example of the use of "pchem" data mining to reveal its properties. Figure 6 of [287] shows clearly both the radial dependence and the angular dependence of the hydrogen bond. Similar evidence is found in later papers; Figure 3 in [483] suggests that hydrogen bonds are stronger when they are both shorter and better aligned. However, the precise relationships between angle and distance can depend on the context, being different in different types of protein structure [32]. Figure 8 of [535] shows a similar relationship between the angle of the hydrogen bond and its distance, derived using protein data. The data in that figure is consistent with a conical restriction on the region of influence of the bond. More recently, the orientation dependence of the hydrogen bond has been revisited. An orientation-dependent hydrogen bonding potential improves prediction of specificity and structure for proteins and protein–protein complexes [355]. All of these involved data mining.

An alternative method for modeling hydrogen bonds is to study their energetics via quantum mechanical calculations and to interpolate the resulting energy surfaces [364, 460]. Close agreement between the orientation dependence of hydrogen bonds observed in protein structures and quantum mechanical calculations has also been reported [283]. Despite the inherent interest in hydrogen bonds, a general model of them has not yet been developed. In particular, hydrogen bonds do not appear as primary bonds in molecular dynamics simulations. Due to the primary importance of the hydrogen bond in protein structure, we will review what is known and not known in Chapter 6.

2.3.5 What is an Answer?

Before we begin to ask questions in earnest, we need to talk about what sort of answers we might expect. In high-school algebra, an answer takes the form of a number or a small set of numbers. In calculus, the answer is often a function. Here, we will often find that the answer is statistical in nature. There appear to be few absolutes in biology, so a probability distribution of what to expect is the best we can hope for.

A probability distribution provides a way to give answers that combine the types of answers you get with high-school algebra and those you get with calculus. An answer that is a number is a Dirac δ-function, whereas a function corresponds to a measure that is absolutely continuous.

This added level of sophistication is especially helpful in a subject where it seems almost anything can happen with some degree of probability.

Mathematics tells us that it is a good idea to have metrics for the space of answers that we expect. Metrics on probability distributions are not commonly discussed. We suggest in Exercise 7.2 one application of metrics on probability distributions to our subject.

In classical physics, problems were often considered solved only when names for the functions involved could be determined. This paradigm is extremely robust and useful. When the names are familiar, they suggest general properties (exponential versus sinusoidal), and they provide a simple algorithm to compute specific values for particular instances. The programming language Fortran was designed specifically to facilitate the evaluation of expressions such as

$$\sin(\log(\tan(\cos(J_1(e^x + \sqrt{\pi x}))))). \tag{2.4}$$

Unfortunately, the classical paradigm is limited by our ability to absorb new names. While the names in (2.4) are familiar to many who have studied Calculus, the list required in practice includes less well-known Bessel functions, Hankel functions, elliptic functions, theta functions, zeta functions, and so on. Moreover, it may be that each new problem requires a new name, in which case the paradigm fails; it is successful only if it provides an abstraction that allows the simplification of the answer. Moreover, strict adherence to this paradigm causes an unnecessary impediment from a computational point of view. All that we may care about is the asymptotic form of a function, or particular values in a certain range, i.e., a plot, or just the point at which it has a minimum.

The newer computational paradigm is not to associate names to solutions, but rather to associate standard algorithms to problems that can be used to provide the information required to understand the mechanism being studied. For example, we may be content if we can specify a well-posed differential equation to be solved to determine numerical values of a function. Thus, we might say that the equation $u' = u$ is a sufficient description of the exponential function. When we discuss quantum mechanics, we will adopt this point of view.

2.4 Multiscale Models

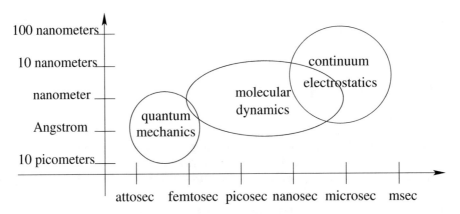

Fig. 2.2 Cartoon of spatial and temporal scales of biomolecular models. See the text for more details.

But why don't we just write down a mathematical model and use it to simulate protein dynamics? This is a reasonable question, and we attempt here to show why such an approach at the moment would not be productive. The difficulty is the particular multiscale aspect of the problem: the temporal scales are huge but the spatial scales overlap, as depicted in Figure 2.2. Of course, existing models are useful in limited contexts. However, we will explain limitations in two such models that must be addressed in order to use them on more challenging simulations.

Models for many systems have components which operate at different scales [236]. Scale separation often simplifies the interactions among the different scales. The differences often occur in both spatial and temporal scales. Scale separation often simplifies the study of complex systems by allowing each scale to be studied independently, with only weak interactions among the different scales. However, when there is a lack of scale separation, interactions among the scales become more difficult to model.

There are three models of importance in protein biochemistry. The different spatial and temporal scales for these models are depicted in Figure 2.2. The smallest and fastest scale is that of quantum chemistry. The model involves continuous variables, partial differential equations and functions as solutions.

The molecular scale is more discrete, described only by the positions of different atoms in space, perhaps as a function of time. The timescale of molecular dynamics is much longer than the quantum scale. But the length scale is comparable with the quantum scale. For example, the Ångstrom can be used effectively to describe both without involving very large or very small numbers.

Finally, the electric properties of proteins are mediated by the dielectric behavior of water in a way that is suitable for a continuum model [110, 447]. But again the length scale is not much bigger than the molecular scale. Many solvated systems are accurately represented using a system in which the size of the solvation layer is the same as the protein dimension. On the other hand, dielectric models are inherently time independent, representing a "mean field" approximation. Thus, there is no natural timescale for the continuum dielectric model, but we have depicted in Figure 2.2 the timescale for so-called Brownian dynamics models which are based on a continuum dielectric model [329].

The lack of physical scale separation, linked with the extreme time separation, in biological systems is the root of some of the key challenges in modeling them. Note that the temporal scales in Figure 2.2 cover fifteen orders of magnitude, whereas the spatial scales cover only three or four orders of magnitude. Many biological effects take place over a timescale measured in seconds, but there may be key ingredients which are determined at a quantum level. This makes it imperative to develop simplified rules of engagement to help sort out behaviors, as we attempt to do here.

We do not give a complete introduction to quantum models, but we do include some material so that we can discuss some relevant issues of interest. For example, molecular-level models utilize force fields that can be determined from quantum models, and this is an area where we can predict significant developments in the future. The hydration structure around certain amino acid residues is complex and something that begs further study. But this may require water models which are currently under development [379], and these models may require further examination at the quantum level.

Multiscale models are most interesting and challenging when there is significant information flow between levels. One of the most intriguing examples is the effect of the electric field on the flexibility of the peptide bond [145]. The electric field is determined by effects at the largest scale and causes a change in behaviors at the smallest scale, forcing a restructuring of the molecular model (Chapter 14).

The Schrödinger equation is a well-accepted model for quantum chemistry. However, it is too detailed for use as a numerical model for large systems. Learning theory [41] is being used to improve standard models used for molecular systems. Recent work [26, 27] suggests that high-dimensional problems like the Schrödinger equation admit accurate low-rank approximations. Density functional theory (DFT) models purport to reduce the complexity of quantum chemistry calculations dramatically, but they require an approximation to the exchange-correlation potential, and at the moment, this remains a heuristic issue, albeit one that is actively studied [92, 256, 434].

Molecular dynamics models are used routinely to simulate protein dynamics, but there are two drawbacks. On the one hand, there are some limitations in the basic theoretical foundations of the model, such as the proper representation of the many-body forces in biomolecules, so the predictions are not fully accurate (cf. Section 14.4). On the other hand, they are still complicated enough that sufficiently long-time simulations, required for biological accuracy, are often prohibitive [12]. The first issue is actively being addressed by improving the interaction potentials used in molecular dynamics simulations [112, 253].

Electrostatic models hope to capture the expected impact of dielectric solvation, but there are limitations here as well. The dielectric coefficient of water is orders of magnitude larger than what would be found inside a large protein. This is a very large jump in a coefficient in a continuum model, and it is prudent to be cautious about any model with such large changes. It is clear that in the neighborhood of the jump in the coefficient, a more complex model might be required [447, 531, 532]. Similarly, attempts are being made to represent the affect of ions more accurately [318].

We can anticipate improvements to the models used for biochemical simulation, and we hope that these will contribute to improved computational techniques in molecular biology. In addition to improvements in models at the various scales, we also anticipate advances in linking models at different scales. While this is an extremely difficult problem, significant advances are being made [121, 128].

2.5 Simple Models

The approach we take in this book is to look for simple models that capture the essential features of protein biophysics. These are sometimes referred to as toy models, or cartoon models, but whatever name is used, the intent is to capture the physical mechanisms with computationally tractable models. A key difficulty of the models discussed in Section 2.4 is that they involve significant cancellations. For example, the molecular dynamics model involves numerous positive and negative charges, yet the net charge of most proteins is small, if not zero. Thus, any inaccuracies in the estimation of charge size or location get amplified by a type of round-off error, since the number of such terms is large. By contrast, we study models where the inherent error may be much larger, such as the dehydron, but there is no cancellation due to competing effects.

2.6 Hydrophobic Interactions

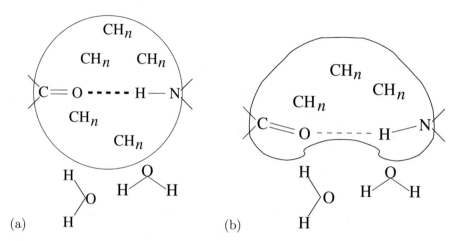

Fig. 2.3 Caricature of wrapping of backbone hydrogen bonds by carbonaceous groups CH_n. (a) Well-wrapped backbone hydrogen bond. (b) Under-wrapped backbone hydrogen bond.

Hydrophobic interactions are sometimes said to be more important than even the hydrogen bond [263]. Although not completely understood, the **hydrophobic force** [48] derives from the **hydrophobic effect** [481]. This effect is one of the central topics of our study. However, the hydrophobic effect has many manifestations in protein behavior.

The hydrophobic effect [48, 481] was proposed in the 1950s [482] as a major contributor to protein structure. However, it is only recently that the detailed nature of hydrophobic forces has been understood. Here, we will characterize a type of intramolecular hydrogen bond, named dehydron, that is strengthened and stabilized as water is removed from its surroundings. Indeed, the dehydron can be viewed as promoting a particular type of hydrophobic effect.

In two recent papers, further understanding of hydrophobic forces have been provided [117, 118]. It was seen that the role of hydrophobic modulation of solvent dielectric is critical to the hydrophobic force [117].

There is a simple view of how hydrophobic forces work. There are certain molecules that are hydrophobic (cf. Section 2.6.2 and Chapter 8), meaning that they repel water. Regions of proteins that have many such molecules, e.g., a protein with a large number of hydrophobic residues on a part of its surface, would tend to prefer association with another such surface to reduce the frustration of having two water–hydrophobe interfaces. It is this simple effect that makes cooking oil form a single blob in water even after it has been dispersed by vigorous stirring.

More precisely, the argument is that the elimination of two hydrophobic surfaces with a water interface is energetically favorable. One could also argue by considering volume changes (cf. Section 5.6) since hydrophobic sidechains take up more volume in water. Recent results show how a hydrophobic force can arise through a complex interaction between polarizable (e.g., hydrophobic) molecules and (polar) water molecules [117, 118]. These arguments are compelling, but they suggest a nonspecific interaction. Indeed, hydrophobic attraction leads to nonspecific binding [176].

But there are other kinds of hydrophobic effects as well. We will show that hydrophobicity plays a central role in a number of electrostatic forces by modulating the dielectric effect of water. In addition, water removal can affect the local polar environment, which can modify the nature of covalent bonds.

2.6.1 Solvent Mediation of Electric Forces

Some bonds become substantially altered in the presence of water. We have already noted that certain ionic bonds (in table salt) are easily disrupted by water. The main bond holding proteins together is the hydrogen bond, and this bond is extremely susceptible to alteration by water interaction since water molecules can each make four hydrogen bonds themselves. So protein survival depends on keeping the hydrogen bond dry in water [146].

Another type of solvent effect that occurs on the quantum level is the rigidity of the peptide bond (Chapter 14) which requires an external field to select one of two resonant states. Such a field can be due to hydrogen bonds (see Section 5.2 and Chapter 6) formed by backbone amide or carbonyl groups, either with other backbone or sidechain groups, or with water. In some situations, water removal can cause a switch in the resonance state to a flexible mode [145].

Another example of a change of electrical properties resulting from differences in the water environment involves a more gross change. Proteins which penetrate a cell membrane go from a fully solvated environment to one that is largely solvent-free (inside the membrane). This can be related to a large-scale change in the secondary structure of the protein conformation that has implications for drug delivery [176].

More generally, solvent mediation can alter any electrostatic force via dielectric effects (Chapter 15). Changes in dielectric properties of the environment can have a substantial impact on any electrical property. Much of our study will be related to the dielectric effect and its modulation by hydrophobic groups. But rather than try to address this by introducing a precise model (see Chapter 15), we prefer to introduce the concept by example. We thus begin by looking at one particular example of hydrophobic modulation of the dielectric behavior of water around hydrogen bonds.

2.6.2 Dehydrons

In [174], a quantifiable structural motif, called **dehydron**, was shown to be central to protein–ligand interactions. A dehydron is a defectively "wrapped" hydrogen bond in a molecular structure whose electrostatic energy is highly sensitive to water exclusion by a third party. Such preformed, but under-protected, hydrogen bonds are effectively adhesive, since water removal from their vicinity contributes to their strength and stability, and thus, they attract partners that make them more viable (see Section 2.6.5 and Chapter 9).

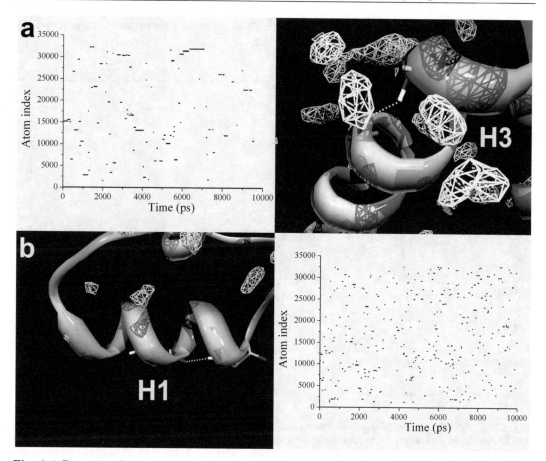

Fig. 2.4 Dynamics of water near hydrogen bonds, adapted from [107, Figure 5]. Plotted on the axes are the residence times of individual waters (numbered as indicated on the vertical axis) as a function of the simulation time (horizontal axis). The orange and yellow polyhedra indicate regions with high residence times for waters. (a) Hydrogen bond in helix H3 is well-wrapped. (b) Hydrogen bond in helix H1 is under-wrapped. The relevant hydrogen bonds are indicated by dotted lines.

A review of protein structure and the role of hydrogen bonds will be presented in Chapter 4. The concept of "wrapping" of a hydrogen bond is based on the hydrophobic effect [48, 481] and its role in modulating the dielectric effect (Chapter 15) through water exclusion. At the simplest level, wrapping occurs when sufficient nonpolar groups (CH_n, $n = 0, 1, 2, 3$) are clustered in the vicinity of intramolecular hydrogen bonds, protecting them by excluding surrounding water [172]. The concept of wrapping of a hydrogen bond is depicted informally in Figure 2.3. A well-wrapped hydrogen bond, Figure 2.3(a), is surrounded by CH_n groups on all sides, and water is kept away from the hydrogen bond formed between the C–O group of one peptide and the N–H group of another peptide (Section 4.1). An under-wrapped hydrogen bond, Figure 2.3(b), allows a closer approach by water to the hydrogen bond, and this tends to disrupt the bond, allowing the distance between the groups to increase and the bond to weaken.

It is possible to identify dehydrons as **under-wrapped hydrogen bonds (UWHB)** by simply counting the number of hydrophobic sidechains in the vicinity of a hydrogen bond. This approach is reviewed in Section 8.3. More accurately, a count of all (nonpolar) carbonaceous groups gives a more refined estimate (Section 8.4). However, it is possible to go further and quantify a force associated with dehydrons which provides a more refined measure of the effect geometry [174] of the wrappers (Section 8.5).

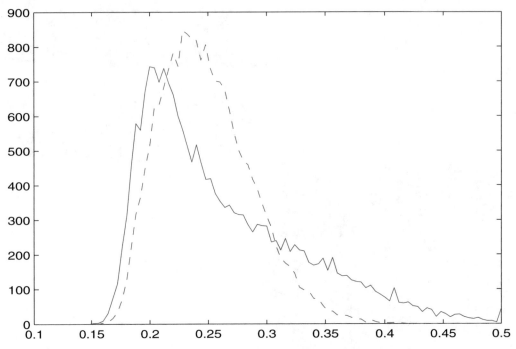

Fig. 2.5 Distribution of bond lengths for two hydrogen bonds formed in a structure of the sheep prion [107]. The horizontal axis is measured in nanometers, whereas the vertical axis represents numbers of occurrences taken from a simulation with 20,000 data points with bin widths of 0.1 Ångstrom. The distribution for the well-wrapped hydrogen bond (H3) has a smaller mean value but a longer (exponential) tail, whereas the distribution for the under-wrapped hydrogen bond (H1) has a larger mean but Gaussian tail.

At the structural level, a significant correlation can be established between dehydrons and sites for protein complexation (Chapter 8). The HIV-1 capsid protein P24 complexed with antibody FAB25.3 provides a dramatic example, as shown in [174, Figure 2] and in a cartoon in Figure 11.1(a).

2.6.3 Dynamics of Dehydrons

The extent of wrapping changes the nature of hydrogen bond [107] and the structure of nearby water [165]. Hydrogen bonds that are not protected from water do not persist [107]. Figure 5 of [107] shows the striking difference of water residence times for well-wrapped and under-wrapped hydrogen bonds. Private communication with the authors of [107] has confirmed that there is a marked difference as well in the fluctuations of the hydrogen bonds themselves. Under-wrapped hydrogen bond lengths are larger (on average) than well-wrapped hydrogen bonds. More strikingly, the distributions of bond lengths as shown in Figure 2.5 are quite different, confirming our prediction based on Figure 2.3 that the coupling of the hydrogen bond characteristics with the water environment would be different.

The H-bond R208–E212 depicted in [107, Figure 5(A)] is well-wrapped, whereas V189–T193 depicted in [107, Figure 5(B)] is a dehydron (see [167, Figure 3a, page 6448]). Well-wrapped hydrogen bonds are visited by fewer water molecules but have longer-lasting water interactions (due to the structuring effect of the hydrophobes), whereas the behavior of dehydrons is more like that of bulk water: frequent rebonding with different water molecules [107].

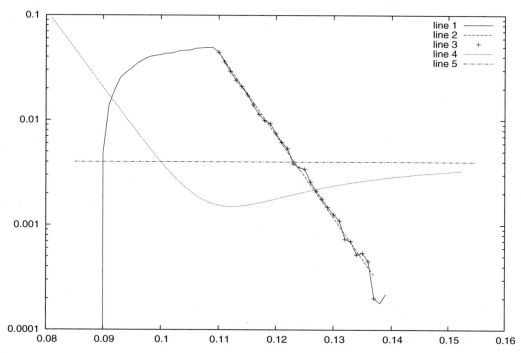

Fig. 2.6 Simulation of a random walk with a restoring force. Shown is the distribution of values x_i defined in (2.5) for 10^5 time steps i, starting with $x_1 = 0.1$, scaled by a factor of 10^{-3}. Also shown is a graph of $\phi + 0.03$ where ϕ is the potential (2.5). The dot-dashed horizontal line provides a reference axis to facilitate seeing where ϕ is positive and negative. The +'s indicate the part of the distribution exhibiting an exponential decay; the dashed line is a least-squares fit to the logarithm of these distribution values. The distribution has been scaled by a factor of 10^{-3} so that it fits on the same plot with ϕ.

The long residence time of waters around a well-wrapped hydrogen bond would seem to have two contributing factors. On the one hand, the water environment is structured by the hydrophobic barrier, so the waters have reduced options for mobility: once trapped, they tend to stay. But also, the polar effect of the hydrogen bond which attracts the water is more stable, thus making the attraction of water more stable. With a dehydron, both of these effects go in the opposite direction. First of all, water is more free to move in the direction of the hydrogen bond. Second, the fluctuation of the amide and carbonyls comprising the hydrogen bond contributes to a fluctuating electrostatic environment. The bond can switch from the state depicted in Figure 2.3(b) when water is near, to one more like that depicted in Figure 2.3(a) if water molecules move temporarily away. More precisely, the interaction of the bond strength and the local water environment becomes a strongly coupled system for an under-wrapped hydrogen bond, leading to increased fluctuations. For a well-wrapped hydrogen bond, the bond strength and water environment are less strongly coupled.

The distance distribution for under-wrapped hydrogen bonds can be interpreted as reflecting a strong coupling with the thermal fluctuations of the solvent. Thus, we see a

Boltzmann-type distribution for the under-wrapped hydrogen bond distances in Figure 2.5. It is natural to expect the mean distances in this case to be larger than the mean distances for the well-wrapped case, but the tails of the distribution are at first more confusing. The distribution in the under-wrapped case exhibits a Gaussian-like tail (that is, an exponential of the distance squared), whereas the well-wrapped case decays more slowly, like a simple exponential. (See Figure 5.13 for a comparison of these distributions.) Thus, the well-wrapped hydrogen bond is sustaining much larger deviations, even though the typical deviation is much smaller than in the under-wrapped case. To explain how this might occur, we turn to a simulation with a simple model.

2.6.4 Simulated Dynamics

The data in Figure 2.5 can be interpreted via a simulation which is depicted in Figure 2.6. This figure records the distribution of positions for a random walk subject to a restoring force defined by

$$x_{i+1} = x_i + \Delta t(f_i + \phi(x_i)) \tag{2.5}$$

with f_i drawn randomly from a uniform distribution on $[-0.5, 0.5]$ and with ϕ being a standard Lennard–Jones potential

$$\phi(x) = (0.1/x)^{12} - (0.1/x)^6. \tag{2.6}$$

The particular time step used in Figure 2.6 is $\Delta t = 0.02$; the simulation was initiated with $x_1 = 0.1$ and carried out for 10^5 steps.

The simulation (2.5) represents a system that is forced randomly with a restoring force back to the stationary point $x = 0.1$, quantified by the potential ϕ in (2.6). Such a system exhibits a distribution with an exponential decay, as verified in Figure 2.6 by comparison with a least-squares fit of the logarithm of the data to a straight line.

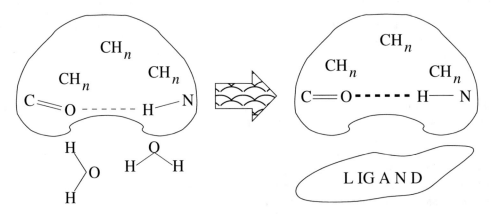

Fig. 2.7 Cartoon showing dehydration due to ligand binding and the resulting strengthening of an under-wrapped hydrogen bond.

2.6.5 Stickiness of Dehydrons

Desolvation of an under-wrapped hydrogen bond can occur when a ligand binds nearby, as depicted in Figure 2.7. The removal of water lowers the dielectric and correspondingly strengthens the hydrogen bond. The resulting change in energy due to the binding effectively means that there is a mechanical force of attraction for a dehydron. This implies that dehydrons are sticky. These concepts are explained in more detail in Chapter 9, and the experimental support for the force of attraction for dehydrons is reviewed as well.

Since dehydrons are sticky, they must be binding sites. Thus, we see in Chapter 7 that they are likely to be found at protein–ligand interfaces. In Chapter 11, we explore several types of protein–ligand interfaces in which dehydrons play a decisive role. Finally, we review in Chapter 13 the role of dehydrons in drug design.

2.7 Dehydron Switch

The strength and stability of hydrogen bonds depend on many factors: the distances between donor and acceptor and other constituents, the angles formed by the constituents, and the local dielectric environment. While we cannot formally quantify the effect of these factors, we can imagine that they combine to form a single variable that describes the 'quality' of the hydrogen bond. Then, the stability and strength of the hydrogen bond depend in some monotonic way on this quality variable.

As wrapping is varied, the quality of a hydrogen bond can vary in a highly nonlinear way, as depicted in Figure 2.8. When a hydrogen bond is extremely underwrapped, the addition of one wrapper could have only a small effect, especially if it appears in a region already wrapped. At the other extreme, addi-

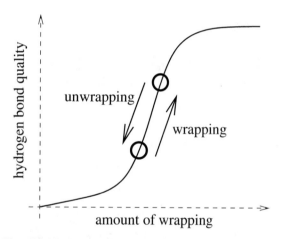

Fig. 2.8 Cartoon depicting the capability of a dehydron to switch on and off when a ligand binds. The binding event may increase wrapping, in which case the hydrogen bond becomes stronger and more stable, or it may decrease wrapping, in which case the hydrogen bond becomes weaker and less stable.

tional wrappers will have diminishing effects on a fully protected hydrogen bond. In the middle, additional wrappers may have the most effect, leading to a sensitive switch.

Binding events can cause the quality of a hydrogen bond to change significantly, due to several factors. Wrapping can increase the quality, and unwrapping can decrease the quality. In addition, large-scale structural changes can occur that may alter the distances and angles that determine the quality of a hydrogen bond. However, even without such gross motion in a protein, the change in the dielectric environment (which has the capability to change by two orders of magnitude) can effect the quality significantly. The former increases the enthalpy of the hydrogen bond and thus causes dehydrons to be sticky in many cases.

Unwrapping a dehydron can effectively switch off the hydrogen bond, eliminating a constraint on the protein. This has the potential to increase the entropy of the overall system because removing the constraint can increase the degrees of freedom of motion. Thus, both effects have the potential to increase free energy upon association.

The variables that determine the quality of a hydrogen bond are not independent. As a hydrogen bond becomes underwrapped, it becomes weaker due just to the change in local dielectric, and this can mean that the structural determinants of the hydrogen bond can change as well. There may be a torque on the hydrogen bond in the original state that is balanced by the hydrogen bond. As the bond becomes weaker, the torque has more effect and the constituents of the hydrogen bond can move. This has the effect of further weakening the hydrogen bond. Thus, the change in dielectric may get amplified by structural changes; the resulting change in quality may be dramatic.

We will discuss experimental evidence in Section 8.1 that a single nonpolar carbonaceous group can have a measurable effect on the strength and stability of a hydrogen bond. However, these results concern formation of basic protein structure (helices) in an isolated solvent environment. In more complex protein environments, the effect of small amounts of wrapping may have less effect due to environmental constraints. These effects can lead to the more dramatic switch behavior depicted in Figure 2.8.

2.8 Exercises

Exercise 2.1. Compare (2.1) with the atomic mass of atoms not listed in Table 2.3. Consult appropriate tables to find out the fraction of different isotopes that occur naturally.

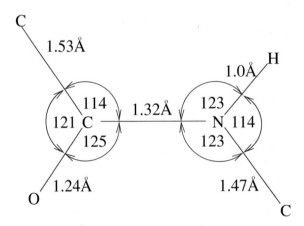

Fig. 2.9 Data reproduced from the figure at the top of page 282 in [382]. Distances are in Ångstroms (indicated by Å) and angles are in degrees. The upper-left and lower-right carbons (C) correspond to sequential C_α carbons in a protein.

Exercise 2.2. Pauling used a simple molecule as a model for the peptide backbone structure in what is called the trans form depicted in Figure 4.2(a). Pauling's data, in the figure at the top of page 282 in [382], is replicated in Figure 2.9. Use this information to compute the distance between C_α carbons according to this model. Explain why you can be sure that these atoms are in a plane (hint: look at the angles). Also use Figure 2.9 to compute the distance between C_α carbons in the cis form as depicted in Figure 4.2(b).

Exercise 2.3. Download a PDB file for a protein and compute the distance distribution between sequential C_α carbons. Instead of plotting the usual distribution, plot the distance on the vertical axis as a function of the order of appearance of C_α carbons in the PDB file. To allow for very small and very large distances, use a log scale for the vertical axis. What causes the outliers? What is the mean of the distribution if you throw out the outliers? Look at PDB file 3SIL and explain why some distances are so small (nearly zero). Look at PDB file 1E08 and explain why one value is significantly smaller than the others. Compare all of this with the distance data in Exercise 2.2.

Exercise 2.4. Download a PDB file for a protein and compute the distance distribution between C_α carbons separated in sequence by k. That is, the sequential neighbors have $k = 1$. How does the mean distance vary as a function of k? Compare the distributions for $k = 3$ and $k = 4$; which has C_α carbons closer together?

Exercise 2.5. Download a PDB file for a protein and compute the N–O distance distribution between the N's and O's that are nearest in sequence (cf. Figure 2.1). That is, for each residue i, there is an O_i and an N_i. Plot the distributions for the distances $|O_i - N_i|$ and $|O_i - N_{i+1}|$.

Exercise 2.6. Download a PDB file for a protein and compute the N–O distance distribution between all pairs of carbonyl and amide groups in the peptide bonds (cf. Figure 2.1). Is there part of the distribution that corresponds to pairs forming local interactions (e.g., a hydrogen bond)? Explain your reasoning. Pick a cutoff distance R and plot the distribution of sequence separation k for all N–O pairs whose distance is less than R. Experiment with R between 3 and 7 Å to see how the distribution changes.

Exercise 2.7. Download a PDB file for a protein and compute the N–O distance distribution between each carbonyl and its nearest amide group with which is not covalently bonded in the peptide bonds (cf. Figure 2.1). That is, for each C–O group, find the nearest N and record the N–O distance. Is there part of the distribution that corresponds to pairs forming local interactions (e.g., a hydrogen bond)? Explain your reasoning. Also record the sequence distance k between the N's and O's and plot a two-dimensional distributions of k versus physical distance d. (Hint: exclude the N's and O's that are near neighbors in the peptide bond backbone.)

Chapter 3
Electrostatic Forces

Peter Joseph William Debye (1884–1966) is known for his explanation
of the role of polar materials leading to the dielectric effect [110]. He is
honored by the unit for a dipole, the Debye.

The fundamental force in biochemistry is electrostatic. Based on electrostatic forces, chemistry distinguishes different **bonds** pairing atoms. In terrestrial biology, water plays a dominant role as a dielectric that modulates different types of electrostatic interactions.

Here, we briefly outline the main types of electrostatic forces as they relate to biology, and especially to proteins and other biomolecular. There are so many books that could be used as a reference that it is hard to play favorites. But the books by Pauling [382, 383] are still natural references.

The order of forces, or bonds [403], that we consider is significant. First of all, they are presented in order of strength, starting with the strongest. This order also correlates with the directness of interaction of the electrons and protons, from the intertwining of covalent bonds to indirect, induced interactions. Finally, the order is also reflective of the effect of solvent interaction to some extent, in that the dielectric effect of solvent is increasingly important for the weaker bonds.

3.1 Direct Bonds

The strongest bonds can be viewed as the direct interactions of positive and negative charges, or at least distributions of charge.

3.1.1 Covalent Bonds

These are the strong bonds of chemistry, and the term molecule simply means atoms paired by covalent bonds. However, their role in structural biology is generally static; they rarely break. They form the backbones of proteins, DNA, and RNA and support the essential linear structure of these macromolecules. Single lines represent single bonds and double (parallel) lines represent double bonds, as depicted in examples in Figure 3.1. The geometry for single and double bonds is often different. Double bonds often confer planar geometry, as shown in

© Springer International Publishing AG 2017
L.R. Scott and A. Fernández, *A Mathematical Approach to Protein Biophysics*,
Biological and Medical Physics, Biomedical Engineering,
DOI 10.1007/978-3-319-66032-5_3

Figure 3.1(a). Certain atoms with only single bonds often confer a tetrahedral geometry, as shown in Figure 3.1(b). The letter R in Figure 3.1(b) stands for "residue" which connotes a complex of from one to eighteen atoms which determine the different amino acid constituents of proteins.

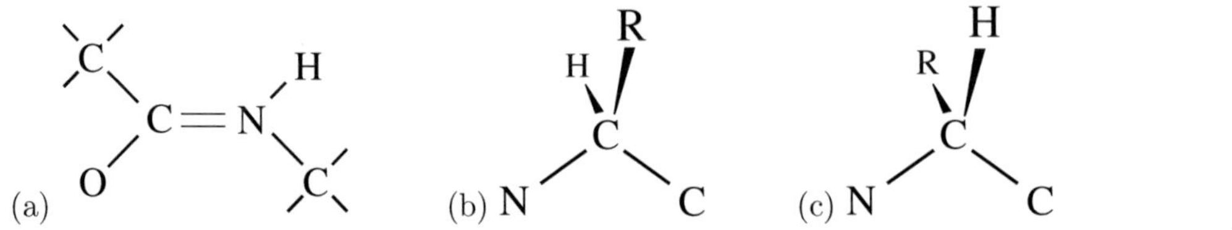

Fig. 3.1 Single and double bonds. The double bond (a) often confers a planar geometry to the atoms; all six atoms in (a) are in the plane of the page. The upper-left and lower-right carbons represent C_α carbons in a peptide sequence and would each have three additional single bonds (not shown). (b) Tetrahedral arrangement of atoms around the central (C_α) carbon in the basic (L-form) peptide unit. The Nitrogen and two Carbons are in the plane of the page, with the Hydrogen lying below the plane, away from the viewer, and the residue R (see text) lying above the plane, toward the viewer. The lower Nitrogen and Carbon would each be double-bonded (not shown) to a Carbon and Nitrogen (respectively) in a peptide sequence, as in (a). (c) The D-form of the peptide base, which occurs (naturally) only infrequently in proteins [348], but plays a significant role in the antibiotic gramicidin [498].

Covalent bonds involve the direct sharing of electrons from two different atoms, as required by the octet rule mentioned in Section 2.1. Such bonds are not easily broken, and they typically survive immersion in water. The octet rule [382, 383] allows the prediction of covalent bond formation through counting of electrons in the outer-most shell (see Table 2.1) of each atom. Explaining further such simple rules for other types of bonds is one of the major goals of this work.

Although covalent bonds are not easily broken in the context of structural biology, their character can be modified by external influences. The most important covalent bond in proteins is the peptide bond (Figure 14.1) formed between amino acids as they polymerize. This bond involves several atoms that are typically planar in the common form of the peptide bond. But if the external electrostatic environment changes, as it can if the amide and carbonyl groups lose hydrogen bond partners, the bond can rotate. We review this effect in Chapter 14.

3.1.2 Ionic Bonds/Salt Bridges

Ionic bonds occur in many situations of biological interest, but it is of particular interest due to its role in what is called a *salt bridge* (Section 4.2.1). Such an ionic bond occurs between oppositely charged sidechains in a protein, as indicated in Figure 3.2. Ionic bonds involve the direct electrostatic attraction between negatively and positively charged groups.

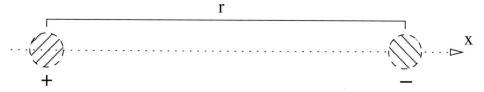

Fig. 3.2 Salt bridges are ionic bonds.

The potential for the electrostatic interaction between two charged molecules, separated by a distance r, is (see Section 3.2)

$$V(r) = z_1 z_2 r^{-1}, \tag{3.1}$$

where z_i is the charge on the i-th molecule. For two molecules with equal but opposite charges, say, $z_1 = 1$ and $z_2 = -1$, the potential is $-r^{-1}$.

We will see that different (noncovalent) bonds are characterized by the exponent of r in their interaction potential. For potentials of the form r^{-n}, we can say that the bonds with smaller n are more long range, since $r^{-n} >> r^{-m}$ for $n < m$ and r large. The ionic bond is thus the one with the longest range of influence.

In addition to being long range, ionic bonds can be stronger in anhydrous environments, such as those occurring in transmembrane proteins. However, the stabilization gained by forming an ionic bond (salt bridge) in a soluble protein is marginal at best. This is so because of the propensity of the ions to be fully hydrated. The hydration is hindered by the formation of the ionic bond. A simple experiment with table salt introduced into a glass of water shows this. Salt forms a stable crystal when dry, but when wet it happily dissolves into a sea of separated ions. The source of attraction between the sodium and chloride ions in salt is the ionic bond.

For all forces of attraction which are of the form r^{-n}, there would be infinite attraction at $r = 0$. However, there is always some other (electrostatic) force of repulsion that keeps the entities from coalescing. Thus, the form of the attractive force is not sufficient to tell us the strength of the bond. However, ionic bonds are often quite strong as well as being long range, second only to covalent bonds in strength.

3.1.3 Hydrogen Bonds

Although weaker than covalent and ionic bonds, hydrogen bonds play a central role in biology. They bind complementary DNA and RNA strands in a duplex structure, and they secure the three-dimensional structure of proteins. However, they are also easily disrupted by water, which is one of the best hydrogen bond makers in nature.

First suggested in 1920, hydrogen bonds were not fully accepted until after 1944 [482]. The detailed structure of hydrogen bonds in biology is still being investigated [283, 355, 464, 490, 516]. Most of the hydrogen bonds of interest to us involve a hydrogen that is covalently bonded to a heavy atom X and is noncovalently bonded to a nearby heavy atom Y. Typically, the heavy atoms X and Y are N, O, or S in protein systems, e.g., NH - - O or OH - - S, etc.; see Table 6.2 for a list. The bond OH − O describes the hydrogen bond between two water molecules.

Hydrogen bonds will be discussed in more detail in Chapter 6.

3.1.4 Cation-π Interactions

Aromatic residues (phenylalanine, tyrosine, and tryptophan: see Section 4.4.6) are generally described as hydrophobic, due to the nonpolar quality of the carbon groups making up their large rings. But their carbon rings have a secondary aspect which *is* polar, in that there is a small negative charge distribution on each side of the plane formed by the rings [103, 200, 538]. This large distribution of negative charge can directly attract the positive charges of cations (e.g., arginine and lysine).

Cation-π interactions will be discussed in more detail in Chapter 12.

3.2 Charge–Force Relationship

We want to talk about the interaction energy (and force) between two charged groups. The units of charge and energy are not the same, and so we need to introduce a conversion factor to allow this.

Suppose we have a charge z at the origin in space $\mathbb{R}^3$. This induces an electric field $\mathbf{e}$ in all of space, and the relationship between the two is

$$\varepsilon \nabla \cdot \mathbf{e} = z\delta, \tag{3.2}$$

where ε is the permittivity and δ denotes the Dirac delta function. Here, $\nabla \cdot \mathbf{e} = \sum_{i=1}^{3} e_{i,i}$ is the divergence operator applied to a vector function $\mathbf{e}$ with components e_i; we have used the "comma" notation to indicate the partial derivative with respect to the i-th variable. The concept of the Dirac delta function is complex but well known: the expression (3.2) means that for any smooth function ϕ that vanishes outside a bounded set

$$-\varepsilon \int_{\mathbb{R}^3} \mathbf{e}(x) \cdot \nabla \phi(x) \, dx = z\phi(0). \tag{3.3}$$

Going between (3.3) and (3.2) is just integration by parts, except that $\nabla \cdot \mathbf{e}$ is not regular enough for this to be justified in a simple way. Thus, (3.3) is taken as definition of (3.2).

When the medium is a vacuum, ε is the permittivity of free space, ε_0. When we write the expression (3.1), we have in mind the permittivity of free space, with appropriate units, cf. (3.6). In other media (e.g., water), the value of ε is much larger. This quantity measures the strength of the dielectric environment. We can now see one example of the lack of duality between hydrophobic and hydrophilic groups mentioned in Section 2.2.1. Hydrophobicity affects the coefficient ε in (3.2), whereas hydrophilic groups would contribute to the right-hand side in the equation.

The exact value for ε_0 depends on the units (Chapter 18) chosen for charge, space, time, etc. The electric field $\mathbf{e}$ does not have units of force. If there is no other charge in the field, no force will be felt. The resulting force on a second charge z' is proportional to the amount of that charge: $z'\mathbf{e}$. So the electric field $\mathbf{e}$ has units of force per unit of charge, whereas z has units of charge. The coefficient ε provides the change of units required by the relation (3.2), cf. (18.2).

The electric field $\mathbf{e}$ can be written as (minus) the gradient of a potential

$$\mathbf{e} = -\nabla V, \tag{3.4}$$

and therefore the potential is related to the charge by

$$-\varepsilon \Delta V = z\delta,$$ (3.5)

where we again have to invoke an interpretation like (3.3) to make proper sense of (3.5). It is not too difficult to verify that a solution to (3.5) is

$$V(r) = \frac{z}{4\pi\varepsilon r}.$$ (3.6)

Note that the units of V are energy per unit charge. We can have a simple representation of the relationship between charge z and its electric potential

$$V(r) = \frac{z}{r},$$ (3.7)

provided we choose the units (Chapter 18) appropriately so that $\varepsilon = 1/4\pi$. The resulting potential energy of a pair of charges z_1 and z_2 is thus given by (3.1).

We will make this simplification in much of our discussion, but it should be remembered that there is an implicit constant proportional to the permittivity in the denominator. In particular, we see that a larger permittivity leads to a smaller potential and related force.

Fig. 3.3 Left: configuration of a dipole consisting of a pair of charged molecules at $(\pm\epsilon, 0, 0)$. Right: abstract representation of a dipole as a vector.

3.3 Interactions Involving Dipoles

A **dipole** is an abstract concept based on a collection of charges, e.g., in a molecule. The simplest example is given by two molecules of opposite charge that are fused together, e.g., by covalent bonds, as depicted in Figure 3.3. Water is a sea of such dipoles, oriented somewhat arbitrarily in response to the ambient electric field. The ability of the water dipole to orient in response to an external field is the basis for the dielectric response.

Mathematically, we imagine that the two charged molecules are placed on the x-axis with the $\pm$ charges at the positions $(\pm\epsilon, 0, 0)$, as shown on the left-hand side of Figure 3.3. We will see that it will be possible (for ϵ small) to think of a dipole as just a vector, as depicted on the right-hand side of Figure 3.3.

We will see many examples of dipoles. One very important one is in the peptide backbone shown in Figure 3.4.

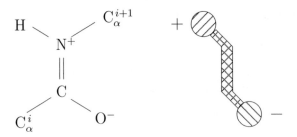

Fig. 3.4 The dipole of the peptide backbone (in the trans enol configuration, cf. Figure 4.2). On the left is the chemical description, and the right shows a cartoon of the dipole. The double bond between the central carbon and nitrogen keeps the peptide bond planar. There is a positive charge center near the hydrogen and a negative charge center near the oxygen.

Many interactions can be modeled as dipole–dipole interactions, e.g., between water molecules. More generally, the use of partial charges (cf. Table 12.1) represents many interactions as dipole–dipole interactions. Forces between molecules with fixed dipoles are often called Keesom forces [196]. For simplicity, we consider dipoles consisting of the same charges of opposite signs, separated by a distance 2ϵ. If the charges have unit value, then the dipole strength $\mu = 2\epsilon$. Interacting dipoles have two orientations which produce no torque on each other.

3.3.1 Single-file Dipole–Dipole Interactions

In the single-file orientation, the base dipole has a unit positive charge at $(\epsilon, 0, 0)$ and a unit negative charge at $(-\epsilon, 0, 0)$; the other dipole is displaced on the x-axis at a distance r: a unit positive charge at $(r + \epsilon, 0, 0)$ and a unit negative charge at $(r - \epsilon, 0, 0)$ (cf. Figure 3.5). Since the potential for two charges is the sum of the individual potentials (i.e., we assume linear additivity), the potential due to the base dipole at a distance $r >> \epsilon$ along the x-axis is

$$
\begin{aligned}
V(r) &= \frac{1}{r - \epsilon} - \frac{1}{r + \epsilon} = \frac{(r + \epsilon) - (r - \epsilon)}{(r - \epsilon)(r + \epsilon)} \\
&= \frac{2\epsilon}{(r - \epsilon)(r + \epsilon)} = \frac{2\epsilon}{r^2 - \epsilon^2} \approx 2\epsilon r^{-2} = \mu r^{-2},
\end{aligned}
\tag{3.8}
$$

where $\mu = 2\epsilon$ is the dipole strength.

We use the expression $f(r) \approx g(r)$ to mean that the expression $f(r)$ is a good approximation to $g(r)$. More precisely, in this case, we mean that the two expressions are asymptotically equal for large r, that is, that

$$
\lim_{r \to \infty} g(r)/f(r) = 1.
\tag{3.9}
$$

In (3.8), $f(r) = 1/(r^2 - \epsilon^2)$ and $g(r) = r^{-2}$, so that $g(r)/f(r) = 1 - \epsilon^2/r^2$, and thus (3.9) follows. Moreover, we can get a quantitative sense of the approximation: the approximation in (3.8) is 99% accurate for $r \geq 10\epsilon$ and even 75% accurate for $r \geq 2\epsilon$.

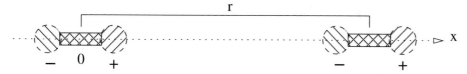

Fig. 3.5 Single-file dipole–dipole configuration consisting of two pairs of molecules with charges ± 1 at $(\pm \epsilon, 0, 0)$ and $(r \pm \epsilon, 0, 0)$.

In the field of the dipole (3.8), the potential energy of a single charge on the x-axis at a distance r is thus μr^{-2}, for a charge of $+1$, and $-\mu r^{-2}$, for a charge of -1. In particular, we see that the charge-dipole interaction has a potential one order lower (r^{-2}) than a charge–charge interaction (r^{-1}). The charge-dipole interaction is very important, but we defer a full discussion of it until Section 10.3.1.

The combined potential energy of two opposite charges in the field generated by a dipole is given by the difference of terms of the form (3.8). In this way, we derive the potential energy of a dipole, e.g., a positive charge at $(r + \epsilon, 0, 0)$ and a negative charge at $(r - \epsilon, 0, 0)$, as the sum of the potential energies of the two charges in the field of the other dipole:

$$\frac{\mu}{(r + \epsilon)^2} - \frac{\mu}{(r - \epsilon)^2}. \tag{3.10}$$

Considering two such charges as a combined unit allows us to estimate the potential energy of two dipoles as follows:

$$
\begin{aligned}
\frac{\mu}{(r + \epsilon)^2} - \frac{\mu}{(r - \epsilon)^2} &= -\mu \frac{(r + \epsilon)^2 - (r - \epsilon)^2}{(r + \epsilon)^2 (r - \epsilon)^2} \\
&= -\mu \frac{4 r \epsilon}{(r + \epsilon)^2 (r - \epsilon)^2} \approx -4 \mu \epsilon r^{-3} = -2 \mu^2 r^{-3}.
\end{aligned}
\tag{3.11}
$$

The negative sign indicates that there is an attraction between the two dipoles in the configuration Figure 3.5.

The electric force field $\mathbf{F}$ is the negative gradient of the potential ∇V. For V defined by (3.8), only the x-component of ∇V is nonzero along the x-axis, by symmetry. Differentiating (3.8), we find that for $r \gg \epsilon$ along the x-axis,

$$
\begin{aligned}
F_x(r, 0, 0) &= -(r - \epsilon)^{-2} + (r + \epsilon)^{-2} \\
&= \frac{-(r + \epsilon)^2 + (r - \epsilon)^2}{(r - \epsilon)^2 (r + \epsilon)^2} = \frac{-4 \epsilon r}{(r - \epsilon)^2 (r + \epsilon)^2} \\
&\approx -4 \epsilon r^{-3} = -2 \mu r^{-3}.
\end{aligned}
\tag{3.12}
$$

The attractive force experienced by a dipole displaced on the x-axis at a distance r, with a positive charge at $(r + \epsilon, 0, 0)$ and a negative charge at $(r - \epsilon, 0, 0)$, is thus (asymptotically)

$$
\begin{aligned}
-\frac{2\mu}{(r + \epsilon)^3} + \frac{2\mu}{(r - \epsilon)^3} &= 2\mu \frac{(r + \epsilon)^3 - (r - \epsilon)^3}{(r + \epsilon)^3 (r - \epsilon)^3} \\
&= 2\mu \frac{6 r^2 \epsilon + 2 \epsilon^3}{(r + \epsilon)^3 (r - \epsilon)^3} \approx 6 \mu^2 r^{-4},
\end{aligned}
\tag{3.13}
$$

which is equal to the derivative of the potential (3.11) as we would expect.

3.3.2 Parallel Dipole–Dipole Interactions

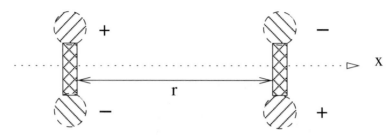

Fig. 3.6 Parallel dipole-dipole configuration with ± 1 charges at $(0, \pm\epsilon, 0)$ and $(r, \mp\epsilon, 0)$.

In the parallel orientation, the base dipole has a positive charge at $(0, \epsilon, 0)$ and a negative charge at $(0, -\epsilon, 0)$; the other dipole is displaced on the x-axis at a distance r: a positive charge at $(r, -\epsilon, 0)$ and a negative charge at $(r, +\epsilon, 0)$ (cf. Figure 3.6).

The potential in the (x, y)-plane due to the base dipole at a distance r along the x-axis is

$$V(x, y) = \frac{1}{\sqrt{(y - \epsilon)^2 + x^2}} - \frac{1}{\sqrt{(y + \epsilon)^2 + x^2}} \tag{3.14}$$

The potential energy of a dipole displaced on the x-axis at a distance r, with a positive charge at $(r, -\epsilon, 0)$ and a negative charge at $(r, \epsilon, 0)$, is thus

$$\left(\frac{1}{\sqrt{(2\epsilon)^2 + r^2}} - \frac{1}{r} \right) - \left(\frac{1}{r} - \frac{1}{\sqrt{(2\epsilon)^2 + r^2}} \right) = -2 \left(\frac{1}{r} - \frac{1}{\sqrt{(2\epsilon)^2 + r^2}} \right)$$

$$= -2 \frac{\sqrt{(2\epsilon)^2 + r^2} - r}{r\sqrt{(2\epsilon)^2 + r^2}} = -2 \frac{\sqrt{(2\epsilon/r)^2 + 1} - 1}{r\sqrt{(2\epsilon/r)^2 + 1}} \tag{3.15}$$

$$\approx - \frac{(2\epsilon/r)^2}{r} = -\mu^2 r^{-3}.$$

Thus, the potential energy of the parallel orientation is only half of the single-file orientation.

The potential $V(x, y)$ in (3.14) vanishes when $y = 0$. Therefore, its derivative along the x-axis also vanishes: $\frac{\partial V}{\partial x}(r, 0) = 0$. However, this does not mean that there is no attractive force between the dipoles, since (by symmetry) $\frac{\partial V}{\partial x}(r, \pm\epsilon) = \pm f(\epsilon, r)$. Thus, the attractive force is equal to $2f(\epsilon, r)$. For completeness, we compute the expression $f(\epsilon, r)$:

$$\frac{\partial V}{\partial x}(x, y) = \frac{-x}{((y - \epsilon)^2 + x^2)^{3/2}} + \frac{x}{((y + \epsilon)^2 + x^2)^{3/2}} \tag{3.16}$$

for general y. Choosing $y = \pm\epsilon$, (3.16) simplifies to

$$\frac{\partial V}{\partial x}(r, \pm\epsilon) = \mp r^{-2} \pm \frac{r}{((2\epsilon)^2 + r^2)^{3/2}} = \mp r^{-2} \left(1 - \frac{1}{((\mu/r)^2 + 1)^{3/2}} \right)$$

$$= \mp \frac{((\mu/r)^2 + 1)^{3/2} - 1}{r^2 ((\mu/r)^2 + 1)^{3/2}} \approx \mp \frac{3\mu^2}{2r^4}, \tag{3.17}$$

for large r/ϵ. The net force of the field (3.17) on the two oppositely charged particles on the right side of Figure 3.6 is thus $3\mu^2 r^{-4}$, consistent with what we would find by differentiating (3.15) with respect to r.

The electric force field in the direction of the second dipole (that is, the y-axis) is

$$\frac{\partial V}{\partial y}(r, y) = \frac{\epsilon - y}{((y - \epsilon)^2 + r^2)^{3/2}} + \frac{\epsilon + y}{((y + \epsilon)^2 + r^2)^{3/2}}. \tag{3.18}$$

At a distance $r \gg \epsilon$ along the x-axis, this simplifies to

$$\frac{\partial V}{\partial y}(r, \pm\epsilon) = \frac{\mu}{(\mu^2 + r^2)^{3/2}} \approx \mu r^{-3}, \tag{3.19}$$

for large r/ϵ. Although this appears to be a force in the direction of the dipole, the opposite charges on the dipole on the right side of Figure 3.6 cancel this effect. So there is no net force on the dipole in the direction of the y-axis.

3.3.3 Dipole Stability

Only the single-file dipole orientation is stable with respect to perturbations. This can be seen as follows. Suppose the dipoles are arranged along the x-axis as above but that they are both tilted away from the x-axis at an angle θ, as shown in Figure 3.7. The connective material between the charge centers has been omitted for simplicity and will be omitted in future drawings. Define θ so that $\theta = 0$ (and $\theta = \pi$) is the single-file dipole configuration and $\theta = \pi/2$ is the parallel configuration. Thus, one dipole has a positive charge at $\epsilon(\cos\theta, \sin\theta, 0)$ and a negative charge at $-\epsilon(\cos\theta, \sin\theta, 0)$. The other dipole is displaced on the x-axis at a distance r: a positive charge at $(r + \epsilon\cos\theta, -\epsilon\sin\theta, 0)$ and a negative charge at $(r - \epsilon\cos\theta, \epsilon\sin\theta, 0)$.

Fig. 3.7 General θ-dependent dipole–dipole configuration. The connective material between the charge centers has been omitted for simplicity.

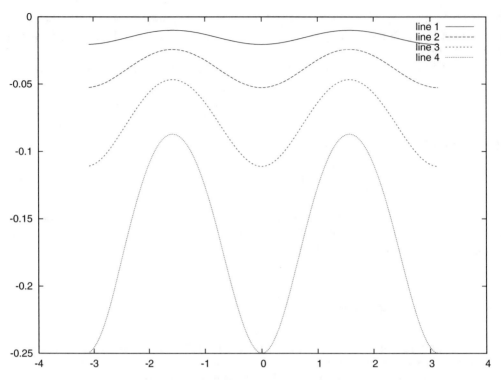

Fig. 3.8 Potential energy variation $v(\rho, \theta)$ as defined in (3.23) (vertical axis) of dipoles as a function of θ (horizontal axis) for the configurations shown in Figure 3.7 for $\rho = 0.02$ (top), $0.05, 0.1, 0.2$ (bottom), where ρ is defined in (3.22).

The potential at the point $(x, y, 0)$ due to the rotated base dipole is

$$V(x, y) = \frac{1}{\sqrt{(x - \epsilon \cos \theta)^2 + (y - \epsilon \sin \theta)^2}} - \frac{1}{\sqrt{(x + \epsilon \cos \theta)^2 + (y + \epsilon \sin \theta)^2}} \tag{3.20}$$

Therefore, the potential energy of the second rotated dipole, with a positive charge at $(r + \epsilon \cos \theta, -\epsilon \sin \theta, 0)$ and a negative charge at $(r - \epsilon \cos \theta, \epsilon \sin \theta, 0)$, is thus

$$
\begin{aligned}
V(r, \theta) &= \frac{1}{\sqrt{r^2 + (2\epsilon \sin \theta)^2}} - \frac{1}{r + 2\epsilon \cos \theta} - \left(\frac{1}{r - 2\epsilon \cos \theta} - \frac{1}{\sqrt{r^2 + (2\epsilon \sin \theta)^2}} \right) \\
&= \frac{2}{\sqrt{r^2 + (2\epsilon \sin \theta)^2}} - \frac{1}{r + 2\epsilon \cos \theta} - \frac{1}{r - 2\epsilon \cos \theta} \\
&= \frac{2}{\sqrt{r^2 + (2\epsilon \sin \theta)^2}} - \frac{2r}{r^2 - (2\epsilon \cos \theta)^2} \\
&= \frac{2}{r} \left(\frac{1}{\sqrt{1 + \rho \sin^2 \theta}} - \frac{1}{1 - \rho \cos^2 \theta} \right) := \frac{2}{r} v(\rho, \theta),
\end{aligned}
\tag{3.21}
$$

where the (non-dimensional) parameter ρ is defined by

$$\rho = (2\epsilon / r)^2. \tag{3.22}$$

This expression

$$v(\rho, \theta) = \frac{1}{\sqrt{1 + \rho \sin^2 \theta}} - \frac{1}{1 - \rho \cos^2 \theta} \qquad (3.23)$$

in (3.21) has a minimum when $\theta = 0$ and a maximum when $\theta = \pi/2$. A plot of v in (3.23) is shown in Figure 3.8 for various values of ρ. When ρ is small, the expression (3.23) tends to the limit

$$\begin{aligned} v(\rho, \theta) &\approx \frac{1}{1 + \frac{1}{2}\rho \sin^2 \theta} - \frac{1}{1 - \rho \cos^2 \theta} \\ &\approx \left(1 - \tfrac{1}{2}\rho \sin^2 \theta\right) - \left(1 + \rho \cos^2 \theta\right) = -\tfrac{1}{2}\rho \left(1 + \cos^2 \theta\right). \end{aligned} \qquad (3.24)$$

Of course, what we have presented is only an indication of the stability and energy minimum of the single-file dipole configuration. We leave a complete proof as Exercise 3.5.

3.3.4 Different Dipoles

So far, we considered dipoles with identical charges and charge distributions (separations). Here, we consider a single-file configuration as in Figure 3.5, but with the dipole on the right consisting of charges $\pm q$ separated by a distance δ, as depicted in Figure 3.9. We consider the potential energy of the right-hand dipole in the potential field (3.8) of the left dipole. Similar to (3.11), we find

$$\frac{\mu q}{(r + \delta)^2} - \frac{\mu q}{(r - \delta)^2} = -\mu q \frac{4r\delta}{(r + \delta)^2 (r - \delta)^2} \approx -4\mu q \delta r^{-3} = -2\mu\nu r^{-3}, \qquad (3.25)$$

where $\nu = 2q\delta$ is the strength of the dipole on the right. Notice that the expression (3.25) is symmetric in the two dipole strengths μ and ν.

Fig. 3.9 Single-file dipole–dipole configuration with different dipole strengths.

3.4 van der Waals Forces

name	constituents	decay rate
Keesom	two fixed dipoles	r^{-3}
Debye	one fixed dipole and one induced dipole	r^{-6}
London dispersion	two correlated dipoles	r^{-6}

Table 3.1 Forces involving dipoles.

Many of the electrostatic forces we consider are induced rather than direct. The best known of these are called van der Waals forces, although this term covers a range of forces known by other names. Keesom forces, which we covered in Section 3.3, are often included in this group, but we will see that there is a qualitative difference in behavior. One prominent Web site went as far as to say "all intermolecular attractions are known collectively as van der Waals forces" but this seems a bit extreme. To clarify names and usages, we summarize in Table 3.1 the different terms used and what they refer to.

We cover van der Waals forces in detail here to clarify that they are electrostatic in nature, and not some new or different type of force. Debye forces and London dispersion forces [196] involve induced dipole–dipole interactions, which we will study using the results derived in Section 3.5. The most significant example is the London dispersion force [196] which results from both dipoles being correlated.

3.4.1 Lennard–Jones Potentials

The van der Waals interactions are often modeled via the Lennard–Jones potential

$$V(r) := \epsilon \left(\left(\frac{\rho}{r} \right)^{12} - 2 \left(\frac{\rho}{r} \right)^6 \right). \tag{3.26}$$

The attractive potential r^{-6} is a precise result of the interaction of a fixed dipole and an induced dipole, which we derive in Section 3.5. In Section 3.5.2, we consider the self-induction of two dipoles. The repulsive term r^{-12} is a convenient model, whereas other terms are more accurate [101].

The minimum of V is at $r = \rho$, with $V(\rho) = -\epsilon$, so we can think of the well depth ϵ as giving the energy scale. The parameter ρ is called the **van der Waals radius** and can be defined as the separation distance at which the force of attraction and repulsion cancel [56]. Typical values for these parameters, from the Amber force field, are shown in Table 3.2. Note that $V(\rho/1.2) \approx -3V(\rho)$, and $V(\rho/2) = -3968V(\rho)$, so the repulsion is quite strong in this model.

The maximal attractive force occurs when $r/\rho = \sqrt[6]{13/7}$ (Exercise 3.8).

3.5 Induced Dipoles

Dipoles can be induced in two ways. Fixed dipoles, such as water molecules, induce a dipole in any polarizable material. Such interactions give rise to what are frequently called Debye forces [196]. More subtly, two molecules can, in a sense, induce dipoles in each other via what are called London dispersion forces [196].

atom	ρ	ϵ	$V(\rho/2)$	D	κ
C (aliphatic)	1.85	0.12	476	1.54	83.1
O	1.60	0.20	794	1.48	33.2
H	1.00	0.02	79	0.74	104.2
N	1.75	0.16	635	1.45	38.4
P	2.10	0.20	794	1.87	51.3
S	2.00	0.20	794	1.81	50.9

Table 3.2 Lennard–Jones parameters from Amber for various atoms involving the van der Waals radius ρ measured in Ångstroms and energy (well depth) ϵ in kcal/mol. For comparison, covalent bond lengths D and strengths [382] κ are given in kcal/mol, together with the repulsion potential energy $V(\rho)$ at the van der Waals radius ρ.

Essentially, all materials are polarizable. This just means that the distributions of electrons can be distorted by an electric field. Table 3.3 gives some typical values of polarizability.

molecule	α_0	α_1	α_2	α_3
benzene		10.66	10.66	4.01
methane	2.62	2.62	2.62	2.62
water	1.49			

Table 3.3 Experimental [10] and derived [500] values for polarizabilities of some molecules. Units are Å³.

3.5.1 Debye Forces

Fig. 3.10 The Debye force: a dipole (left) inducing a dipole in a polarizable molecule (right). The upper configuration (a) shows the dipole and polarizable molecule well separated and the lower configuration (b) shows them closer, with the molecule on the right now polarized.

If a polarizable molecule is subjected to an electric field of strength $\mathbf{F}$, then it is reasonable to expect that an induced dipole μ_i will result, given by

$$\mu_i \approx \alpha\mathbf{F} \qquad (3.27)$$

for small $\mathbf{F}$, where α is the polarizability. This is depicted visually in Figure 3.10, where the upper configuration (a) shows the dipole and polarizable molecule well separated, and the lower configuration (b) shows them closer, with the molecule on the right now polarized.

In general, the electric field $\mathbf{F}$ is a vector and the polarization α is a tensor (or matrix). Also, note that a dipole is a vector quantity: it has a magnitude and direction. In our previous discussion, we considered only the magnitude, but the direction was implicit (the line connecting the two charges). For simplicity, we assume here that α can be represented as a scalar (times the identity matrix), that is, that the polarizability is isotropic.

We can approximate a polarized molecule as a simple dipole with positive and negative charges $\pm q$ displaced by a distance δ, as depicted in Figure 3.9. This takes some justification, but it will be addressed in Chapter 10. There is ambiguity in the representation in that only the product $q\delta$ matters: $\mu = q\delta$.

We derived in (3.12) that the electric force field due to a fixed dipole μ_f has magnitude

$$F_x = 2\mu_f r^{-3}, \tag{3.28}$$

where the x-axis connects the two charges of the fixed dipole. We assume that the molecule whose dipole is being induced also lies on this axis. By combining (3.27) and (3.28), we conclude that the strength of the induced dipole is

$$\mu_i \approx 2\alpha\mu_f r^{-3}. \tag{3.29}$$

From (3.25), we know that the potential energy of the two dipoles is

$$V(r) \approx -2\mu_f\mu_i r^{-3} \approx -4\alpha\mu_f^2 r^{-6}, \tag{3.30}$$

in agreement with the Lennard–Jones model in (3.26).

In Section 3.5.3, we show how a dipole can interact with any matter due to the fact that the positive and negative charges are not colocated. However, we will also see (Section 3.5.3) that this does not provide the expected r^{-6} behavior of the potential energy, but this requires a more detailed argument.

3.5.2 London Dispersion Forces

Suppose now that we start with two nonpolar molecules that are well separated. Due to the long range interaction of the electron distributions of the two molecules, they can become correlated and have a dipole–dipole interaction (attraction), as depicted in Figure 3.11. This property is related to what is known as **entanglement**. To achieve a correct description of the van der Waals force, a quantum mechanics description of the atomic interactions is needed. It can be shown that

$$V(r) \approx c_2 r^{-6}, \tag{3.31}$$

where an expression for the constant c_2 can also be made explicit [6, 88].

(a)

(b)

Fig. 3.11 Correlated polarization of two molecules. The upper configuration (a) shows the molecules well separated, and the lower configuration (b) shows them closer, with the electrons now visibly correlated. Different colors depict instantaneous positions of electrons. In each configuration, favorable dipole-dipole interactions obtain.

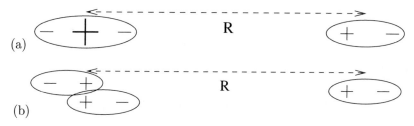

(a)

(b)

Fig. 3.12 (a) Interaction between a dipole (right) and a neutral group (left). The neutrally charged group on the left has charges $+2c$ at the origin and $-c$ located at $(\pm\delta, 0, 0)$, and the dipole on the right has charges ± 1 at $(R \mp \frac{1}{2}\mu, 0, 0)$. (b) Equivalent representation of the interaction between a dipole and a neutral group, in which the neutral group in (a) is written as two dipoles.

3.5.3 Dipole–Neutral Interactions

Suppose we have two charge groups as indicated in Figure 3.12(a). The exact positions of the charges are as follows. We assume that the dipole on the right consists of charges ± 1 located at $(R \mp \frac{1}{2}\mu, 0, 0)$, and the neutrally charged group on the left has charges $-c$ located at $(\pm\delta, 0, 0)$ and $+2c$ at the origin. Then, the potential energy of this system is $cV(\mu, \delta)$ where

$$
\begin{aligned}
V(\mu, \delta) &= \frac{2}{R - \frac{1}{2}\mu} + \frac{1}{R + \frac{1}{2}\mu - \delta} + \frac{1}{R + \frac{1}{2}\mu + \delta} - \frac{2}{R + \frac{1}{2}\mu} - \frac{1}{R - \frac{1}{2}\mu - \delta} - \frac{1}{R - \frac{1}{2}\mu + \delta} \\
&= \frac{2\mu}{R^2 - \frac{1}{4}\mu^2} - \frac{\mu}{R^2 - (\frac{1}{2}\mu - \delta)^2} - \frac{\mu}{R^2 - (\frac{1}{2}\mu + \delta)^2} \\
&= \frac{2\mu}{R^2 - \frac{1}{4}\mu^2} - \frac{\mu}{R^2 - \frac{1}{4}\mu^2 + \mu\delta - \delta^2} - \frac{\mu}{R^2 - \frac{1}{4}\mu^2 - \mu\delta - \delta^2} \\
&= \frac{2\mu}{R^2 - \frac{1}{4}\mu^2} - \frac{2\mu(R^2 - \frac{1}{4}\mu^2 - \delta^2)}{(R^2 - \frac{1}{4}\mu^2 + \mu\delta - \delta^2)(R^2 - \frac{1}{4}\mu^2 - \mu\delta - \delta^2)} \\
&= \frac{2\mu}{\rho + \delta^2} - \frac{2\mu\rho}{\rho^2 - \mu^2\delta^2},
\end{aligned}
\tag{3.32}
$$

where $\rho = R^2 - \frac{1}{4}\mu^2 - \delta^2$. Therefore,

$$
\begin{aligned}
V(\mu, \delta) &= \frac{2\mu}{\rho + \delta^2} - \frac{2\mu\rho}{\rho^2 - \mu^2\delta^2} = \frac{2\mu(\rho^2 - \mu^2\delta^2) - 2\mu\rho(\rho + \delta^2)}{(\rho + \delta^2)(\rho^2 - \mu^2\delta^2)} \\
&= -\frac{2\mu^3\delta^2 + 2\mu\rho\delta^2}{(\rho + \delta^2)(\rho^2 - \mu^2\delta^2)} \approx -\frac{2\mu\delta^2}{R^4},
\end{aligned}
\tag{3.33}
$$

for R large. To check that this result is correct, we could also think of this system as involving two dipoles on the left, of opposite sign and displaced by a distance δ, as depicted in Figure 3.12(b). Since the dipole–dipole interaction is of order R^{-3}, the difference of two such interactions should be of the order R^{-4}.

In Section 10.5, we will see that Figure 3.12 represents the interaction between a dipole and a quadrapole. This is significant because all atoms have distributed charges that appear to some extent like Figure 3.12. That is, there is a positively charged nucleus surrounded by a negatively charged cloud of electrons. The results of this section, and in particular (3.33), show that there is a natural attraction between a dipole and such a distributed charge. We can think of this as being a type of van der Waals force.

3.6 Exercises

Exercise 3.1. Show that the approximation in (3.8) is 96% accurate for $r \geq 5\epsilon$.

Exercise 3.2. Prove that (3.11) is still correct if we use the exact form in (3.8) instead of the approximation $V(r) \approx \mu r^{-2}$.

Exercise 3.3. Prove that (3.13) is still correct if we use the exact form in (3.12), $F_x(r, 0, 0) = -4\epsilon r(r - \epsilon)^{-2}(r + \epsilon)^{-2}$, instead of the approximation $F_x(r, 0, 0) \approx -2\mu r^{-3}$.

Exercise 3.4. Consider the expression in (3.21). Prove that, for any $\rho < 1$, it has a minimum when $\theta = 0$ and a maximum when $\theta = \pi/2$.

Exercise 3.5. Prove that the single-file dipole configuration is stable and an energy minimum. (Hint: derive a formula for the general orientation of two dipoles in three dimensions, cf. Figure 2.2 in [244]. This can be done with one distance parameter and three angular parameters.)

Exercise 3.6. Describe the orientation of the dipoles that corresponds to $\theta = 2\pi$ in Figure 3.8.

Exercise 3.7. Show that the in-line dipole–dipole interaction energy is

$$
-\frac{1}{r - 1} + \frac{2}{r} - \frac{1}{r + 1} \approx -\left(\frac{\partial^2}{\partial r^2}\right)\left(\frac{1}{r}\right) = -\frac{2}{r^3}
\tag{3.34}
$$

as $r \to \infty$ by considering the error in the second-difference operator represented by the left-hand side.

Exercise 3.8. Compute the position of the maximum attractive force for the Lennard–Jones potential (3.26). Determine the magnitude of the force in terms of the parameters of the Lennard–Jones potential.

Chapter 4
Protein Basics

Frederick Sanger (1918–2013) won the Nobel Prize for Chemistry twice, in 1958, "for his work on the structure of proteins, especially that of insulin" and, in 1980, for "contributions concerning the determination of base sequences in nucleic acids."

It is not our intention to provide a complete introduction to the structure of proteins. Instead, we suggest consulting texts [59, 102, 393] for further information. Moreover, we suggest acquiring a molecular modeling set so that accurate three-dimensional models can be constructed. In addition, it will be useful to become familiar with a graphical viewer for PDB files (even the venerable "rasmol" would be useful). We present some essential information and emphasize concepts needed later or ones that may be novel.

4.1 Chains of Amino Acid Residues

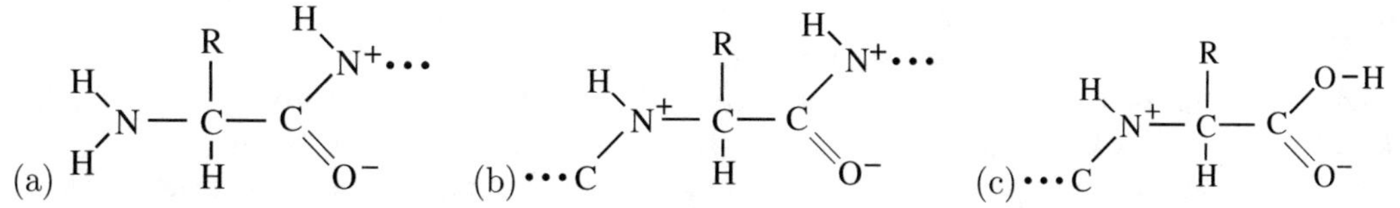

Fig. 4.1 The basic units of a peptide sequence. (a) The initial, or N-terminal, unit. (b) The typical (internal) unit. (c) The final, or C-terminal, unit. The letter R stands for "residue" and can be any of the twenty depicted in Figures 4.3–4.4. The atoms are not all coplanar. In particular, the four bonds around the central C_α Carbon are in a tetrahedral arrangement as shown in Figure 2.1. The dots before the carbons and after the nitrogens indicate the continuation of the peptide sequence.

Proteins are sequences of amino acids which are covalently bonded along a "backbone." The basic units of the backbone are depicted in Figure 4.1. In each unit, there is a residue denoted by R that is a molecule that can vary in size, and it is bonded to the central carbon atom in the unit, called the C_α carbon. The twenty residues R of most interest in biology

© Springer International Publishing AG 2017
L.R. Scott and A. Fernández, *A Mathematical Approach to Protein Biophysics*,
Biological and Medical Physics, Biomedical Engineering,
DOI 10.1007/978-3-319-66032-5_4

are represented in Figures 4.3–4.4. The peptide units bond together to form (arbitrarily long) sequences by forming a double bond between the N-terminus and the C-terminus, as shown in Figure 2.1.

Proteins of biological significance fold into a three-dimensional structure by adding hydrogen bonds between carbonyl and amide groups on the backbone of different amino acids. In addition, other bonds, such as a salt bridge (Section 4.2.1) or a disulfide bond (Section 4.2.2), can form between particular amino acids (cysteine has sulfur atoms in its sidechain). However, the hydrogen bond is the primary mode of structure formation in proteins.

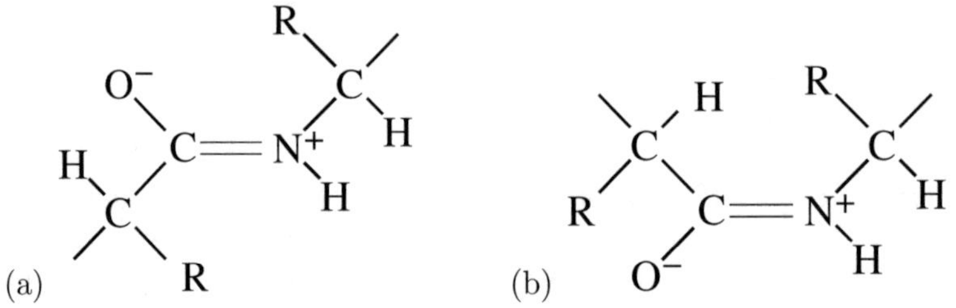

Fig. 4.2 The rigid state of the peptide bond: (a) trans form, (b) cis form [45]. The letter R stands for "residue" and can be any of the twenty depicted in Figures 4.3–4.4. The double bond between the central carbon and nitrogen keeps the peptide bond planar. Compare Figure 14.1.

The basic unit of the peptide group shown in Figure 4.2 comes in two forms that are related by a rotation around the C–N bond. The **trans** form (a) of the peptide bond is the most common state, but the **cis** form (b) has a small but significant occurrence [45, 217, 377]. The occurrences of the cis form are overwhelmingly associated with prolines, but other residues are also involved [245, 264]. The covalent bond linking O-C-N is called a resonance, and it indicated by a polygonal line linking them in Figures 4.1 and 2.1. This particular bond will be considered in more detail in Chapter 14. In Figure 4.2, we represent the C–N bond as a double bond, the resonant state that confers rigidity to the peptide unit.

The peptide chain units are joined at the double bond indicated between the N and the O in Figure 4.2. Thus, we refer to the coordinates of the nitrogen and hydrogen as N^{i+1} and H^{i+1} and to the coordinates of the oxygen and carbon as O^i and C^i.

At the ends of the chain, things are different, as depicted in Figure 4.1(a,c). The **N-terminus**, or **N-terminal end**, has an NH_2 group instead of just N, and nothing else attached, as shown in Figure 4.1(a). In the standard numbering scheme, this is the beginning of the chain. The **C-terminus**, or **C-terminal end**, has a COOH group instead of just CO, and nothing else attached, as shown in Figure 4.1(c). In the standard numbering scheme, this is the end of the chain.

4.1.1 Taxonomies of Amino Acids

There are many ways that one can categorize the amino acid sidechains of proteins. We are mainly interested in protein interactions, so we will focus initially on a scale that is based on interactivity. We postpone until Chapter 7 a full explanation of the rankings, but suffice it to say that we rank amino acid sidechains based on their likelihood to be found in a part of the protein surface that is involved in an interaction.

In the following, we will use the standard terminology for the common twenty amino acids.[1] In Table 4.1, we recall the naming conventions and the RNA codes for each residue. Complete descriptions of the sidechains for the amino acids can be found in Figures 4.3–4.4.

In Table 4.2, we present some elements of a taxonomy of sidechains. We give just two descriptors of sidechains, but even these are not completely independent. For example, all the hydrophilic residues are either charged or polar, and all of the neutral sidechains are hydrophobic. However, some residues have both hydrophilic and hydrophobic regions and are characterized as amphiphilic. For example, the aromatic residues are among the most hydrophobic even though two of them are polar, cf. Section 4.4.6. Even Phenylalanine has a small polarity that can interact with other molecules.

Full name of amino acid	three letter	single letter	The various RNA codes for this amino acid
alanine	Ala	A	GCU, GCC, GCA, GCG
arginine	Arg	R	CGU, CGC, CGA, CGG, AGA, AGG
asparagine	Asn	N	AAU, AAC
aspartate	Asp	D	GAU, GAC
cysteine	Cys	C	UGU, UGC
glutamine	Gln	Q	CAA, CAG
glutamate	Glu	E	GAA, GAG
glycine	Gly	G	GGU, GGC, GGA, GGG
histidine	His	H	CAU, CAC
isoleucine	Ile	I	AUU, AUC, AUA
leucine	Leu	L	UUA, UUG, CUU, CUC, CUA, CUG
lysine	Lys	K	AAA, AAG
methionine	Met	M	AUG
phenylalanine	Phe	F	UUU, UUC
proline	Pro	P	CCU, CCC, CCA, CCG
serine	Ser	S	UCU, UCC, UCA, UCG, AGU, AGC
threonine	Thr	T	ACU, ACC, ACA, ACG
tryptophan	Trp	W	UGG
tyrosine	Tyr	Y	UAU, UAC
valine	Val	V	GUU, GUC, GUA, GUG
stop codons			UAA, UAG, UGA

Table 4.1 Amino acids, their (three-letter and one-letter) abbreviations and the RNA codes for them. For completeness, the "stop" codons are listed on the last line.

We focus here on the properties of individual sidechains, but these properties alone do not determine protein structure: the context is essential. Studying pairs of sidechains that are interacting in some way (e.g., ones that appear sequentially) gives a first approximation of context.

4.1.2 Wrapping of Hydrogen Bonds

A key element of protein structure is the protection of hydrogen bonds from water attack. A different taxonomy amino acids can be based on their role in the protection of hydrogen

---

[1] There are more than twenty biologically related amino acids that have been identified, but we will limit our study to the twenty primary amino acids commonly found.

bonds. We will see in Chapter 7 that this correlates quite closely with the propensity to be at an interface.

Some hydrogen bonds are simply buried in the interior of a protein. Others are near the surface and potentially subject to water attack. These can be protected only by the sidechains of other nearby amino acids. Such protection is provided by the hydrophobic effect (cf. Section 2.6). The hydrophobic effect is complex [48, 481], but suffice it to say that a key element has to do with the fact that certain **nonpolar groups**, such as the **carbonaceous groups** CH_n ($n = 1, 2, 3$), tend to repel polar molecules like water. They are non-polar and thus do not attract water strongly, and moreover, they are polarizable and thus damp nearby water fluctuations. Such carbonyl groups are common in amino acid sidechains; Val, Leu, Ile, Pro, and Phe have only such carbonaceous groups. We refer to the protection that such sidechains offer as the **wrapping of hydrogen bonds**. For reference, the number of nonpolar CH_n groups for each residue is listed in Table 4.2.

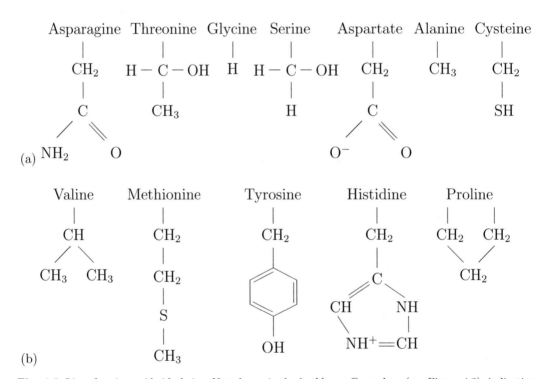

Fig. 4.3 List of amino acid sidechains. Not shown is the backbone C_α carbon (see Figure 4.2), indicating the beginning of the sidechains where the name appears. (a) The smallest, and most likely to be involved in protein–ligand interactions [], ordered from the left (asparagine). (b) The amino acids in the middle ground in terms of interactivity [].

The standard thinking about sidechains has been to characterize them as being hydrophobic or hydrophilic or somewhere in between. Clearly a sidechain that is hydrophobic will repel water and thus protect anything around it from water attack. Conversely, a sidechain that is hydrophilic will attract water and thus might be complicit in compromising an exposed hydrogen bond. In some taxonomies [393], Arg, Lys, His, Gln, and Glu are listed as hydrophilic. However, we will also see that they are indeed good wrappers. On the other hand, Ala is listed as hydrophobic and Gly, Ser, Thr, Cys and others are often listed as "in between" hydrophobic and hydrophilic. And we will see that they are among the most likely to be near underwrapped hydrogen bonds. This is not surprising since they are both polar (see Section 4.4.1) and have a small number of carbonaceous groups.

What is wrong with a simple philic/phobic dichotomy of amino acids is that the "call" of philic versus phobic is made primarily based on the final group in the sidechain (the bottom in Figures 4.3–4.4). For example, Lys is decreed to be hydrophilic when the bulk of its sidechain is a set of four carbonaceous groups. What is needed is a more complete picture of the role of all the groups in the entire sidechain. This requires a detailed understanding of this role, and in a sense that is a major object of this monograph. Thus, it will require some in-depth analysis and comparison with data to complete the story. However in the subsequent chapters this will be done, and it will appear that one can provide at least a broad classification, if

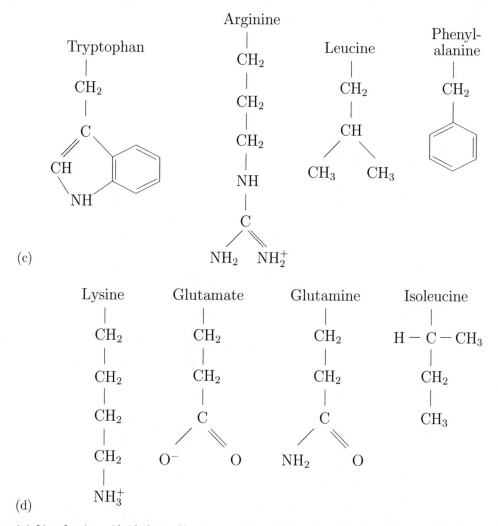

Fig. 4.4 List of amino acid sidechains. Not shown is the backbone C_α carbon (see Figure 4.2), indicating the beginning of the sidechains where the name appears. (c) Less interactive amino acids. (d) The amino acids least likely to be involved in interactions.

not a linear ordering, of amino acid sidechains based on either their ability or propensity to wrap (or not) exposed hydrogen bonds or other electronic bonds.

The ordering of the most interactive proteins is based on a statistical analysis which is described in more detail in Chapter 7. We will also see there that these are likely to be associated with underwrapped hydrogen bonds. On the other hand, it is relatively easy to predict the order for good wrappers based on counting the number of carbonaceous groups. There is not a strict correlation between interactivity and bad wrapping, but a significant trend exists.

4.2 Special Bonds

In addition to the covalent bonds of the backbone and the ubiquitous hydrogen bonds in proteins, there are two other bonds that are significant.

4.2.1 Salt Bridges

Full name of residue	three letter	single letter	water preference	charge or polarity	N	intrinsic pK_a	ΔV
Alanine	Ala	A	phobic	(backbone)	1	NA	−2.6
Arginine	Arg	R	amphi	positive	2	12	+7.9
Asparagine	Asn	N	philic	polar	1	NA	+7.0
Aspartate	Asp	D	philic	negative	1	3.7	+11.9
Cysteine	Cys	C	philic	polar	1	8.5	−1.0
Glutamine	Gln	Q	amphi	polar	2	NA	+1.3
Glutamate	Glu	E	amphi	negative	2	4.2	+8.5
Glycine	Gly	G	NA	(backbone)	0	NA	–
Histidine	His	H	philic	positive	1	6.5	+3.3
Isoleucine	Ile	I	phobic	neutral	4	NA	−2.6
Leucine	Leu	L	phobic	neutral	4	NA	−6.2
Lysine	Lys	K	amphi	positive	3	10.4	+7.6
Methionine	Met	M	amphi	polar	3	NA	+0.7
Phenylalanine	Phe	F	phobic	neutral	7	NA	-0.9
Proline	Pro	P	phobic	neutral	2	NA	−6.2
Serine	Ser	S	philic	polar	0	NA	+1.4
Threonine	Thr	T	amphi	polar	1	NA	+0.3
Tryptophan	Trp	W	amphi	polar	7	NA	+0.6
Tyrosine	Tyr	Y	amphi	polar	6	9.8	−0.3
Valine	Val	V	phobic	aliphatic	3	NA	−3.6

Table 4.2 A taxonomy of amino acids. The code for water interaction is: phobic, hydrophobic; philic, hydrophilic; amphi, amphiphilic. N is the number of CH_n groups in the sidechain. Values of pK_a for ionizable residues are taken from [487] (cf. Table 1.2 of [102]). The indication "backbone" for the polarity of alanine and glycine means that the polarity of the backbone is significant due to the small size of the sidechain. The polar region of Tyrosine and Phenylalanine is limited to a small part of the sidechain, the rest of which is neutral. When histidine is not charged, then it is polar. ΔV is the change in volume in Å^3 [218] of sidechains between protein core and water (Section 5.6).

Certain sidechains are charged, as indicated in Table 4.2, and these can form an ionic bond, as depicted in Figure 3.2 in Section 3.1.2. Depending on the pH level, His may or may not be positively charge, but both Arg and Lys can be considered positively charged in most biological environments. Similarly, Asp and Glu are typically negatively charged. When sidechains of opposite charge form an ionic bond in a protein, it is called a **salt bridge** [99, 122]. Thus there are four (or six, depending on the charge on His) possible salt bridges.

Unmatched charged residues are often found on the surface of a protein where they can be solvated, but not inside a protein core where they would not make contact with water to neutralize the charge.

It is possible to define geometries to characterize salt bridges [122].

4.2.2 Disulfide Bonds

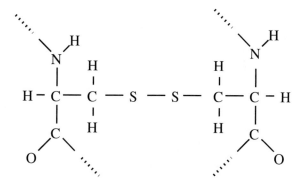

Fig. 4.5 Schematic representation of a disulfide bond. The dotted lines indicate continuation of the peptide sequence. We leave as Exercise 4.11 to complete the drawing including the resonant bond character linking O–C–N as in Figure 4.1.

Proteins are also held together by **disulfide bonds** or **disulfide bridges** which are covalent bonds which form between two sulfurs on cysteines, as depicted in Figure 4.5. Specifically, the hydrogens attached to the sulfur atom on the two Cys sidechains are liberated, and a covalent bond forms between the two sulfur atoms. This is a much stronger bond than a hydrogen bond, but it is also much more specialized. It appears frequently in neurotoxins [209, 371]. These proteins would be highly disordered without the disulfide bridges. There appears to be a strong correlation between the number of disulfide bridges in proteins and their dehydron density (number of dehydrons as a fraction of all hydrogen bonds): an increase in one disulfide bridge per hundred residues bonds correlates with an increase of five dehydrons per hundred backbone hydrogen bonds [163].

Disulfide bonds can also form between two separate proteins to form a larger system. This occurs in insulin and in antibodies. Another example of a covalent bond formed between two proteins is the elastin cross-link which involves two lysines from each chain [496].

4.3 Posttranslational Modifications

Proteins are not quite so simple as the protein sequence might imply. The term **posttrans-lational modification** means changes that occur after the basic sequence has been set. Modifications (**glycosylation, phosphorylation**, etc.) add groups to sidechains and change the function of the resulting protein. A change in pH can cause the ends of some sidechains to be modified, as we discuss in Section 4.5.

Phosphorylation occurs by liberating the hydrogen atom in the OH group of Serine, Threonine and Tyrosine, and adding a complex of phosphate groups (see Section 12.3 for illustrations).

Phosphorylation can be inhibited by the presence of wrappers. Serine phosphorylates ten times more often than Tyrosine, even though the benzene ring presents the OH group further from the backbone.

Phosphorylation is expressed in PDB files by using a nonstandard amino acid code, e.g., SEP for phosphoserine (phosphorylated serine), PTR for phosphotyrosine (phosphorylated tyrosine) and TPO for phosphothreonine (phosphorylated threonine).

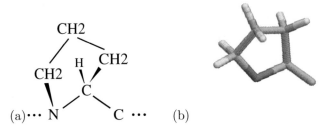

Fig. 4.6 Proline sidechain configuration: (a) cartoon and (b) configuration of Pro165 in the PDB file 1QM0 (the hydrogen attached to the C_α carbon is not shown.

4.4 Special Sidechains

There are many ways that sidechains can be classified, according to polarity, hydrophobicity, and so on. When all such designations are taken into account, each sidechain becomes essentially unique. Indeed, it is advisable to study more complete descriptions of the unique properties of individual sidechains [102]. But there are some special properties of sidechains that deserve special mention here for emphasis.

4.4.1 Glycine (and Alanine)

Glycine is special because it has essentially no sidechain. More precisely, it is the only amino acid without a C_β carbon. As a result, it is appropriate to think of Gly as polar, since the polarity of the backbone itself has a significant impact on the environment near the sidechain. In this regard, alanine can also be viewed to be somewhat polar. Alanine has a C_β carbon, but no other heavy atoms in its sidechain, a feature unique to Ala.

4.4.2 Proline

Proline is unique because it connects twice to the backbone, as depicted in Figure 4.6. This causes a torsional rigidity of the backbone not found with other residues. Proline has substantial impact on the stabilization of loops [14, 211, 213, 229, 332, 386]. Technically, it derives not from an amino acid but rather an imino acid [387]. The backbone nitrogen for proline has no attached hydrogen, due to the extra covalent bond from the sidechain. Thus the proline backbone cannot participate as a hydrogen bond donor. On the other hand, it can be a hydrogen bond acceptor.

There is an extended conformation of protein structures, bringing high exposure of the backbone to water, called PP2 (a.k.a. PPII or PII) which refers to the type of structure that a polyproline strand adopts [183, 455].

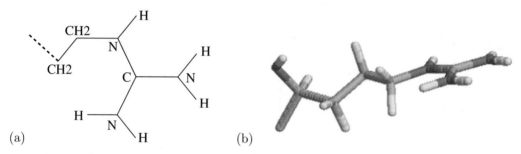

(a)   (b)

Fig. 4.7 The planar structure of the terminal atoms of the Arginine sidechain: (a) cartoon in the plane of the terminal groups, (b) a rotated view of Arg220 in PDB file 1QM0 to show the planar structure from the side.

4.4.3 Arginine

The uniqueness of arginine is highlighted by the fact that the end of its residue closely resembles guanidine (CN_3H_5) and the guanidinium cation (CN_3H_6), cf. Figure 4.8. Compounds of guanidinium (e.g., guanidinium hydrochloride and guanidinium thiocyanate [336, 439]) have the ability to **denature** proteins, that is, to cause them to unfold [127, 543]. Urea (CON_2H_4, a.k.a. carbamide) is also a protein denaturant [71, 433, 440] and resembles guanidine except that the NH group is replaced by an oxygen, cf. Figure 4.8. How the denaturing process occurs is not fully understood [351, 433, 439], although the similarity of urea to the peptide backbone is assumed to play a significant role by intercalating between amide and carbonyl groups on the backbone that would otherwise be making hydrogen bonds [71, 440]. Urea can both denature proteins and dissociate protein complexes [433].

One feature of the arginine residue is that the positive charge at the end of the residue is distributed quite broadly among the atoms at the end of the residue (see Table 8.5). How or why this might have a special effect is not completely explained.

It is very difficult to form natural water structures around the terminal (guanidinium) part of an arginine sidechain [336, 515]. The atoms beyond the C_δ carbon are all in a plane, cf. Figure 4.7. Model building shows that it is impossible for waters attached to the terminal hydrogens (those attached to the three nitrogens) to cohabitate, whereas the terminal N_3 group of Lysine is easily solvated, cf. Exercise 4.2. One can think of the planar structure of the terminal CN_3H_5 group as like a knife blade that cuts through water structures. Similarly, uric acid ($C_5H_4N_4O_3$, the cause of the disease gout) is a planar molecule that is not very soluble in water, despite its relation to urea, which is very soluble.

One property of arginine is that polyarginine is the polypeptide most able to cross a cell membrane without the help of a transporter molecule [347], and compounds rich in Arg have similar behavior [553].

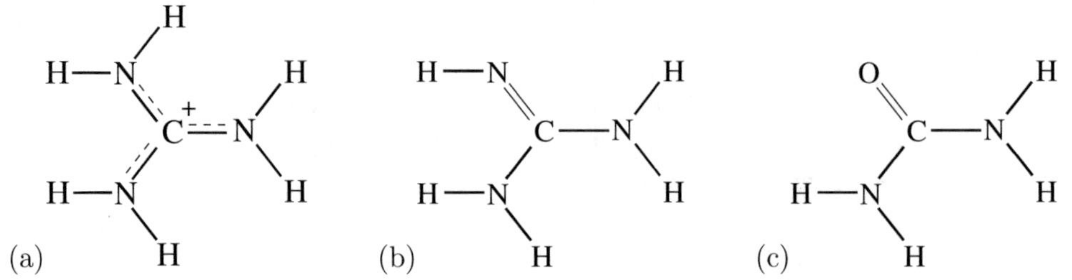

Fig. 4.8 The molecules (a) guanidinium (ion), (b) guanidine and (c) urea. The net positive charge for the guanidinium ion (a) is distributed around the three nitrogen groups, as is indicated by the dashed lines.

4.4.4 Lysine

By contrast, lysine easily solvates, and this feature leads to its ability to function as a sensitive measure of proximity to water [104]. On the other hand, Arginine has almost twice as many hydrogen bonding sites as Lysine, and this may contribute to the greater hydration energy [525, Table I].

4.4.5 Cysteine

What makes cysteine special is the ability of the sidechain to bond with another cysteine sidechain, making a disulfide bridge (Section 4.2.2) [47]. This is the only sidechain that forms a covalent bond with another sidechain.

4.4.6 Aromatic Sidechains

Three sidechains (Tyr, Trp, Phe) have benzene rings as a significant part of their structure. At first, these appear simply hydrophobic, but the electron structure of aromatic rings is complex [123]. There is a doughnut of positive charge located in the plane of the carbon and hydrogen atoms, and the hole of the doughnut contains regions of negative charge extending to both sides of the main positive ring (see Figure 2A of [123]). This makes these sidechains polar. Tyrosine is also polar in a more conventional way at the end of the sidechain due to the OH group there.

Tryptophan deserves special mention for various reasons, not just because of its pop-culture notoriety for sleep induction [362] and other behavioral impact [395]. It is the largest and most complex sidechain, involving two types of rings, the indole ring in addition to the benzene ring.[2] Tryptophan is also one of the least common sidechains, cf. Table 4.3.

4.4.7 Remembering Code Names

Many of the single letter codes for sidechains are obvious (alanine, glycine, histidine, isoleucine, leucine, methionine, proline, serine, threonine, valine), but others require some method to remember. We propose here some nonserious mnemonic devices that may aid in retaining their assignments.

Asp and Glu are the negatively charged residues, and the alphabetic order also corresponds with the size order (Asp is smaller than Glu). The code names are also alphabetical (D and E); the choice of E corresponds to the charge *e* of the extra electron.

Two of the positive sidechains also have special codes. To remember the R for arginine is to think of it as the pirate's favorite sidechain. To "lyse" means to destroy or disorganize, so we can think of lysine as the Killer sidechain.

The biggest sidechains (the aromatic ones) also have letter codes which need special treatment. A way to remember the single letter code for Phe is to misspell it with the Ph changed to F. A way to remember the single letter code for Trp is that it is the Widest sidechain. A way to remember the single letter code for Tyr is to focus on the second letter Y in the name. The sidechain also looks like an inverted Y on top of another Y.

The two remaining proteins are comparable to Asp and Glu, but with nitrogen groups replacing one of the oxygens: asparagiNe and Qlutamine. The emphasis on nitrogen is clear in Asn, since it is the third letter of the code. The letter G is one the most overloaded among first letters in the sidechain names, but Q is a close match for a poorly written G.

4.5 Sidechain Ionization

We will not consider extensively *pH* effects, although these clearly involve a type of modulation of electrical forces. There is significant *pH* variation in different parts of cells, and thus it has a potential role in affecting protein–ligand interactions.

The effects of *pH* are both localized and dynamic in nature, since the number of ions that can be involved in protein–ligand interactions is not large. For example, a well solvated large biomolecule [520] can be modeled dynamically with just over 10^5 atoms, and significantly less than 10^5 water molecules. But at *pH* 7, there is just one hydronium molecule per 5.5508×10^8 water molecules (cf. Section 18.7.1). The number of water molecules in the simulation in [520] used fewer than 55,508 water molecules, and thus would not have included a hydronium ion until the *pH* was less than three. On the other hand, ions cluster around proteins since they have charged and polar residues, so a more complex model is required to account for their effects.

The ionization state of some sidechains can vary depending on the *pH* value of the solvent [102]. The *pH* value (Section 18.7.1) at which the ionization state changes (pK_a) is outside of the range of biological interest in most cases, with the exception of His.

---

[2] In this regard tryptophan shares structure similar to the compound psilocybin which is known to fit into the same binding sites as the neurotransmitter serotonin.

We list the intrinsic pK_a values [102, 375, 487] in Table 4.2 for reference. This value is the pH at which half of the residues would be in each of the two protonation states. For example, for pH below 3.7, Asp would be more likely to be protonated, so that one of the terminal oxygens would instead be an OH group, as shown in Figure 4.9. In this case, it would be appropriate to refer to the residue as aspartic acid. Similarly, for pH below 4.2, Glu would more likely have an OH terminal group, as shown in Figure 4.9, and be called glutamic acid. By contrast, for pH above 8.5, a Cys residue would tend to lose the terminal hydrogen. Correspondingly, the other sidechains with $pK_a > 9$ in in Table 4.2 would also lose a hydrogen.

The relation between pH and pK_a is mathematically simple:

$$
\begin{aligned}
pK_a &= -\log_{10}\frac{[A^-][H^+]}{[AH]} = -\log_{10}\frac{[A^-]}{[AH]} - \log_{10}[H^+] \\
&= -\log_{10}\frac{[A^-]}{[AH]} + pH = \log_{10}\frac{[AH]}{[A^-]} + pH,
\end{aligned}
\tag{4.1}
$$

where $[X]$ denotes the equilibrium number of molecules X and AH denotes the protonated state of molecule A. Thus when $pH < pK_a$, the protonated state AH is more likely than the deprotonated state A^-. However, both states still coexist at all pK_a values; when $pK_a = pH - 1$, $[AH]$ is still one tenth of $[A^-]$. We should note that these statements are correct only for molecules A in pure water.

For simplicity, we refer to the residues in their form that is most common at physiological values of pH. The one sidechain that has a pK_a value near physiological pH (≈ 7) is His [49]. There are two ways in which His can become deprotonated, as indicated in Figure 4.9. The particular choice of site depends on factors similar to those that determine hydrogen location (cf. Section 6.3).

The concept of pK_a is not just a global value. The local pK_a [8, 106, 189, 203, 207, 251, 267, 311, 344, 375, 478, 486] for each sidechain in its own particular local environment can be measured, and these values can be biologically relevant in many cases. For example, the pK_a of Glu35 in hen and turkey lysozyme is just over 6 instead of the nominal value of 4.2 [40]. Thus, the protonation of the carboxyl sidechain of Glu can occur at biologically relevant values of pH in some cases. The particular oxygen that gets protonated will also be dependent on the local environment (cf. Section 6.3). The pK_a can also fluctuate in time, as local changes in chemical environment occur [340].

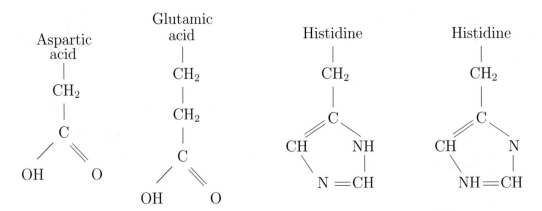

Fig. 4.9 Neutralized sidechains. Either oxygen on Asp and Glu could be protonated. There are two forms of His as shown with either nitrogen being protonated [399].

4.6 Amino Acid Frequencies

We now consider a simple question as an introduction to datamining. We will see that there are two major components to datamining:

- the choice of lens (or filter or metric)
- the choice of dataset.

Here we take a simple lens: we ask about the presence or absence of residues in PDB files. In Table 4.3, we list the frequencies of each residue, as a percentage. To measure this, we count the number of times each one appears in the set of PDB files, we divide this by the total number of residues counted, and then multiply by 100 to get a percentage.

To be precise, if we have a dataset with N different types of characteristics (e.g., $N = 20$ and the characteristics are the different amino acid types), then the **frequency** f_i of the i-th characteristic is defined by

$$f_i = \frac{o_i}{\sum_{j=1}^{N} o_j},\tag{4.2}$$

where o_j is the number of occurrences of the j-th characteristic in the dataset. The f_i's form a probability distribution since they sum to 1. In some cases, frequencies are represented as percentages, in which case we simply multiply by 100 in (4.2) to get $p_i = 100 f_i$. The sum of the p_i's is then 100.

We will extend this later in Section 7.4 when we look at some more sophisticated metrics for examining the likelihood of finding things with a given property.

4.6.1 Choice of Database

To make the exercise more interesting, we look at different datasets, and we see both similarities and differences that are significant. To start with, we have taken for simplicity the abundances published in [62], listed in column two in Table 4.3. We do not claim that the small set of proteins used in [62] provides the optimal reference to measure relative abundance in this setting, but it certainly is a plausible dataset to use. Moreover, the evolution of amino acid abundances studied in [62] is a significant issue in its own right.

The datasets used to generate the third through sixth columns in Table 4.3 were based on the 124,729 protein chains analyzed in the nonredundant PDB dated April 3, 2009 (nrpdb.040309.txt). PDB chains listed as cluster center (group rank 1), using the BLAST pvalue 10^{-7} to determine similarity, were included (there are 14,588 such PDB chains). BLAST is a widely used program to do sequence alignment, and in this case it is used to assure that the sequences used are sufficiently different. Of these, 9,141 chains with flaws in the sidechains (a nonzero value in columns G, H, I, or J in nrpdb.040309.txt) were eliminated, and PDB files having a resolution worse than 2 Ångstroms were also eliminated. Also eliminated were files having no useful temperature factors. This elimination process left a total of 1,989 representative chains. There were 409,730 total residues analyzed, and the corresponding frequencies are listed in column three in Table 4.3.

We will see in Section 5.2 that there are distinctly different parts of proteins based on their secondary structure, which can be described in three motifs: helices, sheets, and loops. The latter is defined to be the complement of the union of the former two classes. Of the total 409,730 residues in the nonredundant dataset, 162,565 are in helices, 91,665 are in sheets, and 155,481 are in loops (19 residues had conflicting structure information). The fourth, fifth, and sixth columns in Table 4.3 list the residue occurrence in secondary structure, helices and sheets or loops. Loops are by definition less structured regions of proteins. Indeed, the latter regions may not have any structure at all, and the corresponding protein sequences are likely underrepresented in the PDB because they frequently cannot form any stable structure, and thus are not imaged via the experimental techniques used to determine the data represented in the PDB [239].

We have a total of five datasets represented in Table 4.3, one for each column. However, the dataset corresponding to the third column is the union of the three datasets representing the final three columns. We would therefore expect that the values in the third column would lie between the min and the max of the three values in the fourth through sixth columns, and indeed they do in all cases. However, the values in the second column come from an unrelated dataset, and it is not surprising that the correlation is less simple. Nevertheless, the values in the second column lie between the values in the fourth through sixth columns 65% of the time, and four of the outliers are Cys, His, Met, and Trp, the four least frequently found sidechains. More importantly, the orderings for the two datasets are similar: both datasets indicate that Leucine is the most prevalent residue overall, and that Cys, His, Met, and Trp are the least common. However, we also see that the prevalence of Leu is based in helices and sheets; its prevalence in loops is about what you would expect by chance (5 percent).

Two other outliers of interest are Gly and Pro; comparing the last four columns, we see that these become dominant in loops, representing more than one-fifth of the residues in loops in the PDB. This suggests that four-fifths of all residues in loops are neighbors of Gly and Pro, if not one of them. Pro makes the structure very rigid nearby, and Gly enhances flexibility. Pro and Gly play the role of bookends for the rigidity–flexibility axis, and their prevalence suggests a concept of structure in loops.

We also see that some residues are more likely to be found in a helix or sheet: Ala and Glu are examples of the former, Ile and Val are examples of the latter.

Residue	[62]	All	Helix	Sheet	Loop
Ala	7.77	8.46	11.4	6.38	6.66
Arg	6.27	4.68	5.41	4.33	4.14
Asn	3.36	4.62	4.17	2.86	6.13
Asp	5.42	5.89	5.80	3.30	7.52
Cys	0.78	1.48	1.22	1.85	1.55
Gln	3.15	3.70	4.55	2.95	3.25
Glu	8.59	6.23	8.06	4.45	5.37
Gly	7.30	8.09	5.21	5.38	12.7
His	1.92	2.26	2.11	2.27	2.42
Ile	6.66	5.50	5.14	9.63	3.43
Leu	8.91	8.48	10.5	9.66	5.70
Lys	7.76	5.79	6.41	4.71	5.78
Met	2.41	1.97	2.27	1.90	1.70
Phe	3.61	4.01	3.91	5.70	3.11
Pro	4.35	4.64	2.66	1.91	8.32
Ser	4.66	6.15	5.80	5.10	7.14
Thr	4.87	5.79	4.79	6.94	6.15
Trp	1.02	1.56	1.58	2.01	1.26
Tyr	3.00	3.71	3.41	5.44	2.99
Val	8.17	6.99	5.66	13.2	4.70

Table 4.3 Amino acid residue frequencies (as percentages) in different datasets. The first column gives the residue three-letter code, and the second column is the frequency reported in [62]. The third through sixth columns are based on the nonredundant dataset described in the text. The third column is the frequency in that dataset for all residues, the fourth column is the frequency in helices, the fifth column is the frequency in sheets, and the sixth column is the frequency in loops.

It is interesting to compare the results presented here with an early study of the distribution of residues found in helices and sheets [518]. In that study, the most likely residues in helices were found to be E, A, L, M, Q, K, R, H (in order of likelihood), and those in sheets, V, I, Y, C, W, F, T. They also estimated the residues most likely in turns to be G, N, P, S, D. The concept of turn (essentially a short loop) has been discontinued in the PBD.

4.6.2 PDB Details

A typical line in a PDB file that we are interested in looks like

```
ATOM    1267  CH2 TRP A 173      -13.763  15.358  47.829  1.00 64.83           C
```

This has been extracted from the PDB file 2ACX, which provides the structural coordinates of G protein-coupled receptor kinase. The first entry indicates this is an atom in the protein, and the second specifies it uniquely as the 1267-th atom. The third and fourth entries identify it as $C_{\eta 2}$ in a tryptophan residue (see Figure 5.6(b)). The fifth and sixth entries identify the atom as being in residue 173 in the A chain. The next three entries are the three-dimensional coordinates of the atom. For now we will stop, except to say that the last column (last entry) confirms that the atom type is C (carbon).

Although the datamining task described here seems simple, dealing with PDB files introduces a serious challenge. The position of the entries in a line in a PDB file is specified precisely in terms of which columns they occupy. We will see that the first entry can include six characters, such as HETATM. The position of the final entry, the atom-type indicator is column 78. We can add the column numbers as follows to see where entries begin and end:

```
ATOM    1267  CH2 TRP A 173      -13.763  15.358  47.829  1.00 64.83           C
123456789 123456789 123456789 123456789 123456789 123456789 123456789 12345678
        1         2         3         4         5         6         7
```

The exact location in physical space of many atoms is not known, and **alternate locations** for them are sometimes listed in a PDB file. Their presence is indicated by the **altloc** indicator in the 17th column of a PDB file. Typical indicators are capital letters at the beginning of the alphabet. So it would be common to see ALYS and BLYS with the same residue sequence number. If it occurred with our example above, it would look like `ATOM 1267 CH2ATRP A 173` ... so parsing has to be done precisely by column. PDB files are not white space delimited. If such occurrences were not flagged, some residues might be overcounted.

It might seem that we could identify residues uniquely by just checking the residue sequence number: columns 23 through 26 in a PDB file. However, this is not sufficient, because there are other reasons that the residue sequence number might be the same for different entries. The **icode** or **insertion code** (column 27) allows for residues to be inserted in a sequence without altering some previously established numbering. Thus, we need to check both the altloc and icode columns to make the correct identification.

PDB files based on NMR data can contain many "models" of a protein complex. Each model is a complete representation of the complex, and there can be dozens of them. We can think of the different models as snapshots at different times. Since the models all represent the same protein(s), we have to consider only one of them in counting frequencies of residues and other features. Each model is preceded by a line in the PDB file of the form

```
MODEL         nm
```

where **nm** represents the model number (it could be just a single digit). In parsing a PDB file, you have to look for these indicators to go by to know which model you are looking at.

A PDB file often represents a complex of several proteins. The individual proteins are in unique **chains**, indicated by a single character in column 22. Often one wants to limit the analysis of frequency data to a specific chain in a PDB file. The **nonredundant PDB** (**NRPDB**) is a database of individual PDB chains that represent to varying degrees unique folds in the PDB. This database can be crucial in terms of avoiding bias in sampling properties of PDB files.

Finally, there is a small challenge in terms of reporting the frequencies of residues in secondary structure. A very small number of residues are ambiguously reported as being in both helix and sheet. This number is small enough that it can typically be ignored, but it can lead to confusing results if one expects the number of residues in helix, sheet, and loop regions to add up exactly.

4.6.3 Hetatoms

In addition to the standard amino acid residues discussed so far, PDB files include additional molecules referred to as **hetatoms**. The most common hetatom is water, with three-letter code HOH. Some hetatoms appear in the same way as regular amino acid residues, such as PTR. This is a version of Tyrosine modified by phosphorylation, the covalent bonding of a phosphate group (PO_4) to the end of the sidechain of Tyrosine; some of these novel residues are listed in Table 4.4. Other hetatoms, like HOH, stand apart, not covalently bonded to other parts of the PDB structure elements; some of these are listed in Table 4.5.

PDB code	sidechain name	analog/comment/formula
ACE	acetyl group	truncated (initial) residue
AIB	alpha-aminoisobutyric acid	truncated Val (C_β removed)
DAL	D-Alanine	D conformation of Ala
DBU	(2e)-2-aminobut-2-enoic acid	$C_4H_7NO_2$
HYP	hydroxyproline	OH group attached to C_γ of Pro
MSE	Selenomethionine	Methionine (S replaced by Se)
PCA	pyroglutamic acid	$C_5H_7NO_3$
PHL	L-Phenylalaninol	Phenylalanine [456]
PTR	phosphotyrosine	phosphorylated tyrosine
SEP	phosphoserine	phosphorylated serine
TPO	phosphothreonine	phosphorylated threonine

Table 4.4 Some common heteroatoms in PDB files which appear as residues in peptide sequences.

PDB code	chemical formula	common name
CA	Ca^+	calcium ion
CL	Cl^-	chloride ion
CCN	C_2H_3N	acetonitrile
GAL	$C_6H_{12}O_6$	beta-d-galactose
HOH	H_2O	water
MOH	CH_4O	methanol
NAG	$C_8H_{15}NO_6$	n-acetyl-d-glucosamine
NGA	$C_8H_{15}NO_6$	N-acetyl-D-galactosamine

Table 4.5 Some common heteroatoms in PDB files which appear as independent molecules.

4.7 Exercises

Exercise 4.1. Draw all the atoms in the tri-peptide GAG, including the C-terminal and N-terminal ends.

Exercise 4.2. Using a model set, build the terminal atoms for Lys and Arg together with some water molecules bound to them. Use this to explain why it is easy to solvate Lys and hard to solvate Arg.

Exercise 4.3. What is the typical distance between the two oxygens at the end of the sidechains of Asp (OD1 and OD2) and Glu (OE1 and OE2)? (Hint: take a small set of PDB files and compute the distribution of distances for all occurrences in those files.)

Exercise 4.4. What is the typical distance between the two heavy atoms at the end of the sidechains of Asn (OD1 and ND2) and Gln (OE1 and NE2)? (Hint: take a small set of PDB files and compute the distribution of distances for all occurrences in those files.)

Exercise 4.5. Three points determine a plane, but there are four atoms in some sidechains that are supposed to lie in a plane (e.g., Asp and Glu). Determine the plane that best fits four points in three dimensions by least squares. Test this on Asp and Glu in several PDB files. How big is the deviation of the four points from the plane (compute the squared deviation)?

Exercise 4.6. Another way to determine how planar four points are in three-dimensional space is to compute the volume of the tetrahedra spanned by these four atoms (as vertices). Test this on Asp and Glu in several PDB files, and contrast this with the corresponding data for Leu (CB, CG, CD1, CD2).

Exercise 4.7. Yet another way to determine how planar four points are in three-dimensional space is to compare angles. Suppose that the four points are $x_0, \ldots, x_3$, and suppose that x_0 is a central point, such as CG in Leu or Asp. Let θ_i be the angle (in radians) formed between x_i and x_0, $i = 1, 2, 3$. If the four points are planar, then $\sum_{i=1}^{3} \theta_i = 2\pi$, and $a = |\sum_{i=1}^{3} \theta_i - 2\pi|$ thus forms a measure of planarity. Test this on Asp and Glu in several PDB files (take x_0 to be the location of CD for Glu), and contrast this with the corresponding data for Leu (CB, CG, CD1, CD2).

Exercise 4.8. Determine how planar the terminal four atom positions (Cx, Cy, Oz1, Oz2) are in Asp (x,y,z=B,G,D) and Glu (x,y,z=G,D,E) in several PDB files. To determine planarity, use the techniques of Exercises 4.5, 4.6 and 4.7. How do the distributions of planarity measure differ? Contrast this with the corresponding data for Leu (CB, CG, CD1, CD2).

Exercise 4.9. Scan a large set of PDB files to see which ones have the largest number of Cys residues. Determine which ones are in a disulfide bond based on proximity of the terminal sulfurs. Plot the distribution of PDB files as a function of numbers of disulfide bonds and the distribution of PDB files as a function of numbers of isolated Cys residues. Redo the calculations plotting instead using numbers of disulfide bonds (respectively, numbers of isolated Cys residues) per hundred residues in each PDB file. What are the proteins (and their function) in the PDB files with the top ten numbers of disulfide bonds (respectively, numbers of isolated Cys residues) per hundred residues in each PDB file? Compare [163].

Exercise 4.10. Determine how long each sidechain is by scanning the PDB. That is, determine the distribution of distances from the C_α carbon to the terminal (heavy) atoms for each residue (amino acid) type (thus ignore Gly). Is the furthest atom always the same for each sidechain? If not, give the distribution for the different atoms. Is the atom furthest from C_α always in the sidechain (hint: check the small residues carefully).

Exercise 4.11. In Figure 4.5, the resonant bond linking O-C-N has been omitted. Re-draw this figure including the rest of the bonds in the peptide unit as done in Figure 4.1.

Exercise 4.12. Repeat the experiment reported in Table 4.3 with a different set of proteins. Pick a set of proteins according to some rule (e.g., all tyrosine kinases) and count the frequencies of the different amino acid residues. Compare with Table 4.3.

Exercise 4.13. Write a code to count the number of residues in a PDB file both by ignoring the altloc indicator and by taking it into account. Test the code on PDB file 3EO6. Verify by another technique that your count is correct.

Exercise 4.14. Write a code to count the number of residues in a PDB file both by ignoring the icode indicator and by taking it into account. Test the code on PDB file 1BIO. Verify by another technique that your count is correct.

Exercise 4.15. Compare the frequency ordering of residues given by columns 2 and 3 in Table 4.3. Determine the extent to which you can cluster groups of residues that are ordered in the same way by both columns. For example, Leu is the most frequent by both columns, and Cys, His, Met, and Trp are the least common according to both. What group of residues is second-most common according to both columns (and so forth)?

Chapter 5
Protein Structure

Rosalind Franklin (1920–1958), according to Francis Crick, provided "the data which really helped us to obtain the structure" of DNA [542]. Her life and work are commemorated by the Rosalind Franklin University in Chicago.

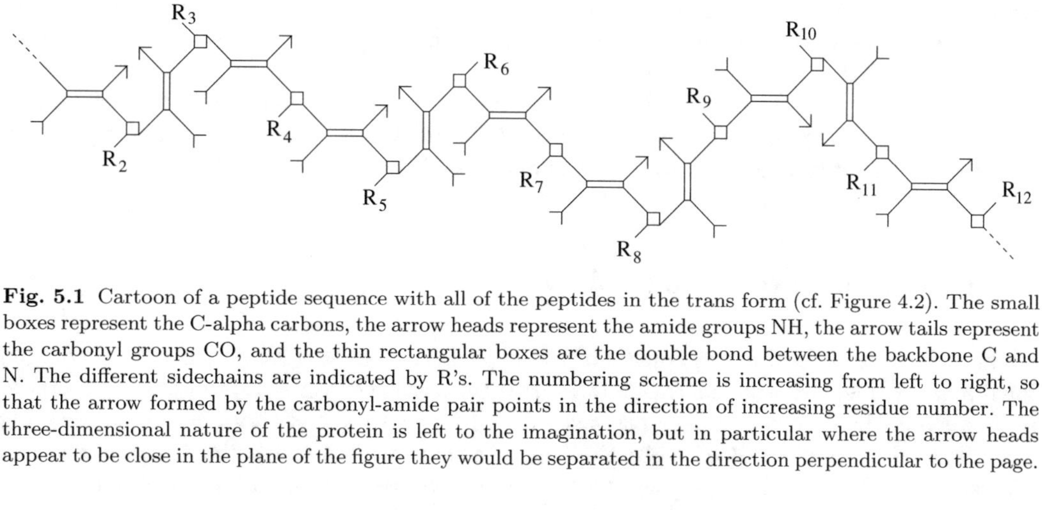

Fig. 5.1 Cartoon of a peptide sequence with all of the peptides in the trans form (cf. Figure 4.2). The small boxes represent the C-alpha carbons, the arrow heads represent the amide groups NH, the arrow tails represent the carbonyl groups CO, and the thin rectangular boxes are the double bond between the backbone C and N. The different sidechains are indicated by R's. The numbering scheme is increasing from left to right, so that the arrow formed by the carbonyl-amide pair points in the direction of increasing residue number. The three-dimensional nature of the protein is left to the imagination, but in particular where the arrow heads appear to be close in the plane of the figure they would be separated in the direction perpendicular to the page.

We now review the basic ideas about protein structure. Much of this chapter is routine, so we review the main ideas quickly, but we also include some novel concepts and points of view.

5.1 Chemical Formula and Primary Structure

An example of a five-residue peptide sequence is given in Figure 2.1. Such an all-atom representation is too busy in most cases, so it is useful to look for a simpler representation. A cartoon of a peptide sequence is depicted in Figure 5.1. Here we retain only certain features,

© Springer International Publishing AG 2017
L.R. Scott and A. Fernández, *A Mathematical Approach to Protein Biophysics*,
Biological and Medical Physics, Biomedical Engineering,
DOI 10.1007/978-3-319-66032-5_5

such as C_α carbon and attached residue, and both the dipole (cf. Figure 3.4) and the double bond of the peptide backbone. This representation is useful in talking about the types of hydrogen bonds that are formed in proteins, as we depict in Figure 5.3. However, in many cases, only the sequence of residues is significant. The representation of proteins as a linear sequence of amino acid residues is called the **primary structure**. More precisely, we can represent it as a string of characters drawn from the twenty-character set A, C, D, ..., W, Y in column three in Table 4.1. For example, the structure of the six-residue DRYYRE [180] is discussed in Section 10.5.3.

The primary structure representation of proteins is fundamental, but it does not directly explain the function of proteins. Essentially all proteins function only in a three-dimensional structure. That structure can be described in a hierarchical fashion based on structural subunits, as we now explain.

5.2 Hydrogen Bonds and Secondary Structure

Proteins can be described using a hierarchy of structure. The next type after the primary, linear structure is called **secondary structure**. Many components of secondary structure have been identified, but the main ones may be categorized by two primary types: alpha helices and beta sheets (a.k.a., α-helices and β-sheets). These form the basic units of secondary structure, and they can be identified in part by the pattern of hydrogen bonds they make, as depicted in Figure 5.3. These units combine to form "domains" or "folds" that are characteristic structural patterns that can be viewed as widgets used to build protein structure, and presumably govern its function. Structural relationships among these widgets form interesting networks, as will be described in Section 5.7.

5.2.1 Secondary Structure Units

Alpha helices are helical arrangements of the peptide complexes with a distinctive hydrogen bond arrangement between the amide (NH) and carbonyl (OC) groups in peptides separated by k steps in the sequence, where primarily $k = 4$ but with $k = 3$ and $k = 5$ also occurring less frequently. An example of a protein fragment that forms a rather long helix is given in Figure 5.2(c). What is shown is just the "backbone" representation: a piecewise linear curve in three dimensions with the vertices at the C_α carbons. The hydrogen bond arrangement in a helix is depicted in Figure 5.3(a) between two such peptide groups. The helices can be either left-handed or right-handed, but protein structures are dominated by right-handed helices [260, 368, 467].

Beta sheets represent different hydrogen bond arrangements, as depicted in Figure 5.3: (b) is the antiparallel arrangement and (c) is the parallel. Both structures are essentially flat, in contrast to the helical structure in (a).

Both alpha helices and beta sheets can be distinguished based on the angles formed between the protein backbone units, as described in Section 5.3. It is also possible to find distinctive patterns in the distribution of residues found in helices and sheets as shown in Table 4.3.

Other structural units include turns and loops (or coils). The former involve short (three to four) peptide units, whereas the latter can be arbitrarily long. The PDB classification eliminated "turn" as a special classification in 2009. The term **loop** is now taken to mean the complement of α-helix and β-sheet regions in protein sequence.

There is a characteristic alternation of hydrophobic and hydrophilic sidechains in helices and sheets [130]. Not surprisingly, the frequency of alternation is approximately two in beta sheets, so that one side of the sheet tends to be hydrophobic and the other side hydrophilic. In alpha helices, the period closely matches the number k of residues between the donor and acceptor of the mainchain hydrogen bonds.

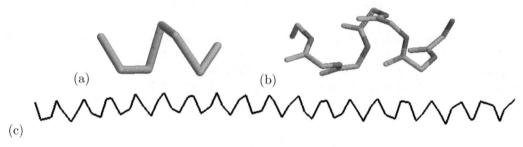

Fig. 5.2 Two helices (one short, one long) in chain B in PDB file 1HTM, which depicts the Influenza Hæmagglutinin HA2 chain. (a) and (b) represent the helix formed by the seven residues B-Asn-146 through B-Ile-152: (a) shows only the backbone and (b) indicates all of the backbone heavy atoms, showing in particular the direction of the carbonyl (C-O) orientation, pointing toward the four nitrogen residues ahead in the sequence. (c) is the backbone representation of the helix formed by the sixty-six residues B-Ser-40 through B-Gln-105.

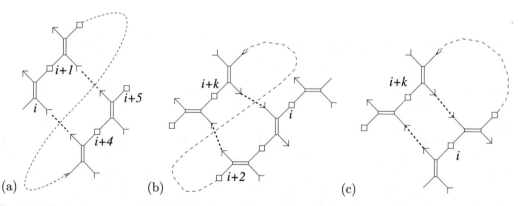

Fig. 5.3 The hydrogen bond (dashed line) configuration in (a) α-helix, (b) parallel β-sheet, and (c) antiparallel β-sheet. The dotted lines indicate how the backbone is connected. The amide (N-H) groups are depicted by arrow heads and the carbonyl (O-C) groups are depicted by arrow tails, thus indicating the dipole of the backbone.

5.2.2 Folds/Domains

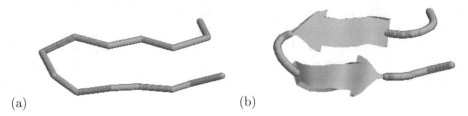

Fig. 5.4 Two views of the peptide in PDB file 1K43 which forms an antiparallel β-sheet: (a) backbone, (b) ribbon cartoon.

A **fold** or **domain** is a collection of basic structural units, as defined in Section 5.2.1, together with topological information on relations among the basic units [523] (Figure 5.4). The Structural Classification of Proteins (or SCOP) database [9] provides a classification of these folds [523]. A hypothetical example is depicted in Figure 5.5.

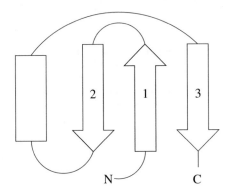

Fig. 5.5 Example of a simple fold with three beta sheets and one alpha helix (in the primary sequence between sheets 2 and 3). The letters N and C indicate the N-terminal and C-terminal ends. Proximity of the edges of the beta sheets indicates hydrogen bonding.

The topological representation of folds can be viewed as a type of language based on the basic units (α and β) as characters. This set of characters can be extended to include other units (e.g., turns). This linguistic approach to structure has been used as the basis of approaches to secondary structure prediction [16, 72, 129]. The I-sites library of a dozen or so structural units has been used [72] to facilitate the prediction. Thus, we might think of folds (domains) as representing words in a linguistic representation of structure. With this view, proteins represent phrases in the language with particular significance. A single protein can consist of a single fold, or it may be made up of several different folds.

SCOP release 1.73 (November 2007) divided proteins into seven classes of protein structure groupings, together with four additional groupings of special cases. These are listed in Table 5.1, which shows the number of folds for each class. Not counting the last four classes, there are 1086 folds represented (cf. [524]). Thus, we can think of the set of known folds as a small dictionary of the words formed in the language based on the characters of secondary structural elements.

Combinations of folds interact [315] to form a variety of structures; the shape that they adopt is called the **tertiary structure**. The combination of several proteins in a unified (functional) structure is called a **protein complex**. The shape that a protein complex adopts is called its **quaternary structure**.

Name of fold classes	number of folds in class
All alpha-helix proteins	258
All beta-sheet proteins	165
Alpha and beta proteins a/b	141
Alpha and beta proteins a+b	334
Multidomain proteins alpha and beta	53
Membrane and cell surface proteins and peptides	50
Small proteins	85
Coiled coil proteins	7
Low-resolution protein structures	26
Peptides	120
Designed proteins	44

Table 5.1 The major classes in SCOP release 1.73. The a/b class consists of mainly parallel beta sheets, whereas the a+b class consists of mainly antiparallel beta sheets. The last four lines of the table are not considered "true classes" of protein folds.

5.3 Mechanical Properties of Proteins

The Protein Data Bank (PDB) supports a simple mechanical view of proteins. The positions of the backbone and sidechain atoms are specified, together with the positions of some observed water molecules and other atoms. This basic information allows the derivation of extensive additional information, as we will explain subsequently. But for the moment, we simply recall some information on the static description of proteins.

5.3.1 Conformational Geometry of Proteins

We recall the basic ingredients of the peptide group from Figure 4.2. If x is a given residue, then $N(x)$, $H(x)$, $C(x)$, and $O(x)$ denote the position vectors of the corresponding atoms in the peptide group. For the remaining atoms, the standard notation from the PDB is as follows:

$$C_\alpha(x), C_\beta(x), C_\gamma(x), C_\delta(x), C_\epsilon(x), C_\zeta(x), C_\eta(x)$$

are the $\alpha, \beta, \gamma, \delta, \epsilon, \zeta, \eta$ carbons (denoted in plain text in the PDB by CA, CB, CG, CD, CE, CZ, CH) in the sidechain structure of residue x. Most of these can also appear with subscripts, e.g., C_{γ^i} for $i = 1, 2$ in Ile and Val. Correspondingly, $N_{\delta^i}(x), N_{\epsilon^i}(x), N_{\eta^i}(x)$ are the i-th δ, ϵ, η nitrogens, denoted in plain text in the PDB by NDi, NEi, NHi for $i = 1, 2$. Notation for oxygens is similar. Unfortunately, the plain text descriptor OH for O_η in Tyr is a bit confusing, since this oxygen has an attached hydrogen. The PDB descriptors for histidine and tryptophan are depicted in Figure 5.6.

We can view $C_\alpha(x), N_{\delta^i}(x)$, etc., as three-dimensional vectors, using the corresponding coordinates from the PDB. For amino acids x_i, x_{i+1} which are adjacent in the protein sequence, the *backbone vector* is defined as

$$\mathcal{B} = C_\alpha(x_{i+1}) - C_\alpha(x_i). \tag{5.1}$$

The *sidechain vector* $\mathcal{S}(x)$ for a given amino acid x, defined by

$$\mathcal{S}(x) = C_\beta(x) - C_\alpha(x), \tag{5.2}$$

will be used to measure sidechain orientation. $\mathcal{S}$ involves the direction of only the initial segment in the sidechain, but we will see that it is a significant indicator of sidechain conformation. For $x = Gly$, we can substitute the location of the sole hydrogen atom in the residue in place of C_β. For each neighboring residue pair x_i, x_{i+1}, the sidechain angle $\theta(x_i, x_{i+1})$ is defined by

$$\cos \theta(x_i, x_{i+1}) = \frac{\mathcal{S}(x_i) \cdot \mathcal{S}(x_{i+1})}{|\mathcal{S}(x_i)||\mathcal{S}(x_{i+1})|}, \tag{5.3}$$

where $\mathcal{B}$ is defined in (5.1), and $A \cdot B$ denotes the vector dot product.

It is not common to characterize the secondary structures (helix and sheet) by θ, but θ is strongly correlated with secondary structure, as shown in Figure 5.7, and it gives a simple interpretation. Values $70 \le \theta \le 120$ are typical of α-helices, since each subsequent residue turns about 90 degrees in order to achieve a complete (360 degree) turn in four steps (or 72 degrees for five steps, or 120 degrees for three steps). Similarly, $140 \le \theta \le 180$ is typical of β-sheets, so that the sidechains are parallel but alternate in direction, with one exception. Some β-sheets have occasional "spacers" in which θ is small, in keeping with the planar nature of sheets.

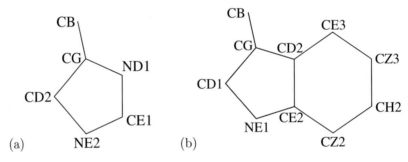

Fig. 5.6 PDB notation for (a) histidine and (b) tryptophan.

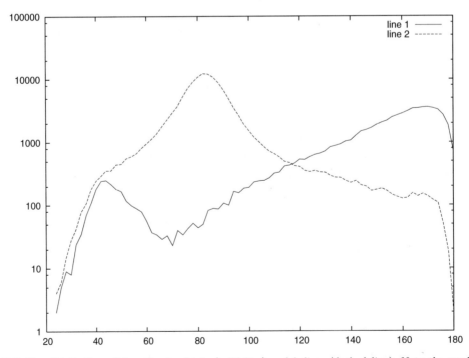

Fig. 5.7 The distribution of θ angles in sheets (solid line) and helices (dashed line). Note the small but significant distribution for sheets in the region $35 \le \theta \le 60$.

The distribution of the θ angle peaks roughly at 44, 82, and 167 degrees as shown in Figure 5.7. The peptide bond makes it difficult for θ to be much less than 50 degrees, thus the smaller peak corresponds to a motif where the sidechains align as closely as possible. A small number of these occur in beta sheets, but the majority of them may constitute an independent motif whose properties deserve further study.

5.3.2 Dihedral Angles

We also recall the standard mainchain **dihedral** or **torsion** angles. Given a sequence of four points A, B, C, D in $\mathbb{R}^3$, the **dihedral angle** represents the angle between the planes spanned by subsequent triples of points, namely between the planes spanned by A, B, C and by B, C, D, as depicted in Figure 5.8. For two intersecting planes, there are in general two angles between them. Let μ denote the smaller (positive) angle, in radians; the other angle is $\pi - \mu$ radians. (Of course, both angles could be $\pi/2$.) Choosing which angle to call the dihedral angle is a convention, although it is possible to define it in a consistent way so that it depends continuously on the points A, B, C, D. The dihedral angle can be determined by the orientation of the normal vectors, with the normal determined in a consistent way. In Figure 5.8, the normal vectors are determined using the left-hand rule.

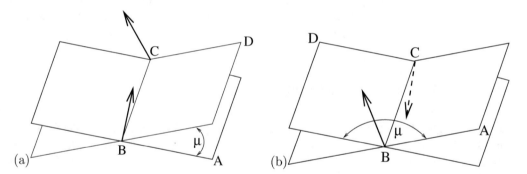

Fig. 5.8 The geometry of the dihedral angle $[A, B, C, D]$ in two different configurations: (a) cis form [45], (b) trans form. The normals are defined using the left-hand rule.

The angle can be defined in terms of the angle between the normal vectors for the two planes. The normal to the plane spanned by A, B, C is proportional to $n_1 = (A-B) \times (C-B)$, where $\times$ denotes the vector "cross" product in $\mathbb{R}^3$. Similarly, the normal to the plane spanned by B, C, D is proportional to $n_2 = (B-C) \times (D-C)$. The angle μ between these two normals is given by

$$\cos \mu = \frac{n_1 \cdot n_2}{|n_1| \, |n_2|}, \tag{5.4}$$

where $|n_i|$ denotes the Euclidean length of n_i. Let $[A, B, C, D] = \mu$ denote the dihedral (or torsion) angle between the planes defined by the points A, B, C and B, C, D. The dihedral angle can be viewed as a rotation about the line $\overline{BC}$ at the intersection of the two planes.

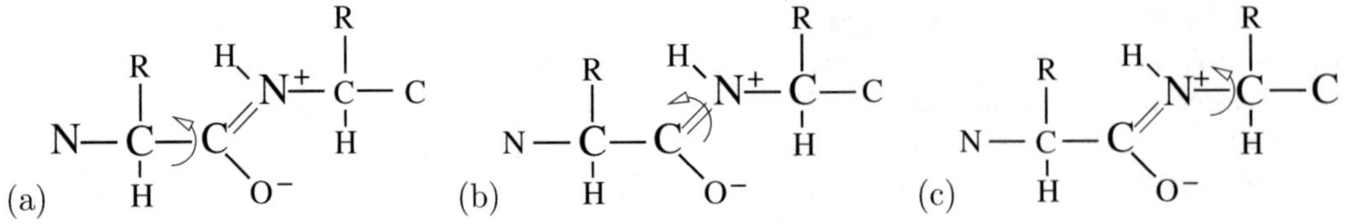

Fig. 5.9 The primary dihedral angles: (a) ψ, (b) ω, (c) ϕ. The four points used in each definition are in a larger font. The arrows indicate the axes of rotation.

Then the ψ, ω, and ϕ angles (Figure 5.9) are defined by

$$\psi(x_i) = [N(x_i), C_\alpha(x_i), C(x_i), N(x_{i+1})]$$
$$\omega(x_{i+1}) = [C_\alpha(x_i), C(x_i), N(x_{i+1}), C_\alpha(x_{i+1})] \qquad (5.5)$$
$$\phi(x_{i+1}) = [C(x_i), N(x_{i+1}), C_\alpha(x_{i+1}), C(x_{i+1})].$$

It might appear that there is a possible degeneracy, if three of the consecutive points are collinear. Since this cannot happen for the three angles in (5.5), we do not worry about this case. The reason that the positions of N-C-C are not collinear is the tetrahedral structure depicted in Figure 3.1.

In Chapter 14, we study the effect of a polar environment on the flexibility of peptide bond rotation in the ω angle. The typical trans form of the peptide bond as depicted in Figure 4.2(a) is associated with $\omega = \pi$ radians. The cis form of the peptide bond as depicted in Figure 4.2(b) is associated with $\omega = 0$ radians.

5.3.3 ϕ, ψ Versus ψ, ϕ: The Role of θ

The pair of angles ϕ_i, ψ_i captures the rotation of the peptide chain around the i-th C_α carbon atom. The θ angle measures the rotation that corresponds with comparing angles ψ_i, ϕ_{i+1} in successive peptides (cf. Exercise 5.9). This correlation has recently been observed to have significant predictive power [185].

The conformations of ϕ_i, ψ_i are typical of different secondary structures, such as α-helix or β-sheet. The Ramachandran plot [237] depicts the distributions of angles that are commonly adopted (cf. Exercise 5.12). Conformational ϕ, ψ space is a torus, but only restricted regions are visited by the residues in proteins, even in loops, due to steric hindrance [237].

5.3.4 Sidechain Rotamers

The sidechains are not rigid, so the geometric description of a sidechain requires more information than ϕ, ψ and so forth. As we use the ϕ, ψ angles to define the positions of the backbone relative to the position of the N-terminal end, we can also use dihedral angles to define the positions of the sidechains. For example, we can define

$$\chi_1 = [N, C_\alpha, C_\beta, X_\gamma], \quad \chi_2 = [C_\alpha, C_\beta, X_\gamma, Y_\delta], \quad \text{and so forth,} \qquad (5.6)$$

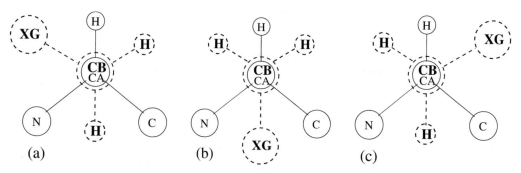

Fig. 5.10 The primary sidechain rotamer conformations (a) gauche+, (b) gauche-, and (c) trans, corresponding to χ_1 values of (a) -60 degrees, (b) +60 degrees, and (c) 180 degrees. The view is oriented so that the C_α and C_β atoms are aligned perpendicular to the plane of the page. The closer (and therefore larger) atoms are indicated with dashed lines and bold letters. The hydrogens for C_β are indicated. The atom marked "XG" corresponds to either a C_γ or an O_γ atom.

for all sidechains that have a C_β atom attached to an additional "heavy" atom (not Hydrogen). This excludes glycine and alanine, but includes all the other sidechains. For valine, there are two candidates for the X_γ (CGi, $i = 1, 2$) atom, but the angle is (or should be, according to standard models) the same for either choice due to the planar structure of terminal groups. For other sidechains, there are similar ambiguities in the definition of γ_2.

Libraries of angular orientations of the different segments have been developed [323, 401]. The possible orientations are not uniformly distributed in many cases, but rather show a strong bias for a few discrete orientations. For example, carbon chains typically orient so that the hydrogen atoms are in complementary positions. In Figure 5.10, the three primary conformations are shown for sidechains with C_β and C_γ constituents. These conformations are known as **gauche+**, **gauche-**, and **trans**, corresponding to mean χ_1 values of -60 degrees, +60 degrees, and 180 degrees, respectively.

However, the distributions can change depending on local neighbor context [317].

5.4 Ramachandran Plot

The **Ramachandran plot** defines all *a priori* possible conformations of a single residue in a protein chain without making any assumption on the structure in which the residue may appear. These conformations are described in terms of the (φ^i, ψ^i) torsional angles for the i-th residue. Glycine and proline are outliers since they are the extreme points in terms of conformational freedom, so these have their own special Ramachandran plots. But the other 18 residues have a common Ramachandran plot.

Not surprisingly, the observed (φ, ψ) values for the most common structural motifs fall well within the allowed regions of the Ramachandran plot. The three connected components [333, Figure 10, page 4627] of the allowed region contain angles that are observed for α-helices, β-sheets, and left-handed helices, respectively. The program PROCHECK [296] is used to analyze the (φ, ψ) values for a given PDB file and is used to validate PDB structures.

Ideal protein backbone helical structures have the same values of dihedral angles φ and ψ from unit to unit and thus can be uniquely determined by a pair (φ, ψ). The two main types of protein helical structure, α- and 3_{10}-helices, are stabilized by hydrogen bonding patterns involving residues $i \to i + 4$ and $i \to i + 3$, respectively, and the dihedral angle pairs $(-63°, -43°)$ and $(-60°, -25°)$ characterize the most common helices within each group, respectively.

Sasisekharan was the first to study steric constraints, published in 1962 [436]. Further developments, done with his graduate adviser G. N. Ramachandran, appeared in 1965 [411]. A description of the derivation of the Sasisekharan–Ramachandran plot is given in [333, Figure 10, page 4627]. One conclusion is that the location of the preferred (φ, ψ) angles do *not* lie on the boundary of the allowed region. Thus, the typically observed (φ, ψ) values are not determined by steric hindrance alone.

Sasisekharan treated atoms as simple impenetrable spheres and found three major *normally allowed* (φ, ψ) regions for standard radii of peptide atoms and *outer limit* regions for the smallest radii that can be still considered plausible. The φ, ψ plot, with allowed regions marked, received the name of Ramachandran plot. The outer limit region is important because the 3_{10}-helices belong to it and not to the normally allowed region, as shown in [333, Figure 9, page 4627].

The original map was derived before any protein structures had been solved, but it has been tested on known secondary structures such as α-helix and β-sheet and on small peptides (see [410, Figure 6, page 54]).

The question of what exactly determines the observed dihedral angle pairs $(-63°, -43°)$ and $(-60°, -25°)$ is of significant interest [373]. Two hypotheses are examined in [373], namely that the observed angles are determined by

- minimizing the hydrogen bond length, the distance between hydrogen and oxygen, and
- aligning the hydrogen bond angle made between the N-H vector and the C-O vector.

Surprisingly, neither of these objectives, or any combination of them, appear to characterize the observed angles. By contrast, it is found that the observed dihedral angle pairs correlate with maximizing hydration of the backbone [373].

5.5 Pair Frequencies in Proteins

We expand upon Section 4.6 which looked at amino acid residue frequencies in different secondary structures. Instead we now ask what *pairs* of residues are found frequently in different structural motifs. More precisely, we focus on pairs found in sequence, that is, (R_i, R_{i+1}), where R_i denotes the residue type (Asn, Asp, etc.). Thus, we are looking for any possible effects of neighbors, either attraction or exclusion. For example, we might think it unlikely for sequence–neighbor residues would have like charge, due to the resulting repulsion, or that opposite-charged residues would be likely sequence neighbors. This requires us to develop some technology to examine such issues precisely. It also introduces some complexities in terms of analyzing PDB files.

5.5.1 Pair Frequency Definitions

In addition to looking at the frequencies of individual residues, one can also look at the frequencies of pairings. A standard tool for doing this is the **odds ratio**. Suppose that f_i is the frequency of the i-th amino acid in some dataset, and suppose that C_{ij} is the frequency of the pairing of the i-th amino acid with the j-th amino acid. Then the odds ratio O_{ij} is defined as

$$O_{ij} = \frac{C_{ij}}{f_i f_j} \tag{5.7}$$

and has the following simple interpretation. If the pairing of the i-th amino acid with the j-th amino acid was random and uncorrelated, then we would have $C_{ij} = f_i f_j$, and thus $O_{ij} = 1$. Therefore, an odds ratio bigger than 1 implies that the pairing is more common than would be expected for a random, uncorrelated pairing, and conversely, less likely if it is less than 1.

Let us break down the process in more detail. Suppose that the pairs i, j are observed in an amount n_{ij}. Then by the standard definition of frequency (4.2), the pair frequency is

$$C_{ij} = \frac{n_{ij}}{N}, \qquad \text{where} \quad N = \sum_{i,j} n_{ij}. \tag{5.8}$$

If the pairs are not ordered, then the frequency of occurrence of residue i in a pair would be given by

$$f_i = \sum_j C_{ij}, \tag{5.9}$$

by the following reasoning. The quantify

$$o_i = \sum_j n_{ij}, \tag{5.10}$$

is the number of occurrences of the i-th residue in any of the pairs. Thus by (4.2),

$$f_i = \frac{o_i}{\sum_j o_j} = \frac{\sum_j n_{ij}}{N} = \sum_j C_{ij}, \tag{5.11}$$

proving (5.9). Therefore

$$O_{ij} = \frac{n_{ij}}{N f_i f_j} = N \frac{n_{ij}}{\left(\sum_k n_{ik}\right)\left(\sum_k n_{jk}\right)}, \tag{5.12}$$

providing a way to compute the odds ratio directly from the occurrences n_{ij}.

The **log odds ratio** is often defined by simply taking the logarithm of the odds ratio. This has the benefit of making the more likely pairings positive and the less likely pairings negative. Moreover, since the odds ratio is a multiplicative quantity, the logarithm linearizes it so that the quantities less than 1 are not compressed compared with the ones greater than 1. In [208], a quantity G_{ij} is defined by multiplying the log odds ratio by a numerical factor of 10.

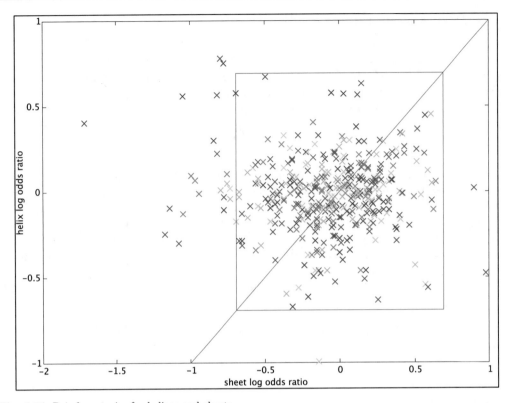

Fig. 5.11 Pair frequencies for helices and sheets.

In the case of counting residue pairs in secondary structure in proteins, a further complication occurs: The order of the pair matters. That is, the frequency of the pair where A occurs just before C can be different from the frequency of the pair where C comes before A. This also means that the simple question of how often A appears becomes more complex. Do we mean how often in appears in a pair AX or in a pair XA? Thus, we can define C_{ij} to be the frequency of the pair where the amino acid i is followed by j. Then

$$f_i^b = \sum_j C_{ij} \tag{5.13}$$

is the frequency of occurrence of residue i before another residue in a sequence, and

$$f_j^a = \sum_i C_{ij} \tag{5.14}$$

is the frequency of occurrence of residue i after another residue in a sequence. Thus, the correct way to write the odds ratio for the occurrence of i before j is

$$O_{ij} = \frac{C_{ij}}{f_i^b f_j^a} = N \frac{n_{ij}}{\left(\sum_k n_{ik}\right)\left(\sum_k n_{kj}\right)}. \tag{5.15}$$

If the matrix of occurrences n_{ij} was symmetric, then this distinction would not be necessary.

5.5.2 Application to Protein Sequences

We now apply these ideas to residue pairs in different secondary structure. We saw in Section 4.6 that there are different frequencies for residues in different secondary structure, and we might expect the pair frequencies to have an even stronger differentiating behavior. In Table 4.3, we see that individual frequencies can vary by almost an order of magnitude. However, we will see that pair frequencies are more complicated to interpret.

There are several residue pairs that have extreme probabilities of occurrence in helices versus sheets, as indicated in Table 5.2 and Figure 5.11. The range of odds ratios in Table 5.2 is an order of magnitude, as we found for residue frequencies in Table 4.3. But the extreme odds ratios in Table 5.2 are dominated by pairs with Cys, which is itself very rare. Thus, high values of the odds ratio occur in part because of the small denominator.

The box in the middle of Figure 5.11 segregates pairs for which the odds ratio is either greater than 2 or less than $\frac{1}{2}$. We can take the criterion to define what we mean by highly likely or highly unlikely. The pairs inside the box are not more (or less) likely by a factor of 2 than random, and almost all 400 pairs fall within this box. On the other hand, over 5% of the pairs are unlikely (by a factor greater than 2 over random) to be found in sheets. By contrast, no pair is that unlikely to be found in helices.

One potential application of the residue pair analysis might be to help identify secondary structure by sequence analysis [477]. However, as Figure 5.11 shows, the odds ratios are similar for most of the residue pairs for helices and sheets. There are only two pairs that are both highly likely to be in helices and highly unlikely to be in sheets, according to our factor of 2 standard.

helix	sheet	residue pair
2.18	0.45	His Cys
2.12	0.46	Cys Pro
1.50	0.18	Pro Pro
0.37	0.87	Cys Met
0.62	2.66	Lys Cys
1.02	2.46	Cys Lys
1.88	1.16	Cys Cys

Table 5.2 Amino acid pairs with extreme values for odds ratios. Listed are the two highest odds ratio pairs for helices and sheets as well as the lowest odds ratio pairs for both. In addition, the duplex pair of Cys residues is included for reference.

5.5.3 Like-Charged Pairs

Conventional wisdom would suggest that like-charged residues would not likely be found adjacent in proteins, since they repel each other [321]. Thus, it is interesting to look at the occurrences of like-charged pairs in sequence. In Table 5.3, the frequencies of such pairs are given for residues that appear in helices. Surprisingly, some pairs actually occur more frequently than chance would predict.

For like-charged pairs, we might have expected odds ratios much less than one, whereas all of the like-charged pairs involving Asp, Glu, Arg, Lys are no less than 0.89, and many are greater than 1. It should be remembered that His is somewhat of a wild card in that its protonation state is strongly dependent on pH in the normal physiologic range. Moreover, the highest odds ratio is not for a putative salt bridge (residues of opposite sign) but rather for the repulsive, like-charged pair Arg-Arg.

The low odds ratios for oppositely charged sequence–neighbor pairs is somewhat misleading in that most salt bridges are formed by pairs that differ in sequence by 3 or 4 [122].

residue	Asp	Glu	Arg	His	Lys
Asp	0.97	1.03	0.88	1.06	0.88
Glu	0.96	1.13	1.04	1.07	1.06
Arg	1.12	1.16	1.36	1.06	0.98
His	0.65	1.00	0.63	1.08	0.78
Lys	1.07	1.21	0.89	0.96	1.03

Table 5.3 Odds ratios for charged amino acid sequence–adjacent pairs in helices.

5.6 Volume of Protein Constituents

The number of atoms in amino acid sidechains varies significantly. Although atoms do not fill space in the way we would imagine from a terrestrial point of view, it is possible to associate an exclusion volume [52] with them that is useful in comprehending them. In 1975, Chothia initiated a study of the size of sidechains and the change in size in the core of proteins, an early use of data mining in the PDB. That study was later revisited [218], and subsequent studies have further refined estimates of sidechain the exclusion volume, including sizes of individual atom groups [494]. For reference, we have summarized in Tables 5.4–5.6 some estimates [494] of the sizes of the constituent groups inside the core of proteins. The numbers presented represent the dispersion in mean values over the different datasets used [494]. For reference, a water molecule occupies about 30Å^3 (see Section 18.7.3).

The definition of "volume" of sidechain or atom-group is important. It might be best to think of this as an "excluded volume" in the sense described subsequently. The approach of [218, 494] uses a Voronoi decomposition of space based on vertices at the location of the heavy atoms in proteins (excluding hydrogens). We will give the essentials of this approach but describe certain important details only briefly.

The basic Voronoi decomposition involves polytopes defined for each point in the input set S, defined as follows. The polytope P_s for each $s \in S$ is defined as the closure of the set of points in space closer to s than to any other point $t \in S$. Thus, for $s \neq t$ in S, the interiors of P_s and P_t will not intersect. What is somewhat magic about the set of polytopes is that they form a decomposition of space (there are no voids). But this is just because, for every point in space, there is a closest point in S. Some of the polytopes are infinite, and these are not useful in the application to proteins. In [494], a modified procedure was used that involves moving faces in the interior, and presumably adding faces at the boundary, based on van der Waals radii determined separately for each atom. The volume of a residue is defined as the sum of volumes of its atoms. In Figure 5.12, we have depicted only the interior of a Voronoi diagram, excluding the boundary where such infinite domains would occur.

The faces of each polytope are perpendicular to lines between nearby vertices and are equidistant from the two vertices. Every vertex (heavy atom) is associated with a unique polytope (that contains no other heavy atoms). The heavy atom need not be at the center of the polytope. In Figure 5.12, we show an extreme case in which the points in S can

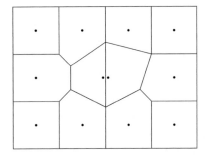

Fig. 5.12 Voronoi diagram for a set of points in two dimensions.

get arbitrarily close to the boundary of the enclosing P_s. In [218, 494], a slight modification is made to the basic Voronoi definition by moving the faces along the line joining the two atoms based on the relative van der Waals radii of the two atoms.

At first blush, it might seem that the volumes occupied, e.g., by the carbonaceous groups in Table 5.4 are too small, because we know that oil floats on water. But a water molecule is twenty to fifty percent heavier than a CH_n group (for $n = 0, \ldots, 3$), so the size estimates are quite in line with what we would expect. It also might seem that there is a very big variation in volume of different sidechain constituents. However, if we think in terms of the size of a box (or ball) of comparable volume, then the side of the box (or diameter of the ball) does not vary quite so much. For reference, in Table 5.7 we give the relevant volumes for boxes of various sizes, ranging from 2Å to 3.4Å on a side, and we see that this range of linear dimensions accounts fully for the range in volumes observed in Tables 5.4–5.6.

We have omitted sizes for some atom groups from the tables; for example, the volume of most sidechain oxygens is 14.9–19.2Å^3, with the exception of $O(\gamma^1)$ in threonine, whose volume is about 10Å^3. There are some obvious trends, e.g., in Table 5.4 it is clear that the volume of the carbonaceous groups CH_n is an increasing function of n, but the increase in size is greater than linear. On the other hand, although there is a similar trend for the nitrogen-based groups NH_n, the increase is less pronounced; in particular, the volume of the NH_3 group in lysine is only 20.6–21.9Å^3, smaller than the NH_2 groups of arginine.

But there is also a more subtle size issue that is solvent dependent. One of the significant conclusions [218] is that hydrophobic residues occupy less volume inside the core of a protein than they do in bulk water. Similarly, hydrophilic residues occupy more volume inside the core of a protein than they do in bulk water. We have listed the change in volume of the various sidechains in Table 4.2. Given our general understanding of the hydrophobic effect, the changes in mean exclusion volume are consistent with our understanding of the hydrophobic effect.

In typical proteins, the increase in volume due to burying hydrophilic residues is compensated by the decrease in volume due to burying hydrophobic residues. That is, the net volume change upon folding is typically quite small. However, for other systems, such a balance does not seem to be so close. For example, cell membranes are made of lipid layers composed substantially of hydrophobic chains. Thus, simple pressure tends to keep such cell membranes intact. To break apart, the cell membranes constituents would have to undergo a substantial increase in volume and thus induce a significant increase in pressure.

volume (Å^3)	n	t	carbon groups
9.2—10.1	0	12	$D(\gamma)$, $E(\delta)$, $F(\gamma)$, $H(\gamma)$, $N(\gamma)$, $Q(\delta)$, $R(\zeta)$, $W(\gamma, \delta^2, \epsilon^2)$, $Y(\gamma, \zeta)$
14.1—14.8	1	3	$I(\beta)$, $L(\gamma)$, $V(\beta)$
20.0—21.7	2	16	$H(\delta^2, \epsilon^1)$, $F(\delta^{1,2}, \epsilon^{1,2}, \zeta)$, $W(\delta^1, \epsilon^3, \zeta^{2,3}, \eta^2)$, $Y(\delta^{1,2}, \epsilon^{1,2})$
22.5—24.2	2	25	$E(\gamma)$, $I(\gamma^1)$, $K(\gamma, \delta, \epsilon)$, $M(\gamma)$, $P(\delta)$, $Q(\gamma)$, $R(\gamma, \delta)$ plus 15 β's
25.2—25.8	2	2	$P(\beta, \gamma)$
35.7—38.5	3	9	$A(\beta)$, $I(\gamma^2, \delta^1)$, $L(\delta^{1,2})$, $M(\epsilon)$, $T(\gamma^2)$, $V(\gamma^{1,2})$

Table 5.4 Volume of carbonaceous atom groups CH_n in protein sidechains [494]. The group sized 22.5—24.2 also includes all C_β carbons with the exception of Ala, Ile, Pro, and Val which are listed in other size groups. t is the number of carbonaceous groups in the category and n is the number of hydrogens in these groups.

volume (Å^3)	n	t	nitrogen groups
14.8—17.0	1	4	$H(\delta^1, \epsilon^2)$, $R(\epsilon)$, $W(\epsilon^1)$
20.6—23.9	2-3	5	$R(\eta^{1,2})$, $K(\zeta)$, $N(\delta^2)$, $Q(\epsilon^2)$

Table 5.5 Volume of nitrogen atom groups NH_n in protein sidechains [494]. t is the number of carbonaceous groups in the category, and n is the number of hydrogens in these groups.

atom	group of 17	alanine	glycine	proline
C	8.4—8.9	8.8—8.9	9.5—9.8	8.7—8.8
O	15.7—16.3	16.0—16.3	16.1—16.5	15.8—16.3
N	13.3—14.1	13.8—14.0	14.5—14.9	8.5—8.8
CA	12.9—13.5	14.0—14.1	23.3—23.8	13.8—14.0

Table 5.6 Volume of protein backbone atom groups (Å^3) [494]; "group of 17" refers to the residues other than Ala, Gly, and Pro.

The volumetric cost of burying hydrophilic residues makes one wonder why they appear inside proteins at all. However, without them the electrostatic landscape of the protein would be far less complex. Moreover, if proteins had only hydrophobic cores, they would be harder to unfold. Both of these effects contribute to an understanding for why charged and polar residues are found in protein cores.

The volumes for the sidechain atoms plus the backbone atoms is about the same in the two papers [218, 494], although the apportionment between backbone and sidechain differs systematically. This would lead to different values for the sidechain volume in Table 4.2, but they would just be shifted by a fixed amount. Thus, the relative size change between hydrophobic residues and hydrophilic residues would remain the same.

r	2.1	2.2	2.3	2.4	2.5	2.6	2.7	2.8	2.9	3.0	3.1	3.2	3.3	3.4
r^3	9.26	10.6	12.2	13.8	15.6	17.6	19.7	22.0	24.4	27.0	29.8	32.8	35.9	39.3

Table 5.7 Relation between volume r^3 and length r in the range of lengths (in Ångstroms) relevant for proteins.

5.7 Fold Networks

We have followed a natural progression in the hierarchy of structure of proteins. A natural next step is to look at interactions between different structural units. To study interactivity from a more global viewpoint, we need some new mathematical technology. A natural representation for interactions is to use graph theory. We will see that appropriate concepts from this theory provide useful ways to compare interactions quantitatively.

Relations among proteins can be determined by various means, and here we look at relations between the basic tertiary structural units of proteins. This will provide both a baseline of what to expect in terms of the graph theory of protein interactions as well as an application of some basic concepts of tertiary structure of proteins. The notion of "fold" or "domain" characterizes the basic unit of tertiary structure of proteins, as described in Section 5.2.2. A fold consists of basic units of secondary structures together with relations among them. Secondary structure consists of different types of helices and beta sheets and other motifs, as described in Section 5.2. These structural subunits form different groupings (folds) that have characteristic shapes that are seen repeatedly in different proteins.

Although most proteins consist of a single domain, a significant number contain multiple domains [506]. The distribution of multiple domains is known [523] to be exponential. More precisely, the probability $p(k)$ of having k domains was found [523] to be closely approximated by

$$p(k) \approx 0.85e^{0.41(k-1)}. \tag{5.16}$$

Such a distribution implies [523] "the evolution of multidomain proteins by random combination of domains."

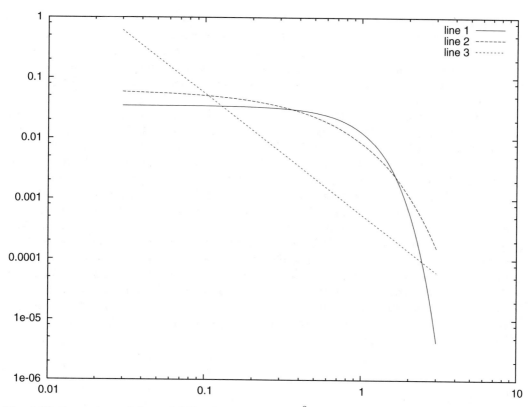

Fig. 5.13 Comparisons of the probability distributions $c_1 e^{-x^2}$ (solid line), $c_2 e^{-2x}$ (dashed line), and $c_3 x^{-2}$ (dotted straight line) on the interval $[0.03, 3]$. The constants c_i are chosen to make each distribution integrate to one on $[0.03, 3]$.

One way to form a relation among proteins is to define the vertices of a graph to be the protein domains and the edges of the graph to be pairs of domains found in a single protein [529]. The resulting graph of protein domain connectivity has many nodes with low degree with just a few highly connected nodes, in agreement with a power-law distribution [407, 529]. More precisely, if we let $p(k)$ be the probability of having k neighbors in this graph, we find

$$p(k) \approx ck^{-\gamma}. \tag{5.17}$$

A network with a distribution (5.17) is often called a **scale free network** [38]. This provides a useful baseline regarding what to expect in terms of protein interactions.

The definition of domain interaction requires some explanation. If all three-dimensional structures were available in the PDB, this would be simplified significantly. However, the current fraction of proteins available in the PDB is tiny, and it is expected to remain so for a long time due to the cost and technical obstacles of structure determination. In [529], interaction was determined from different databases devoted to the characterization of protein domains, including both structural means and other techniques such as sequence alignment which do not require structural information. One of the databases used in [529] was Pfam [184]. The objective of "Pfam is a comprehensive collection of protein domains and families, represented as multiple sequence alignments and as profile hidden Markov models" [184]. For a description of profile hidden Markov models, see [129].

Knowing the sequence description of a domain allows the determination of proteins with common domains via sequence alignment. If one of the sequences is represented in the PDB, then a fold can be assigned to all the similar sequences. It may seem odd that it would be possible to predict what is essentially structural information (a fold) for a large number of proteins, but one view suggests that relatively few PDB structures are needed [504]. On the other hand, sequence similarity does not always imply fold similarity [212].

The distribution of folds in different species is not uniform. Quite the contrary, the distribution appears to provide a distinct signature for the fourteen different species studied in [523] (twenty species were studied in [407]). Thus, it is natural to consider the connectivity of folds separately for each species, and the distribution of connectivity of folds for each species was found [407, 529] to follow a power-law with a distinctive decay rate for each species, with "those of smaller genomes displaying a steeper decay" [407], that is, a larger γ in (5.17).

We have described two characteristic probability distributions here, the exponential (5.16) and the power-law (5.17). These are both quite different from the more familiar Gaussian (normal) distribution. Since all of these are being fit to a finite number of data points, it is a reasonable question to ask whether these distributions are really different enough to distinguish them with just a few points. To address this question, Figure 5.13 presents a comparison of the Gaussian distribution with the power-law distribution (5.17) and the exponential distribution (5.16). One thing revealed by Figure 5.13 is that data that appear to fall along a linear curve in a log–log plot much more closely fits a power-law distribution than it does either a Gaussian or an exponential distribution.

5.8 Exercises

Exercise 5.1. Suppose that we create a string of letters from an alphabet A as follows. We start with an empty string σ. With probability p, we pick a letter $x \in A$ and define a new string $\sigma \leftarrow \sigma + x$ where $+$ means "append to the string." With probability $q = 1 - p$, we stop. Then the probability of having a string of length zero is q, and the probability of having a string of length one is pq. Continuing in this way, show that the probability of having a string of length k is $p^k q$. Prove that

$$\sum_{k=0}^{\infty} p^k q = 1, \tag{5.18}$$

and that $f(k) = p^k q = ce^{-\gamma k}$ for some c and γ. Explain why this supports the statement that the evolution of multidomain proteins can be modeled by the "random combination of domains" [523].

Exercise 5.2. Use protein constituent volume data [218, 494] (cf. Tables 5.4–5.6) to estimate the density of typical proteins. What is the mass per residue of typical proteins? Compute the mass of alpha helices, beta sheets, and loops separately, as well as estimating the mass of complete proteins. (Hint: use Table 4.3 to estimate the relative abundance of each residue.)

Exercise 5.3. Plot a Gaussian distribution and an exponential distribution as in Figure 5.13 but on a log-linear plot. That is, use a logarithmic scale on the vertical scale but an ordinary (linear) scale on the horizontal axis. Explain how to distinguish these distributions by this device.

Exercise 5.4. Determine whether the choice of atom X_γ in (5.6) matters for computing χ_1 for Val by examining high-resolution structures. For each Val, compute the difference of dihedral angles for the two choices of terminal atoms, and plot the resulting histogram for all Val residues in your dataset. Similarly, give a scatter plot for the distribution of angle pairs, in degrees. (Use the tools from Exercise 4.8 to know if the groups of atoms are planar or not.)

Exercise 5.5. Determine whether the choice of atom Y_δ in (5.6) matters for computing χ_2 for Asp, Asn, and Leu by examining high-resolution structures. For each residue type, compute the difference of dihedral angles for the two choices of terminal atoms, and plot the distribution of angle pairs, in degrees. (Compare with Exercise 4.8.)

Exercise 5.6. Draw figures analogous to Figure 5.10 for the residues Thr and Ser, assuming that they adopt the same basic gauche± and trans configurations. How many possible orientations of the atoms are there? Check to see whether the rotameric states you propose are probable [323].

Exercise 5.7. Draw figures analogous to Figure 5.10 for the residues Asp and Asn, assuming that they adopt the same basic gauche± and trans configurations. Include the possible orientations of the terminal pairs (OD1 and XD2) of atoms (where X=O for Asp and X=N for Asn). How many possible orientations of the atoms are there? Check to see whether the rotameric states you propose are probable [323]. (Hint: recall the planarity of the terminal four atoms, Exercise 4.8.)

Exercise 5.8. Explain how Figure 5.3(a) represents either a right-handed or left-handed helix depending on whether the dashed line goes above or below the indicated hydrogen bonds, that is, toward or away from the viewer. Recall that a helix is right-handed if it turns clockwise as it moves away from you. (Hint: cut a narrow strip of paper and mark donors and acceptors for the hydrogen bonds at regular intervals. Experiment by twisting the paper into both right-handed and left-handed helices.)

Exercise 5.9. In typical peptide bonds, the ω angle is constrained to so that the peptide bond is planar (cf. Figure 14.1). In this case, there is a relationship imposed between θ, ϕ, and ψ. Determine what this relationship is.

Exercise 5.10. Proteins are *oriented*: there is a C-terminal end and an N-terminal end. Determine whether there is a bias in α-helices in proteins with regard to their *macrodipole* μ which is defined as follows. Suppose that a helix consists of the sequence $p_i, p_{i+1}, \dots, p_{i+\ell}$ where each p_j denotes an amino acid sidechain. Let $\mathcal{C}(p)$ denote the charge of the sidechain p, that is, $\mathcal{C}(D) = \mathcal{C}(E) = -1$ and $\mathcal{C}(K) = \mathcal{C}(R) = \mathcal{C}(H) = +1$, with $\mathcal{C}(p) = 0$ for all other p. Define

$$\mu(p_i, p_{i+1}, \dots, p_{i_\ell}) = \sum_{j=0}^{\ell} \mathcal{C}(p_{i+j}) \left(j - \tfrac{1}{2}\ell\right) \tag{5.19}$$

Plot the distribution of μ over a set of proteins. Compare with the peptide dipole, which can be modeled as a charge of $+0.5$ at the N-terminus of the helix and a charge of -0.5 at the C-terminus of the helix. How does this differ for left-handed helices versus right-handed helices? What happens if you set $\mathcal{C}(H) = 0$? (Hint: the PDB identifies helical regions of protein sequences. The peptide dipole in our simplification is just ℓ, so μ/ℓ provides a direct comparison.)

Exercise 5.11. Consider the definition of macrodipole introduced in Exercise 5.10. Explain why the α-helical polypeptide $Glu_{20}Ala_{20}$ would be more stable than $Ala_{20}Glu_{20}$.

Exercise 5.12. Determine the Ramachandran plot [237] for a set of proteins. That is, plot the ϕ_i and ψ_i angles for all peptides in the set. Use a different symbol or color for the cases where the i-th peptide is said to be a helix or sheet in the PDB file.

Exercise 5.13. From Figure 4.2, we would conclude that the distances between two C-alpha carbons is much smaller when the peptide bond is in the cis form. Compare the C-alpha distances with the ω angle.

Exercise 5.14. Correlate the distance between C_α carbons separated in sequence by k with the secondary structure call in a given PDB file. That is, give the (three, separate) distributions of distances as a function of k for helices and sheets and loops (i.e., for residues not in either structure).

Exercise 5.15. Consider the distribution of distances between C_α carbons separated in sequence by $k = 2$ in a given PDB file. Explain the bimodal character in PDB file 2P9R. Correlate the distance with ϕ and ψ angles.

Chapter 6
Hydrogen Bonds

Linus Carl Pauling (1901–1994) was awarded the Nobel Prize in Chemistry
in 1954 and the Nobel Peace Prize in 1962, making him the only person
to be awarded two unshared Nobel Prizes. He promoted the regular use
of large dose of vitamin C, and later in life the consumption of significant
doses of alcohol, to support good health.

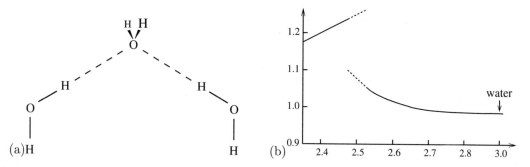

Fig. 6.1 (a) Cartoon of the configuration of three water molecules and the hydrogen bonds formed, depicted
by the dashed lines. The oxygen atoms accept two hydrogen bonds, whereas the hydrogen atoms are involved
in donating only a single bond. (b) Cartoon of the bifurcation of O–H..O hydrogen bonds from a symmetric
arrangement to an asymmetric arrangement, based on Figure 4 of [397]. The horizontal axis is the O–O
distance and the vertical coordinate is the O–H distance (both in Ångstroms). The upper-left segment is the
symmetric arrangement. Note that the water–water hydrogen bond is well away from the symmetric region.
The dashed curves indicate where data have been found in both states.

Hydrogen bonds are central to biology, yet they are not fully understood. Thus we present
a detailed review of its physical and geometric characterization, specialized to a biomolecular
context.

6.1 Introduction to Hydrogen Bonds

The concept of the hydrogen bond was established by 1920 [297], and possibly earlier [98, 272].
Hydrogen bonds are the most important bonds in biochemistry, so we need to understand
them in some depth. Unfortunately, there are several challenges. First of all, although hydro-

© Springer International Publishing AG 2017
L.R. Scott and A. Fernández, *A Mathematical Approach to Protein Biophysics*,
Biological and Medical Physics, Biomedical Engineering,
DOI 10.1007/978-3-319-66032-5_6

gen bonds in proteins have been considered extensively [32, 287, 483], they are not yet fully understood and are still actively studied [249, 250]. Secondly, in most PDB files, hydrogens are not listed at all, due to the difficulty of locating them by typical imaging techniques. Thus, an initial step is to place hydrogens in the appropriate places. We will explain how this is done in simple cases and what the problems are in the difficult cases. We describe how their locations can be inferred starting in Section 6.3. Many different techniques are in use, and comparisons of the different techniques have been made [188].

Hydrogen bonds can be viewed as our simplest example of a switch that is crucial to protein biophysics. Although detailed models do not yet exist to quantify the behavior of a hydrogen bond as a switch, we will see qualitatively that their geometry determines their quality, and that a change in geometry effectively can reduce their viability. Their dependence on distance is not a simple one like what we have seen so far for point charges and dipoles. Although hydrogen bonds are often modeled as dipole–dipole interactions, it is known that this model is insufficient. They have a stronger angular dependence and they also involve induction, somewhat like van der Waals interactions.

The general hydrogen bond is of the form XH - - Y where X and Y are "heavy atoms" such as F, N, O, S, or even C in some cases. The X atom is called the **donor** of the bond, and the Y atom is called the **acceptor** of the bond. An example based on the interaction of water molecules is given in Figure 6.1(a).

Acceptor	Donor	System	$R(\text{Å})$	$\Delta E(\text{kcal})$
HF	NH_3	$HF–HNH_2$	3.45	1.3
H_2O	NH_3	$H_2O–HNH_2$	3.41	2.3
H_3N	NH_3	$H_3N–HNH_2$	3.49	2.7
HF	H_2O	$HF–HOH$	3.08	3.0
H_2O	H_2O	$H_2O–HOH$	3.00	5.3
H_3N	H_2O	$H_3N–HOH$	3.12	5.8
HF	HF	$HF–HF$	2.72	9.4
H_2O	HF	$H_2O–HF$	2.75	11.7
H_3N	HF	$H_3N–HF$	2.88	4.6

Table 6.1 Distances and energies of various hydrogen bonds. R is the distance (in Ångstroms) between the donor and acceptor (heavy) atoms. The energy ΔE of the hydrogen bond is given in kcal/mole. Note that R "is primarily a function of the degree of positive charge on the hydrogen in the H bond" [277].

6.2 Types of Hydrogen Bonds

Hydrogen bonds differ based on the heavy atoms that are involved. The variation in bond distance and strength is illustrated in Table 6.1 which has been extracted from [277]. What is clear from this data is that the donor type (the side of the bond that includes the hydrogen) is the primary determinant of the hydrogen bond strength (and length) in these cases. This is interpreted to mean that the dipole of the donor is the determining factor [277]. In some sources (including Wikipedia), the electronegativity of the constituents is given as the key factor. But according to [277], "the ability of proton donors and acceptors to form hydrogen bonds (X-H ... Y) is more closely related to their respective acidity or basicity than to the electronegativities of X and Y."

One basic question about hydrogen bonds is whether the hydrogen is in a symmetric position between the donor and acceptor, or whether it favors one side (donor) over the other.

The answer is: yes and no [397]. Both situations arise in nature, and there is an intriguing bifurcation between the two configurations, as depicted by the caricature in Figure 6.1(b). Depicted is a curve that was fit [397] to extensive data on bond lengths of OH - - O hydrogen bonds. The horizontal axis is the distance between the oxygen centers, and the vertical coordinate is the (larger) distance between oxygen and hydrogen. The upper-left segment, where the O–H distance is exactly half of the O–O distance, is the symmetric arrangement. In this configuration, you cannot distinguish one of the oxygen atoms as the donor; both are donor and acceptor.

The dashed parts of the curves indicate where data have been found in both configurations. But what is striking is the void in the O–H distance region between 1.1 Å and nearly 1.2 Å. Thinking in bifurcation terms, one can stretch the O–O distance in the symmetric configuration, but at a certain point it loses stability and has to jump to the asymmetric configuration in which the hydrogen has a preferred partner. Moreover, as the O–O distance continues to increase, the (smaller) O–H distance *decreases*, as the influence of the other oxygen decreases with increasing distance. Note that the O–O distance (for waters) reported in Table 6.1 is 3.0 Å, thus clearly in the asymmetric regime (almost off the chart in Figure 6.1(b)).

6.2.1 Hydrogen Bonds in Proteins

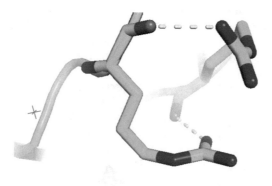

Fig. 6.2 Example of two sidechain–mainchain hydrogen bonds pairing residues Arg 836 and Arg 858 in the oncogenic mutant L858R in EGFR kinase (PDB file 2ITV). The dotted yellow lines connect the heavy atoms (NH1 in Arg 836 to the backbone O in Arg 858, and NH1 in Arg 858 to the backbone O in Arg 836) in the hydrogen bonds. The backbone oxygen in Arg 836 is in the foreground.

As indicated in Table 6.1, hydrogen bonds vary in character depending on the donor and acceptor. In proteins, there are two classes of donors and acceptors, mainchain (or backbone) and sidechain. The mainchain–mainchain hydrogen bond was depicted schematically in Figure 2.3. In Figure 6.2, we show two sidechain–mainchain hydrogen bonds indicated by dashed yellow lines, between the backbone carbonyls and sidechain N–H groups in Arg 836 and Arg 858. The protein depicted is the oncogenic **mutant** L858R in EGFR kinase (PDB file 2ITV). The term mutant means that one residue is changed from what is considered to be the native form of the protein. In this case, Lys 858 is changed to Arg 858 in PDB file 2ITV, indicated by the notation L858R.

All backbone nitrogens (with the exception of proline, unless it is N-terminal) can act as donors of hydrogen bonds, and all backbone oxygens can be acceptors of hydrogen bonds.

These donors and acceptors were represented as outgoing and incoming arrows in Figure 5.1. In addition, many of the standard sidechains can act as donors or acceptors, as listed in Table 6.2. Note that certain atoms can be both donors and acceptors, as are the oxygens in water.

Given two classes of contributors, mainchain (M) and sidechain (S), there are four classes of bond pairs: M–M, M–S, S–M, and S–S. We have differentiated between S–M and M–S depending on whether the donor or acceptor is M or S, but in some cases these two classes are lumped into one class.

Given the rigidity of the backbone and the flexibility of the sidechains, it would be reasonable to assume that S–S bonds were the most common and M–M the least. Curiously, it is just the opposite [470]. In [446], we see that mainchain–mainchain are much more common. By simple counts in a database of 1547 nonredundant structures, the number of M–M bonds is nearly four times the number of mainchain–sidechain (M–S and S–M) bonds combined, and it is seven times the number of sidechain–sidechain bonds. On the other hand, one finds a significant number of potential sidechain–water hydrogen bonds in many PDB files. These include apparent water bridges [394, 535]. It is not clear how fully waters in PDB files are reported, but their importance to protein structure is significant.

Full name of amino acid	three letter	single letter	Donors (PDB name)	Acceptors (PDB name)
Arginine	Arg	R	NE, NH1, NH2	—
Asparagine	Asn	N	ND2	OD1
Aspartate	Asp	D	—	OD1, OD2
Cysteine	Cys	C	SG*	SG
Glutamine	Gln	Q	NE2	OE1
Glutamate	Glu	E	—	OE1, OE2
Histidine	His	H	ND1, NE2	ND1, NE2
Lysine	Lys	K	NZ	—
Methionine	Met	M	—	SD
Serine	Ser	S	OG	OG
Threonine	Thr	T	OG1	OG1
Tryptophan	Trp	W	NE1	—
Tyrosine	Tyr	Y	OH	OH

Table 6.2 Donors and acceptors for sidechain hydrogen bonds. *If a Cys is involved in a disulfide bridge, it cannot be a hydrogen bond donor.

Full name of nonstandard residue or molecule	PDB three letters	Donors	Acceptors
Acetyl group	ACE		O
Glycerol	GOL	O1, O2, O3	O1, O2, O3
Nitrate Ion	NO3		O1, O2, O3
Phosphotyrosine	PTR	N, O2P‡, O3P‡	O, OH, O1P, O2P, O3P
Pyroglutamic acid	PCA	N†	O, OE
Phosphono group	PHS		O1P, O2P, O3P
Phosphate Ion	PO4		O1, O2, O3, O4
Sulfate Ion	SO4		O1, O2, O3, O4

Table 6.3 PDB codes for donor and acceptor atoms in some nonstandard residues and molecules. Key: † Only N-terminus. ‡ In case that the hydrogens PHO2, PHO3 exist in the PDB files.

Typical hydrogen bond donors would make only one hydrogen bond, whereas typical oxygen acceptors can make two hydrogen bonds. However, more complex patterns are possible; see the figures on page 139 of [250]. In some cases, the network of hydrogen bonds can be complicated, as shown in Figure 6.3.

Fig. 6.3 Three serines (52, 54, and 56) from chain H the PDB file 1DQM come together to form a complex of hydrogen bonds. The amide groups on the peptide base of H-Gly55 and H-Ser56 provide donors to the oxygens on the end of the sidechain of H-Ser52. The sidechain–sidechain hydrogen bonds among the terminal OH groups on the serines depend on the hydrogen placements.

6.2.2 Hydrogen Bond Strength

Assessing the strength of hydrogen bonds remains a significant challenge [298, 473]. There is a strong angular dependence for the energy of the hydrogen bond [364, 483]. Moreover, the nature of the hydrogen bond can depend on the context: even backbone–backbone hydrogen bonds can be different in alpha helices and beta sheets [275, 295, 483].

One might hope that modeling the hydrogen bond as a simple dipole–dipole interaction (Section 10.2.1) would be sufficient to capture the angular dependence. But a purely partial charge model of hydrogen bonds is not sufficient to capture the angular dependence of the energy: "At the distances where H bonding occurs, the dipole moment approximation is a poor one and higher multipoles must be considered" [277], as we confirm in Section 10.2.1.

Attempts have been made to model hydrogen bonds via more sophisticated interactions. In addition to partial charges, dipole, quadrapole and higher representations of the donor, and acceptor groups have been used [67, 68]. A model of this type for the hydrogen bonds in water has been proposed that includes terms for the polarization of water [46, 346]. The difficulty with models of this type is that the multipole expansion converges rapidly only for large separation of the donor and acceptor. Thus, these models provide very accurate representations of the asymptotic behavior of the interaction for large separation distances R, but for values of R of close to native separation distances it is only slowly converging. As a remedy to this, distributed multipole expansions, in which the representation involves partial charges (and dipoles, etc.) at many positions, have been proposed [472, 474].

A model to represent the angular and distance dependence of the energy of the hydrogen bond, based only on the atomic distances among the primary constituents, has been proposed [364], in which the dominant term appears to be a strong repulsion term between the like-charged atoms. Such a model is simple to implement because it uses exactly the same data as a dipole model, but with a more complex form and with additional data derived from ab initio quantum chemistry calculations.

More recently, a model based on distance and angular data has been proposed [86] that has been fitted to approximation quantum (DFT) computations and empirical data on hydrogen bonds at interfaces. This model uses a reduced representation based on only one distance and two angles, and it is similar to a model used earlier [105]. The earlier model was based on one distance and three angles, with the angle not used in [86] playing only a minor role.

The accurate simulation of one of the simplest hydrogen bonds, in the water dimer, has been of recent interest [271], even though this computation has been carried out for several decades [277]. The fact that this simple interaction is still studied is an indicator of the difficulty of determining information about general hydrogen bonds. Models of the water trimer, tetramer, and hexamer have also received recent attention [210, 265, 319].

For more information on hydrogen bond models, see [113].

6.3 Identification of Hydrogen Positions

Many PDB files do not include locations of hydrogens. Only the heavier atoms are seen accurately in the typical imaging technologies. In general, hydrogen placement is a difficult problem [188, 483]. However, in many cases, the positions of the missing hydrogens can be inferred according to simple rules [353]. We begin with the situations that are the simplest to predict and then consider the more complex ones.

Most hydrogens can be located uniquely. In particular, the Appendix in [353] depicts the locations of such hydrogens, as well as providing precise numerical coordinates for their locations. However, other hydrogen positions are not uniquely determined.

The position of most hydrogens can be modeled by the bond lengths and angles given in [353]. A program called HBPLUS [339] was developed based on this information to provide hydrogen positions in a PDB format. More recently, sophisticated software has emerged to provide estimates of hydrogen positions, including decisions about sidechain ionization (cf. Section 4.5) [188, 203]. The program Reduce is commonly used today [300, 527]. Reduce can be accessed via the website MolProbity.

6.3.1 Fixed Hydrogen Positions

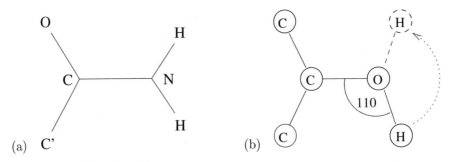

Fig. 6.4 Hydrogen placement for three sidechains. (a) Asn and Gln. Shown is the terminal group of atoms for the sidechains. The atom marked C' denotes the preceding carbon in the sidechain, viz., CB for Asn and CG for Gln. (b) Tyr. The hydrogen is in the plane of the aromatic ring, with the angle between C–O and O–H being 110 degrees. Both positions are possible for the terminal hydrogen.

The position of the hydrogen that is attached to the mainchain nitrogen (see Figure 3.1a) can be estimated by a simple formula. The C–O vector and the H–N vector are very nearly parallel, so one can simply take

$$H = N + |C - O|^{-1}(C - O) \tag{6.1}$$

since the N–H distance is approximately one Ångstrom. We leave as an exercise (Exercise 6.1) to make the small correction suggested by the figure on page 282 in [382], replicated in Figure 2.9; also see the more recent corrections to estimates of bond lengths and angles in [247].

The bond length $|N - H|$ in the formula (6.1) is taken to be 1.0 Å in accordance with [382]. However, examination of all PDB files incorporating hydrogens [449] indicates that, in practice, different lengths are chosen. There are significant numbers of PDB file structures derived by NMR having mean $|N - H|$ lengths (in Å) of 0.98, 1.00, 1.01, and 1.026, many of these with standard deviations significantly less than 0.01, indicating that a particular length was chosen as a model. Similarly, there are significant numbers of PDB file structures derived by X-ray techniques having mean $|N - H|$ lengths (in Å) of 0.86, 0.98, 1.00, and 1.015, again with many of these having standard deviations significantly less than 0.01, indicating the use of a model to set the bond length within each individual PDB file. The lengths 0.86 and 1.02 are obtained by using Reduce with two different options. The former is obtained by specifying the electronic center and the latter corresponds to the nuclear position. More precisely, for the PDB file 1abb, one obtains N–H distances of 0.860 ± 0.002 for the electronic option and 1.020 ± 0.002 for the nuclear option. Thus in any analysis in which the bond length $|N - H|$ is a critical variable, care must be taken to view the sensitivity of the results on $|N - H|$. See [449] for more details.

As another simple example, the position of the hydrogens that are attached to the terminal nitrogen in Asn and Gln can also be estimated by a simple formula. The terminal $O-C-NH_2$ group of atoms are all coplanar, and the angles formed by the hydrogens around the nitrogen are all 120 degrees, as depicted in Figure 6.4(a). The angle between the C–N and the C–O vectors is very close to 120 degrees [353], so the C–O vector and one of the N–H vectors are very nearly parallel. So one can again take

$$H^1 = N + |C - O|^{-1}(C - O) \tag{6.2}$$

as the location for one of the hydrogens attached to N, since again the N–H distance is approximately one Ångstrom. For the other hydrogen bond, the direction we want is the bisector of the C–O and C–N directions. Thus, the second hydrogen position can be defined as

$$H^2 = N + \tfrac{1}{2}\left(|O - C|^{-1}(O - C) + |N - C|^{-1}(N - C)\right) \tag{6.3}$$

We leave as an exercise (Exercise 6.2) to make the small corrections suggested by [353, Figure 13].

The hydrogen attached to the terminal oxygen in the tyrosine sidechain has two potential positions. The hydrogen must be in the plane of the aromatic ring, but there are two positions that it can take. This is depicted in Figure 6.4(b). The one which makes the stronger H bond with an acceptor is presumably the one that is adopted.

6.3.2 Variable Hydrogen Positions

The terminal OH groups in serine and threonine are even less determined, in that the hydrogen can be in any position in a circle indicated in Figure 6.5(a). A Cys sidechain that is not engaged in a disulfide bond would be similar, as shown in Figure 6.5(b). The expected orientations of the terminal OH groups would presumably be determined by the local electrostatic environment.

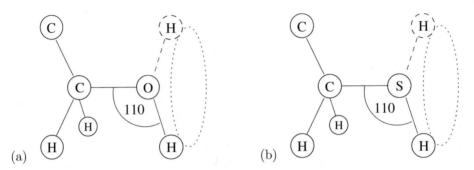

Fig. 6.5 Hydrogen placement for (a) Ser and Thr and (b) Cys: anywhere on the dotted circle. The angle between C–O and O–H being 110 degrees, but the hydrogen is otherwise unconstrained.

6.3.3 Ambiguous Hydrogen Positions

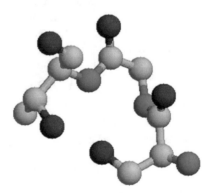

Fig. 6.6 Ambiguous hydrogen placement for serine-28 (lower right)—threonine-30 (upper left) sidechain–sidechain hydrogen bond involving the terminal O–H groups; from the B chain in the PDB file 1C08. The sidechain of isoleucine-29 has been omitted but the backbone atoms are shown connecting the two residues. Only the oxygen atoms in the terminal O–H groups are shown. The distance between Ser28 OG and Thr30 OG1 is 2.57 Å.

An interesting example of the ambiguity of the assignment of the hydrogen location for serines and threonines occurs in the PDB file 1C08. In chain B, Thr30 and Ser28 form a sidechain–sidechain hydrogen bond involving the terminal OH groups, as indicated in Figure 6.6. But which is the donor and which is the acceptor cannot be differentiated by the data in the PDB file in a simple way. Model building shows that both are possible, and indeed, there could be a resonance (Section 14.1) between the two states. One state may be forced by the local environment, but without further determining factors both states are possible. It is possible to critique the detailed geometry by considering the quality of the corresponding dipole–dipole interaction (see Section 10.2.1). According to this metric, Thr30 is the preferred donor.

6.4 Geometric Criteria for Hydrogen Bonds

The simplest approach to approximating the angular dependence of the hydrogen bond is to use angular limits, as well as distance limits, in the definition. Each hydrogen bond can be defined by the geometric criteria (Figure 6.7) based on those used in [32], as we now enumerate:

1. Distance between donor and acceptor $|D - A| < 3.5\,\text{Å}$
2. Distance between hydrogen and acceptor $|H - A| < 2.5\,\text{Å}$
3. Angle of donor–hydrogen–acceptor $\angle DHA > 90°$
4. Angle of donor–acceptor–acceptor antecedent $\angle DAB > 90°$
5. Angle of hydrogen–acceptor–acceptor antecedent $\angle HAB > 90°$

To be declared a hydrogen bond, all five criteria must be satisfied.

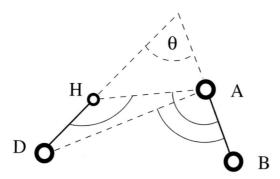

Fig. 6.7 Geometric model for hydrogen bonds: D is the donor atom, H the hydrogen, A the acceptor, B acceptor antecedent (i.e., an atom one covalent bond away from the acceptor).

In [250], data are presented for hydrogen bonds. In particular, [250, Figure 2] depicts how $\angle DHA$ interacts with the distances for a particular class of hydrogen bonds, including water–water hydrogen bonds. It is clear that using default limiting distances of $|D - A| = |H - A| + 1$ is not appropriate in many cases, unless $\angle DHA = \pi$. If $\angle DHA < \pi$, then $|H - A| > |D - A| - |D - H|$. In particular, if $\angle DHA = \pi/2$, then $|H - A| = \sqrt{|D - A|^2 - |D - H|^2}$. For most of the hydrogen bonds considered here, the donor–hydrogen distance $|D - H| = 1$, so if we are to allow angles as small as $\angle DHA = \pi/2$, we should allow hydrogen–acceptor distances as large as $\sqrt{|D - A|^2 - 1}$. Thus, if our limit is $|D - A| \leq 3.5$, the appropriate limit for the hydrogen–acceptor distance is $|H - A| \leq 3.3541$. In general, you can (Exercise 6.13) use the law of cosines to give express the distance $|H - A|$ in terms of $|D - H|$, $|D - A|$, and $\angle DHA$.

It is surprising that there is no widely accepted model for hydrogen bonds other than these geometric constraints. There is no reliable quantitative estimate of the strength or stability of hydrogen bonds based on the angle and distance parameters used in the geometric definition even when the context of the bonds is specified (e.g., the hydrogen bonds in particular secondary structures). Note that the five parameters of this model are not independent; only four independent parameters are required (see Exercise 6.13).

An additional angle, denoted by θ in Figure 6.7, is also sometimes useful to characterize the quality of a hydrogen bond. It is clearly dependent on the other three angles used in the definition of the hydrogen bond. It can be easily computed from the vectors $H - D$ and $A - B$ via

$$\cos \theta = \frac{(H - D) \cdot (A - B)}{|H - D| \, |A - B|}. \tag{6.4}$$

Here we are not assuming that A, B, D, and H lie in a plane.

It is possible to examine classes of hydrogen bonds in specialized environments. For example, we can look at hydrogen bonds that appear in the basic secondary structural elements of proteins. The paper [32] has surveyed these extensively, and we just give some highlights to indicate what has been found. It Table 6.4, we see hydrogen bond distance and angle data for hydrogen bonds in various types of secondary structures.

The differences in geometry for the different structures lead to slightly different distance and angle data. But a more refined metric [15, 32] decomposes $\angle_{HOC}$ into two components β and γ shown in Figure 6.8. More precisely, let P denote the plane of the backbone including the C, O, N, and C_α atoms. Then γ is the out of plane angle formed by the OH vector, and β is the angle formed between the CO vector and the projection of the OH vector onto the plane P.

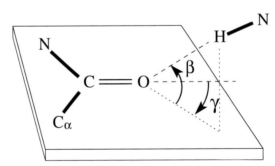

Fig. 6.8 Angles β (out of plane) and γ (in plane) were defined in [15, Figure 11, page 751] and later used in [32, Figure 22, page 143]. Our figure is fashioned after the original [15, Figure 11, page 751] which was reproduced in the subsequent reference.

structure	N_T	d_{ON}	d_{OH}	$\angle_{NHO}$	$\angle_{HOC}$	β	γ
α helix	14	2.99±0.14	2.06±0.16	155±11	147±9	+27	−18
3_{10} helix	17	3.09±0.14	2.17±0.16	153±10	114±10	+55	−25
$\parallel\beta$ sheet	16	2.92±0.14	1.97±0.15	161±9	155±11	0	−20
$\nparallel\beta$ sheet	16	2.91±0.14	1.96±0.16	160±10	150±12	−10	20

Table 6.4 Average hydrogen bond quality data for hydrogen bonds in different secondary structures: α helices, 3_{10} helices, ($\parallel\beta$) parallel beta sheets, and ($\nparallel\beta$) antiparallel beta sheets. N_T indicates the table number in [32] from which the data is drawn. The designators d_{AB} indicate the distance between atoms A and B in Ångstroms; the designators $\angle_{ABC}$ indicate the angle (in degrees) formed by the three atoms A, B, and C. Values of β and γ have been estimated by comparing data from [15] and [32].

6.5 Special Classes of Hydrogen Bonds

Hydrogen bonds can occur in a wide variety of situations. Here we gather some cases that might otherwise be obscure.

Electrostatic interactions between positively charged residues, such as Lys and Arg, and negatively charged residues, Asp and Glu, can be characterized ambiguously as both salt bridges and hydrogen bonds in many cases [289, 290]. Also, His can be positively charged in some cases.

The constituents of a salt bridge have significant energy cost for desolvation. That is, each constituent prefers to be solvated rather than forming an ionic bond. Thus, interactants for which none of the heavy atoms were less that 4 Å apart were considered destabilizing [289]. A 4 Å has been used [290] as a cutoff distance for distinguishing between different salt bridge interactions. There is some statistical evidence to support this [187].

It is possible to define geometries relevant for salt bridges [122] that can potentially be used to differentiate between true salt bridges and ones that only form hydrogen bonds or other polar interactions.

Under suitable conditions, the terminal groups of Asp and Glu can become protonated. The resulting OH group can then form hydrogen bonds with oxygens, including the ones in the terminal groups of other Asp and Glu residues [522]. These are referred to as carboxyl–carboxylate hydrogen bonds. Although these bonds would be expected in low pH environments [438], they have been found to be critical elements of ion channels [356]. In typical PDB structures, the hydrogen in a carboxyl–carboxylate hydrogen bond would not be visible. Thus, it could be associated with either oxygen unless further information is available to reveal the association.

In general, any negatively charged entity might provide an acceptor for a hydrogen bond. We will see that the faces of the aromatic rings in Phe, Tyr, and Trp provide appropriate negative charge regions that act as acceptors of hydrogen bonds. We will postpone the discussion of them until a more detailed analysis of the aromatic sidechains (Chapter 12).

In some cases, carbonaceous groups (CH) can act as donors for hydrogen bonds [255, 334].

6.6 Exercises

Exercise 6.1. Refine the formula (6.1) to give a more precise location for the hydrogen attached to the nitrogen in the peptide bond, e.g., following the figure on page 282 in [382], replicated in Figure 2.9, or the more recent corrections to estimates of bond lengths and angles in [247, 488].

Exercise 6.2. Refine the formulas (6.2) and (6.3) to give a more precise location for the hydrogens attached to the terminal nitrogen in the residues Asn and Gln, using the data in [353, Figure 13].

Exercise 6.3. Use the model for the energy of a hydrogen bond in [364] to estimate the strength of hydrogen bonds. Apply this to antibody–antigen interfaces to investigate the evolution of the intermolecular hydrogen bonds at the interfaces.

Exercise 6.4. Hydrogen positions can be inferred using neutron diffraction data, because hydrogen is a strong neutron scatterer. There are over a hundred PDB files including neutron diffraction data. Use this data to critique the models for hydrogen locations presented in this chapter.

Exercise 6.5. Helical secondary structure is formed by amide–carbonyl hydrogen bonding between peptides i and j, where $3 \leq |i - j| \leq 5$. Determine how frequent it is to have $i - j = k$ for the different possible values of $k = -5, -4, -3, 3, 4, 5$. Are there instances where amide–carbonyl hydrogen bonding between peptides i and j, where $|i - j| = 2$ or $|i - j| = 6$?

Exercise 6.6. The C–O (carbonyl) groups in the peptide backbone can make upto two hydrogen bonds (typically), whereas the N–H (amide) group usually forms only one hydrogen bond. How common is it for carbonyl groups to make two backbone hydrogen bonds in helical secondary structures? In β-sheet structures? As a first step, you can define helical carbonyls to be ones where there is bonding between peptides i and j, where $3 \leq |i-j| \leq 5$, but determine how many double bonds there are for each value of $k = i - j$. At the second step, use the secondary structure information in the PDB as a guide to determining which residues you should consider for interrogation regarding hydrogen bonds. Do the similar study of numbers of hydrogen bond partners for backbone N–H (amide) groups.

Exercise 6.7. The C–O (carbonyl) groups in the peptide backbone can make upto two hydrogen bonds (typically), whereas the N–H (amide) group usually forms only one hydrogen bond. Determine the frequency of backbone carbonyl groups making k bonds ($k = 0, 1, 2, 3$) in helical secondary structures (and, respectively, in β-sheet structures) including sidechain–mainchain bonds. Do the same study for backbone donor amide groups. Use the secondary structure information in the PDB as a guide to specify residues you should consider regarding potential hydrogen bonds.

Exercise 6.8. Determine the angular dependence of the mainchain–mainchain hydrogen bond. What is the distribution of O–H distances and C–O, N–H angles? Consider the different classes of bonds separately: those in (1) parallel and (2) antiparallel sheets, and those in helices of separation $k = \pm 3, \pm 4, \pm 5$ (cases 3-8). How does the bond distance and angle correlate? What is the mean distance and angle in each case?

Exercise 6.9. Write a code to compute the distances $d_i = |O_i - N_{i+1}|$ for the consecutive backbone oxygens and nitrogens in a PDB file. Here $i \geq 1$ denotes the occurrence of each backbone O and N in the PDB file, not the residue number. Plot d_i as a function of i for the PDB file 2ACX and one other PDB file of your choice. By looking to see where they occur in 2ACX, explain the large values (> 3 Å) that you see. After eliminating these outliers, compute the mean value of d for 2ACX. For the PDB file of your choice, do a similar process if needed (that is, if there are outliers). Explain why O_i and N_{i+1} cannot form a hydrogen bond even though the distance d_i appears to be favorable.

Exercise 6.10. The pairing of Gln (i) and Asp ($i+4$) in alpha helices can make a sidechain–sidechain hydrogen bond in addition to the mainchain–mainchain hydrogen bond [240]. What is the odds ratio of this pairing? How favorable is the geometry for this sidechain–sidechain hydrogen bond (determine the distribution of angles and distances for these bonds).

Exercise 6.11. The pairing of Gln (i) and Asp ($i+4$) in alpha helices can make a sidechain–sidechain hydrogen bond in addition to the mainchain–mainchain hydrogen bond [240]. What other sidechain pairs can lead to sidechain–sidechain hydrogen bonds in alpha helices or beta sheets?

Exercise 6.12. Using the frequency data in Table 4.3, determine the relative frequencies of nonmainchain donors and acceptors in a typical protein based on the data in Table 6.2.

Exercise 6.13. Use the law of cosines to express the distance $|H - A|$ in terms of $|D - H|$, $|D - A|$, and $\angle DHA$ in Figure 6.7.

Exercise 6.14. Use the law of cosines to express the distance $|D - H|$ in terms of $|A - H|$, $|D - A|$, and $\angle DHA$ in Figure 6.7. Use Reduce to generate hydrogen positions and Wrappa to determine hydrogen bond distances and angles for a set of PDB files. Use your formula to determine resulting the distribution of N–H distances in protein backbones.

Chapter 7
Composition of Protein–Protein Interfaces

Dorothy Mary (Crowfoot) Hodgkin (1910–1994) won the Nobel Prize in Chemistry in 1964 "for her determinations by X-ray techniques of the structures of important biochemical substances."

We now turn to a key question: What factors are most influential in protein–ligand binding? We review attempts to answer this question both to give a sense of the historical development and also to emphasize key aspects of the data mining techniques used. Later in the book, we will clarify the role of dehydrons in this process, but for now we proceed naïvely to get a sense of how the ideas developed. Although we ultimately want to consider general ligands, we work only with ones that are themselves proteins in this chapter.

Protein associations are at the core of biological processes, and their physical basis, often attributed to favorable pairwise interactions, has been an active topic of research [55, 91, 195, 257, 320, 352, 417, 511]. A common belief has been that hydrophobic interactions play a central role in protein–ligand binding. According to [493], the "hydrophobic effect [...] is believed to play a dominant role in protein–protein interactions." The guiding principle of [208] was that in the past "interfaces were found to have more hydrophobic residues than the rest of the protein surface. Recent large-scale studies of protein complexes have confirmed the importance of hydrophobicity in protein–protein interactions." However, more precise comments could also be found. According to [535], "The prevailing view holds that the hydrophobic effect has a dominant role in stabilizing protein structures." Whether or not hydrophobic effects are important for protein–ligand binding, one might also expect that the sort of bonds that help proteins form their basic structure would also be involved in joining two different proteins together.

7.1 Direct Bonds

Both hydrogen bonds and salt bridges play a significant role at protein interfaces. The mean density of hydrogen bonds between two different proteins at an interface is reported to be $0.474/nm^2$ [535], about one per 2.11 square nanometers, corresponding to a grid of size 14.5 Å on a side. The mean size of a protein–protein interaction patch was found to be 1558 Å^2; data were based on the 46 single-patch interfaces listed in Table I of [81]. For reference, the

© Springer International Publishing AG 2017
L.R. Scott and A. Fernández, *A Mathematical Approach to Protein Biophysics*,
Biological and Medical Physics, Biomedical Engineering,
DOI 10.1007/978-3-319-66032-5_7

interfacial area per atom reported in [81] has a mean of 9.25 $Å^2$; this corresponds to 10.81 atoms per nm^2, or about one atom per grid block with a grid size of about 3.05 Å. Grids of these respective sizes are indicated in Figure 7.1; a typical patch would consist of a square composed of about 13 grid steps on each side.

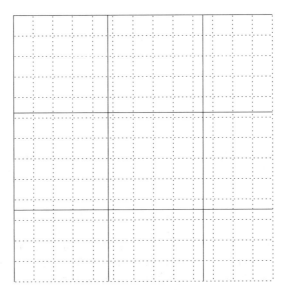

Fig. 7.1 Meshes indicating the average density of atoms and hydrogen bonds at typical protein–protein interfaces, based on the data from [81, 535]. The finer, dotted mesh indicates the atom density (one per square), and the coarser, solid mesh indicates the hydrogen bond density (one per two squares). A typical interaction patch consists of about 13 grid steps on each side, corresponding to the entire figure.

The average number of hydrogen bonds per interface studied in [535] is about ten. Thus, there is a small discrepancy between the data in [535] and [81], as is typical for different studies of interfaces. On the other hand, the average number of salt bridges per interface is only two [535]. Disulfide bonds play a more limited and specialized role.

It might be that the story of protein–protein interactions ends here, with the intermolecular hydrogen bonds and salt bridges being the whole story. However, three of the 54 high-resolution structures studied in [535] have no hydrogen bonds or salt bridges, and another dozen have no salt bridges and five or fewer hydrogen bonds. Not surprisingly, we will begin to see indications of the role of intramolecular hydrogen bonds that become enhanced upon binding, as we depicted in Figure 2.7.

One factor that complicates the picture of protein–protein interactions is the appearance of water molecules that play a structural role, as opposed to simply mediating interactions via dielectric effects. In the protein interfaces studied in [535], polar atom pairs bridged by water across the interface with hydrogen bonds were potentially more numerous than direct hydrogen bond pairs, with each water molecule connecting 3.8 cross-chain atom pairs on average. This unusually high number is not fully explained, but the prospect was suggested that "water molecules are most likely to allocate their hydrogen bonds to the protein atoms dynamically, in the so-called flip-flop mechanism" [535].

	area	length
atoms	$9.25\,\text{Å}^2$	$3.05\,\text{Å}$
hydrogen bonds	$211\,\text{Å}^2$	$14.5\,\text{Å}$
salt bridges	$\approx\!800\,\text{Å}^2$	$28.3\,\text{Å}$
interface patch	$\approx\!1600\,\text{Å}^2$	$40\,\text{Å}$

Table 7.1 Density of atoms, hydrogen bonds and salt bridges in proteins according to [81, 535]. Areas associated with a single entity; density is the inverse of this area. The length in column 3 is just the square root of the area.

7.2 Hydrophobic Interactions

There have been several attempts to define a hydrophobicity scale for protein sidechains as a guide to protein–ligand binding [52, 130, 292, 344, 349, 369, 402]. A similar, but distinct, concept is that of **lipophilicity** [335] which measures the extent to which substances dissolve in a nonpolar solvent. Although the scales are designated as hydrophobicity measures, they are really intended to be proxies for the local dielectric environment [344]. One characteristic feature of most hydrophobicity measures is that the scale attempts to balance hydrophobicity with hydrophilicity, in such a way that amphiphilic residues tend to be in the middle of the scale. However, a hydrophilic residue does not cancel the hydrophobic effect in a simple way, at least regarding its impact on the local dielectric. A hydrophilic residue surrounded by hydrophobic groups will not have a strong effect on the dielectric environment. As we will see in Chapter 15, it is both the abundance and the mobility of water molecules that contribute to the dielectric effect. A small number of (confined) water molecules hydrogen bonded to a singular polar group on a protein sidechain will not cause a significant increase in the local dielectric. We will present here a scale for sidechains based on data mining protein interfaces that turns out to correlate closely with the amount of wrapping.

We review one attempt [208] to discover hydrophobic interactions by examining protein–protein interfaces. The hypothesized form of the interactions in such studies determined the basic choices that guided the data mining. In particular, a definition of "sidechain interaction" using only the proximity of C_β carbons was used [208] to quantify interactivity; it does not take into account individual sidechain features. This is depicted in Figure 7.2 in which two sidechains on the right have close C_β carbons, but two on the left are not so close even though they form a hydrogen bond. Such a definition is appropriate for postulated hydrophobic interactions which are generally nonspecific, but it is not designed to detect more subtle relationships. By a re-examination of the data, the studies actually provide confirmation that hydrophobic–hydrophobic interactions are not prominent in the interfaces studied.

Indeed, hydrophobic–polar pairings at protein–protein interfaces are frequent and challenge the commonly held view regarding hydrophobic interactions. The prediction and explanation of binding sites for soluble proteins require that we quantify pairwise energy contributions. But also we must concurrently explain the extent to which surrounding water is immobilized or excluded from the interactive residue pairs. As proteins associate, their local solvent environments become modified in ways that can dramatically affect the intramolecular energy [21, 143, 160, 164, 182, 361, 511].

Water removal from hydrophobic patches on protein surfaces has a high thermodynamic benefit [55, 91, 195, 257, 320, 352, 417, 511], due to an entropic gain by the solvent. The water next to hydrophobic patches lacks interaction partners (hydrogen bond partners), and in moving to a bulk environment it gains hydrogen bonds. Thus, hydrophobic patches are possible binding regions provided there is a geometric match on the binding partner. However, most protein surfaces have ratios of hydrophilic to hydrophobic residues reportedly ranging from

7:1 to 10:1 [174]. Moreover, hydrophobic patches involved in associations at an interface are often paired with polar groups [465]. We will ultimately explain how this can be energetically favorable, but we begin with some small steps to understand better the nature of protein interfaces.

7.3 Amino Acids at Protein–Protein Interfaces

We begin with a simple use of data mining applied to the understanding of amino acid tendencies at interfaces. There are different questions that one can ask, and of course it is natural that amino acids get ranked in different orders accordingly. For simplicity, we contrast just two that have appeared in the literature, but we also review others in Section 7.5.3. The data here are drawn primarily from [62, 181, 208]. The two basic questions we address are the following.

- What residues are most likely to be *found* at an interface?
- What residues are most likely to be *interacting* at an interface?

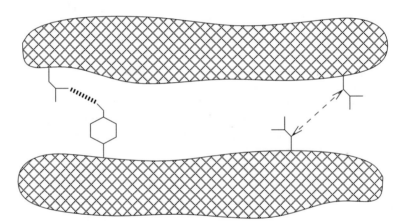

Fig. 7.2 Cartoon showing a protein–protein interface and some possible ways to characterize interactions between residues. On the right, two residues (Asp and Asn fit the description, but Gln, Glu, or Val are similar) have their C_β carbons close, the criterion used in [208]. On the left, a hydrogen bond between Tyr and Gln is depicted, but with the C_β carbons of each at a greater distance.

It is important to realize how these questions differ and how they drive the resulting data mining experiments.

Unfortunately, there is no universal way to define a protein–protein interface. The basic idea is that it is the region of the proteins in which there are two sidechains, one from each protein, having a specified relationship. For example, two such sidechains are said to be "interacting" in [208] if the distance between their C_β carbons (C_α for Gly) was less than 6 Ångstroms. A shortcoming of this definition is indicated in Figure 7.2, in which two sidechains making a hydrogen bond across the interface will not be counted as interacting, whereas two others may be counted even though there is no obvious type of interaction.

On the other hand, the JAIL database [215, 507] defines an interface by requiring two atoms from sidechains from separate parts of protein structures to be within 4.5 Å, in which case the two entire residues are considered to be interacting. This definition further requires at least 5 C_α atoms to be within the interaction zone [507]. Furthermore, distinctions are made between **obligate interactions**, between two fold domains within a single protein, and transient (nonobligate) interactions, between different proteins [331, 507].

Another approach to define an interface uses the concept of **solvent-accessible surface**. This has a mathematically rigorous definition, although it does not describe accurately the solvation of a protein. It should really be called a **sphere-accessible surface**. To define this surface, we first define a domain that could be called the van der Waals domain of a protein:

$$W = \cup_j \left\{ x \mid \text{distance}(x, a_j) \le r_j \right\}, \qquad (7.1)$$

where a_j is the location of the j-th atom in the protein and r_j is the van der Waals radius (Table 3.2) of the j-th atom. We then define a set of points which are centers of (open) spheres (of radius $\lambda > 0$) contained in the exterior of W:

$$X_\lambda = \left\{ x \mid \text{distance}(x, W) > \lambda \right\}. \qquad (7.2)$$

The solvent-accessible surface S_λ is the boundary of X_λ:

$$S_\lambda = \partial X_\lambda. \qquad (7.3)$$

The physical justification of this definition is not that water is roughly spherical, but rather that the locations of the points in a water molecule, for arbitrary possible orientations of a water molecule, should be represented by a spherical volume. Thus, we can say that (with the appropriate λ), X_λ is a set of points where we can be sure that a single water molecule could be accommodated (without steric conflict). The set of points where water could actually go is potentially larger [539], since it is not a spherical object. On the other hand, if we considered the hydrophobicity of a patch of protein, it might be that water would not be very likely to get as close as S_λ might suggest. Thus, a different notion of **solvent exclusion surface** has been proposed [539]. Finally, and most importantly, water must form a network of hydrogen bonds, so estimating the closest approach of a single water molecule is not very relevant. Instead, a more global approach should be used to assess the set of possible water locations. So S_λ gives only an approximation to an actual solvation surface, and the errors in prediction could be on either side of the surface.

The contact area for a protein–protein interface may be estimated by comparing the area of S_λ in (7.3) (or other estimated solvation surface area) for the two proteins separately with that of the joined unit. In [81], the sidechains contributing to this discrepancy in area are counted as being in the interface. Another approach to identify interfaces uses threading techniques originally used to predict protein folding [324]. Since there is little agreement in how interfaces are defined, we will not attempt to give the details in each case. Differences in conclusions could be largely affected by differences in definitions.

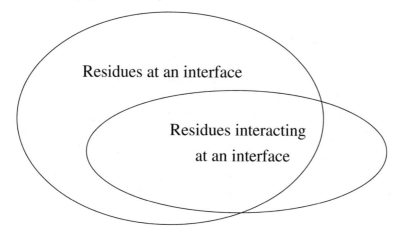

Fig. 7.3 Cartoon showing possible relationship between two data sets. Because the characterization of the interface and of interactions can differ, data sets do not necessarily satisfy obvious inclusion relations.

The site specificity of protein–protein interactions has been widely studied due to its central biological significance [111, 208, 238, 257–259]. Hydrophobic residues such as Leu and Val are more abundant at protein–ligand interfaces. As a result, the role of hydrophobic residues in the removal of water surrounding the protein surface has been assumed to be a dominant factor for association [172, 463]. But it is also true that such residues are more abundant in proteins in general (see Tables 4.3 and 7.3).

The first question [181] we consider is about the amino acid composition of protein–protein interfaces. This can be done by simply counting, once an identification has been made regarding which amino acids are at an interface. However, simple frequencies are misleading: Leu is the most common residue at interfaces, but it is also overwhelmingly the most common residue in most proteins. Thus, one has to normalize by the natural frequencies of amino acids in proteins [62].

The second question [208] is about the amino acid composition for pairs of amino acids at interfaces *that are interacting*. There are many ways to define interaction, but proximity [208] is a natural metric. Thus, two residues are defined [208] as interacting if their C_β coordinates differ by at most 6 Å (with C_α used for Gly). This notion is simplistic in that the C_β atom is only the first in the sequence. The key feature of this definition of interactivity is that it does not discriminate *how* sidechains might be interacting. Moreover, it might be that two sidechains are forming a bond and yet their C_β atoms are further apart than 6 Å.

In view of Section 7.1, it would be reasonable to pre-filter potential interactions by known interactions, such as hydrogen bonds, salt bridges, and so forth (e.g., cation-π interactions and dehydrons) [93]. But this approach was not taken in [208].

Let us compare and contrast the two questions. The first question seeks to determine clues for protein–protein association by investigating all residues, suitably normalized. The second question assumes that proximity of sidechain pairs is a significant factor in protein–protein association and thus looks for consequences of restricting to such pairs. Not surprisingly, each question returns different answers regarding the relative significance of different residues. In Figure 7.3, we depict the difference between the two data sets. We allow for the fact that being "at the interface" may be differently defined in each case, leading to the possibility that neither set contains the other.

The distribution of amino acid composition in proteins displays evolutionary trends [62], and this can require extra care to reveal subtle relationships. Here we limit our investigations to fairly strong trends for simplicity. However, the precise numerical data presented would differ if different databases were chosen for the primary data being used.

7.4 Interface Propensity

A common belief is that hydrophobic residues on the surface of proteins are likely candidates to support interfaces in protein–protein association. In Section 7.5, we present evidence that supports this case with suitable clarifications. However, [181] presents data with a distinctively different conclusion, by focusing on all residues found at an interface and normalizing the relative abundance of residues at the interface by their overall abundances. The residues with the highest relative propensity [181] to be at interfaces are, in decreasing order of frequency, the group depicted in Figure 4.3(a). None of these residues is distinctively hydrophobic. This is quite a surprising result, and it demands an explanation.

To begin with, let us clarify the basic notions. If we have a data set with N different types of characteristics (e.g., $N = 20$ and the characteristics are the different amino acids), recall that the frequency f_i of the i-th characteristic is defined by (4.2), where o_j is the number of occurrences of the j-th characteristic in the data set.

If we have two data sets with the same characteristics (e.g., amino acid names), with frequencies f_i and g_i, respectively, as defined in (4.2), then one can define a **relative frequency**

$$r_i = f_i/g_i \qquad (7.4)$$

of the characteristics between the two data sets. There are some problems with this measure of occurrence. First of all, it might happen that $g_k = 0$ for some k, making the interpretation difficult. Related to this is the need for normalization in order to be able to relate two different comparisons. In [181], the following approach was taken.

Define a normalized **relative propensity** via

$$R_i = \frac{r_i}{\sum_{j=1}^{N} r_j}. \qquad (7.5)$$

These relative propensities sum to one, so we can think of them like ordinary frequencies. Similarly, we multiply by 100 in (4.2) to convert to percentages as the unit of "frequency."

3-letter code	1-letter code	Nonpolar Carbons	Interface Rel. Prop.	Dehydron Rel. Prop.	Hydro-pathy	Levitt s value
Asn	N	1	+1.28	+1.63	−3.5	0.2
Thr	T	1	+1.10	+1.41	−0.7	−0.4
Gly	G	0	+0.99	+1.42	−0.4	0.0
Ser	S	0	+0.60	+0.80	−0.8	0.3
Asp	D	1	+0.34	+0.76	−3.5	2.5
Ala	A	1	+0.29	+0.60	1.8	−0.5
Cys	C	0	+0.25	+0.24	2.5	−1.0
Val	V	3	+0.20	−0.31	4.2	−1.5
Met	M	1	+0.10	+0.10	1.9	−1.3
Tyr	Y	6	+0.10	+0.10	−1.3	−2.3
His	H	1	−0.25	−0.25	−3.2	−0.5
Pro	P	2	−0.25	−0.25	−1.6	−1.4
Trp	W	7	−0.33	−0.40	−0.9	−3.4
Arg	R	2	−0.35	−0.40	−4.5	3.0
Leu	L	4	−0.35	−1.10	3.8	−1.8
Phe	F	7	−0.40	−0.40	2.8	−2.5
Lys	K	3	−0.42	−0.38	−3.9	3.0
Glu	E	2	−0.50	−0.11	−3.5	2.5
Gln	Q	2	−0.62	−0.60	−3.5	0.2
Ile	I	4	−0.70	−0.92	4.5	−1.8

Table 7.2 Amino acids ranked according to their likelihood of being found at protein–protein interfaces. The third column indicates the number of nonpolar carbon groups in the sidechain (see Table 8.2). Interface and dehydron relative propensity (Rel. Prop.) is given as $R_i − 5$ as in (7.5). Dehydron propensity is also presented as frequency $f − 5$; 5% is the average propensity to be at an interface or engaged in a dehydron. The hydropathy scale of Kyte et al. [292] and the hydrophobicity values s of Levitt [307] are included for reference.

If we apply this approach to data sets of proteins, and the characteristics are the different amino acid constituents, then we obtain the scheme used in [181]. In this case, the sum of the relative propensities (in percentage units) is one hundred, so the mean is five. In Table 7.2, data from [181] is presented in terms of the deviation of these relative propensities from the mean of five. That is, the data represent $100R_i - 5$.

The unusual ranking of residues in Table 7.2 was explained in [181] by noting that it correlates closely with the propensity to be engaged in under-wrapped backbone hydrogen bonds, among amino acids acting as either proton donors or acceptors for mainchain hydrogen bonds. These data are presented in the fifth column in Table 7.2, and the correlation is striking. Such bonds, in turn, are determinants of protein–protein associations, as discussed subsequently.

Since we expect a significant number of intermolecular hydrogen bonds (and some salt bridges) at interfaces, we might expect residues capable of making them (cf. Table 6.2) to be more likely at interfaces. But these residues are uniformly distributed in Table 7.2, not clustered near the top. If anything, the charged residues are clustered near the bottom. This implies that another factor determines the propensity to be at an interface, as suggested in [181], namely, the amount of wrapping a residue can provide.

As noted in [181], the five of the seven residues in Figure 4.3(a), with the highest propensity for being engaged in under-desolvated hydrogen bonds, also have at most one torsional degree of freedom in their sidechain. Thus, we expect the entropic loss resulting from the conformational hindrance of the sidechains upon protein association is minimal with these sidechains, so that the energetic benefit of intermolecular protection of pre-formed hydrogen bonds is most beneficial. The only purely hydrophobic residue that has an appreciable propensity to be in an interface is Val (cf. Figure 4.3(b)), with only one sidechain rotameric degree of freedom. Therefore, its conformational hindrance upon binding also entails minimal loss in conformational entropy.

Consideration of the residues ranked at the bottom of Table 7.2 demonstrates that hydrophobic residues on the protein surface are infrequent relative to their overall abundance. This implies that they are negatively selected to be part of binding regions, and thus, they must play a secondary role in terms of binding.

Note that the polar residues (Asn, Asp, Ser, Cys, and Thr) with a minimal distance from their polar groups to the backbone are likely to be engaged in dehydrons, according to Table 7.2. It is presumed [181] that this arises not only because they have minimal nonpolar carbonaceous groups, but also because the relative proximity of their polar groups to a backbone hydrogen bond may limit further clustering of hydrophobic groups around the bond. Gly is itself the greatest under-wrapper and can even be thought of as polar due to the fact that the polar environment of the peptide bond is exposed; Ala is the penultimate under-wrapper and may also exhibit some of the polar qualities of Gly (cf. Section 4.4.1).

7.5 Amino Acid Pairs at Interfaces

We now return to the second question raised at the beginning of the chapter regarding the amino acid composition for interacting pairs of amino acids at interfaces. We recall the warm-up exercise regarding the pair frequencies of amino acid residues in secondary structure presented in Section 5.5. This gives a baseline for what to expect regarding pair frequencies in general. We review the results in [208] which attempt to explain the common belief that hydrophobic interactions should dominate. We show in fact that a re-examination of the the results in [208] indicates something contrary.

7.5.1 Pair Interactions at Interfaces

We review the results in [208] which use proximity as an interaction metric in which two residues are defined as interacting if their C_β coordinates differ by at most 6 Å. We utilize the basic technology regarding pair frequencies developed in Section 5.5. In [208], a quantity G_{ij} is defined by multiplying the log odds ratio by a numerical factor of ten. We recall that the hypothesis of [208] was that hydrophobic interactions would be the most prevalent. In this setting, some dominant residue pairs are indeed hydrophobic, although it is pointed out in [208] that they "occurred more often in large contact surfaces, while polar residues prevailed in small surfaces," anticipating the subsequent discussion regarding "core" versus "rim" residues [81]. We present in Table 7.3 the residues and their relative propensies, as defined in (7.5), in decreasing order, as determined in [208].

Two of the residues in Table 7.3 with greatest relative propensity, namely Trp and Pro, are distinctively hydrophobic, as expected in the basic hypothesis of [208]. However, these are also two of the most unique residues, as discussed in Section 4.4. Trp–Pro can be involved in a what is called a "sandwich" [408], so this is indeed a result in line with the basic hypothesis. But there are other sidechains which are purely hydrophobic and have few other interesting features, such as Val, Leu, Ile, and Phe. If the hypothesis of [208] were correct, these sidechains would be dominant among the leading pairs. Instead, other high-ranking residue pairs in Table 7.3 involve residues ranked at the top in Table 7.2. Indeed, one might wonder why there would be any differences in the two tables. The differences between Table 7.3 and Table 7.2 reflect the fact that we are now asking about residues which are in proximity of a specific residue.

Res. Code	Pairing Rel. Prop.	Pairing Rel. Freq.	Pairing Freq. [208]	Total Abundance [62]	Interface Rel. Prop.	Rim/Core freq. [81]
Cys	5.4	2.40	1.87	0.78	+0.25	0.45
Trp	1.9	1.60	1.63	1.02	−0.33	0.32
Pro	1.7	1.55	6.74	4.35	−0.25	1.24
Ser	1.5	1.50	7.01	4.66	+0.60	1.04
Asn	1.3	1.46	4.90	3.36	+1.28	1.19
Thr	1.1	1.41	6.87	4.87	+1.10	1.19
His	0.76	1.33	2.56	1.92	−0.25	0.52
Tyr	0.32	1.23	3.70	3.00	+0.10	0.67
Gly	0.11	1.18	8.59	7.30	+0.99	1.16
Ala	0.11	1.18	9.18	7.77	+0.29	0.95
Phe	−0.15	1.12	4.02	3.61	−0.40	0.33
Gln	−0.33	1.08	3.41	3.15	−0.62	1.03
Met	−0.72	0.99	2.38	2.41	+0.10	0.54
Asp	−0.98	0.93	5.06	5.42	+0.34	1.48
Val	−1.2	0.87	7.12	8.17	+0.20	1.09
Leu	−1.6	0.79	7.05	8.91	−0.35	0.82
Ile	−1.8	0.75	5.00	6.66	−0.70	0.76
Arg	−1.9	0.71	4.46	6.27	−0.35	1.19
Glu	−2.6	0.55	4.71	8.59	−0.50	1.87
Lys	−2.9	0.48	3.73	7.76	−3.9	2.16

Table 7.3 Amino acids which occur in pairs at interfaces and their relative abundances as determined in [208]. Primary data is taken from the indicated references. Relative propensity is defined in (7.5) and relative frequency is defined in (7.4). Interface relative propensity from Table 7.2 is included for comparison. The rim versus core frequencies ratio indicates the ratio of frequencies of residues found in the rim versus core of a protein interface [81], as depicted in Figure 7.4. Column 4 is computed from (7.6) using C_{ij} data from [208]. Column 3 is computed using (7.4) using the frequency data in columns 4 and 5. Column 2 is derived from (7.5) applied to columns 3.

Since Table 7.3 does not provide relative abundances directly, we need to say how these have been derived. The fundamental data in Table 7.3 are Table II on page 93 in [208], which lists the "contact" matrix C_{ij}. This is a matrix that counts the number of times that residue i contacts (is within the proximity radius of) residue j. Summing a column (or row) of C_{ij} and normalizing appropriately give the total frequency F_i of the i-th amino acid involved in such pairings. More precisely, to report frequencies as a percentage, define

$$F_i = 100 \frac{\sum_{j=1}^{20} C_{ij}}{\sum_{i,j=1}^{20} C_{ij}} \tag{7.6}$$

to be the amino acid pairing frequency, shown in the column entitled "Pairing Freq. [208]" in Table 7.3.

The abundance of each amino acid in such pairings needs to be normalized by an appropriate measure. Here we have taken for simplicity the abundances published in [62] which are reproduced in the column entitled "Total Abundance [62]" in Table 7.3. We do not claim that this provides the optimal reference to measure relative abundance in this setting, but it certainly is a plausible data set to use. The data shown in the column entitled "Pairing Rel. Freq." in Table 7.3 represent the ratio of F_i, defined in (7.6), to the abundances reported in [62].

The fact that Cys appears to have the highest relative abundance in pairs at interfaces reflects the simple fact that when Cys appears paired with another residue, it is unusually frequently paired with another Cys to form a disulfide bond (Section 4.2.2), as confirmed in [208].

Res. Pair	log odds ratio	odds ratio
Cys-Cys	0.626	1.87
Trp-Pro	0.351	1.42
Asp-His	0.220	1.25
Arg-Trp	0.205	1.23
Asp-Ser	0.202	1.22
Asp-Thr	0.191	1.21
Cys-Ser	0.181	1.20
Asp-Gln	0.174	1.19
Met-Met	0.145	1.16
Cys-His	0.136	1.15

Table 7.4 Highest log-odds ratios reported in Table III in [208]. Note that the numbers listed in that table are the log-odds ratios inflated by a factor of $A = 10$.

It is noteworthy that the odds ratios indicated in Table III of [208], the largest of which we have reproduced in Table 7.4, are all between one half and two. That is, there are no pairs which occur even as much as twice as frequently as would be expected randomly (or half as frequently). The pair in [208] with the highest odds ratio (1.87) is Cys-Cys, a disulfide bridge. Although Cys is uncommon, when it does appear we can expect it to be involved in a disulfide bridge. The next highest odds ratio pair is Trp–Pro (1.42), which pairs two of the most unique sidechains (Sections 4.4.2 and 4.4.6). The lack of rotational freedom in proline may be significant since there is no entropic loss in the pairing, but the story is likely more complex, e.g., Trp–Pro can be involved in a sandwich [408].

The subsequent four pairs in [208] with the next highest odds ratios involve charged residues:

Asp–His (1.25), Arg–Trp (1.23), Asp–Ser (1.22) and Asp–Thr (1.21).

The first of these pairs is a salt bridge, and the second is a charge–polar interaction known as a cation-π interaction [103, 200, 538] (see Section 3.1.4 and Section 12.3) based on the special polarity of aromatic residues (Section 4.4.6). The latter two pairs are charged and polar residues as well, and their interactions could well be based on hydrogen bonds. The next four pairs in ranking of odds ratio are Cys–Ser (1.20),

Asp–Gln (1.19), Met–Met (1.16) and Cys–His (1.15).

These pairings show a similar mix of polar interactions, not the expected hydrophobic–hydrophobic interactions. In particular, why are Phe–Phe, or Phe–Leu, or the other purely hydrophobic residue pairs not found as commonly at interfaces? In [208], the typical configuration of Arg–Trp is pictured, a likely cation-π interaction, and similar polar pairings are highlighted, such as Lys–Lys, with an odds ratio 0.81, which is a fairly high odds ratio for pairs which repel each other.

There is no absolute scale on which to measure odds ratios, and the significance of any deviation from one is context dependent. But it is notable that the pair frequencies reported in [208] are much smaller than found for alpha helices or beta sheets as indicated in Figure 5.11. We interpret that to mean that the hydrophobic pairs involved in interfaces are more nearly random and uncorrelated than pair occurrences in helices and sheets. This does not mean that the approach of [208] was flawed, rather that the hypothesis was proved to be incorrect. If the dominant interactions had been hydrophobic, these techniques would have discovered them. On the other hand, it could be possible to pre-filter the data to eliminate identifiable interactions, and this might provide interesting new data.

7.5.2 Core Versus Rim

When we add the further analysis in [81] which differentiated the prevalence of core versus rim residues in protein interfaces, the picture is clarified. In [81], interface topology was characterized in detail, and it was found that interfaces could typically be described in terms of discrete patches of about 1600 Å^2 in area. For each patch, the boundary (rim) residues were identified versus the interior (core) residues, each of which constituted about half of the total.

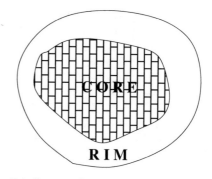

Fig. 7.4 Cartoon showing a protein–protein interface subdivided into rim versus core regions.

The statistics for amino acid preferences for the rim versus the core are reproduced in Table 7.3. There is a strong correlation between being charged or polar and preferring the rim, as indicated in Table 7.5.

Similarly, it is noteworthy that the variance in relative propensities is much greater for pairs of interacting residues at interfaces (Table 7.3) than it is for all (unrestricted) residues at interfaces (Table 7.2). This is not surprising because we have selected for a particular subset of pairs (instead of including all pairs). Combining the previous two observations, we can say that interacting pairs at the core of interfaces are more likely to involve a hydrophobic residue, but the pair compositions involving hydrophobes are nearly random.

To illustrate the sensitivity of results depending on the database chosen, we review the results in [28] which is very similar in spirit to [81], the difference being the use of homodimers for the study of interfaces. In Table 7.5, we present this data, with the residues reordered to give the rim/core preferences in order for the data in [81] to facilitate comparison with the data in [28]. What we see is the same general trend, namely that charged and polar residues prefer the rim, but with changes in the particular rankings among the different groups. However, there is a significant reversal in the roles of arginine and valine [28].

The dissection trilogy is completed in [29] in which an attempt is made to determine amino acid distributions for "nonspecific" interactions. This is intended to be a proxy for any surfaces which might bind however briefly to other protein surfaces. The data set is determined by looking at crystal contact surfaces in the PDB. We leave as an exercise to compare the data for these surfaces with the other data presented here. See [29] for a comparison with the data in [81] and [28].

Res. Code	Rel. Prop.	Rel. Prop.	Pair Freq.[208]	Total Abundance [62]	Rim/Core freq. [81]	Homodimer Rim/Core [28]
Lys	−2.9	0.48	3.73	7.76	2.16	2.19
Glu	−2.6	0.55	4.71	8.59	1.87	1.48
Asp	−0.98	0.93	5.06	5.42	1.48	1.61
Pro	1.7	1.55	6.74	4.35	1.24	1.51
Asn	1.3	1.46	4.90	3.36	1.19	1.49
Thr	1.1	1.41	6.87	4.87	1.19	1.16
Gly	0.11	1.18	8.59	7.30	1.16	1.38
Arg	−1.9	0.71	4.46	6.27	1.19	0.85
Val	−1.2	0.87	7.12	8.17	1.09	0.83
Ser	1.5	1.50	7.01	4.66	1.04	1.15
Gln	−0.33	1.08	3.41	3.15	1.03	1.22
Ala	0.11	1.18	9.18	7.77	0.95	0.93
Leu	−1.6	0.79	7.05	8.91	0.82	0.61
Ile	−1.8	0.75	5.00	6.66	0.76	0.55
Tyr	0.32	1.23	3.70	3.00	0.67	0.58
Met	−0.72	0.99	2.38	2.41	0.54	0.68
His	0.76	1.33	2.56	1.92	0.52	0.85
Cys	5.4	2.40	1.87	0.78	0.45	0.81
Phe	−0.15	1.12	4.02	3.61	0.33	0.40
Trp	1.9	1.60	1.63	1.02	0.32	0.60

Table 7.5 Amino acids which occur in pairs at interfaces and their relative abundances. All data are the same as in Table 7.3 except the last column, but the order of the rows has been shuffled to conform to the Rim/Core frequency order. Primary data is taken from the indicated references.

7.5.3 Further Studies

We have made several observations based on analyzing existing data sets. These conclusions should be viewed as preliminary since these data sets must be viewed as incomplete. Our primary intent was to introduce a methodology for exploring such data sets and to indicate the type of results that can be obtained.

Our basic analysis of pairwise interaction data was taken from [208]. However, the methodology is quite similar to that of the earlier paper [493], although there are differences in the way the interior (and non-interior) sidechains in the interaction zone are defined. That is, the classification of rim and core residues in the interface [208] is different in definition from exposed and interior residues in the interface in [493], although similar in spirit. Figure 3B of [493] shows how the residues that are interacting (proximate) in an interface are very similar in composition to ones in the interior of proteins.

Protein–ligand interfaces differ in function, and interfaces with different function can have different composition. In [257], basic differences between protein–antibody and enzyme–inhibitor pairs, as well as others, are explored. Using more extensive data sets available more recently, this approach has been refined to allow classification of interface type based on amino acid composition [370].

The aromatic sidechains do play a special role in protein interfaces through what is called a cation-π interaction [200] (see Section 12.3). The special polar nature of the aromatic residues (Section 4.4.6) provides the opportunity for interaction with positively charged (cation) residues (Lys, Arg, His), however other types of bonds can be formed as well, including a type of hydrogen bond [308]. The cation-π motifs play a special role in protein interfaces [103, 538]. The cation-π interaction also has a significant role in α-helix stabilization [454].

A study of the role of evolution on protein interface composition can be found in [74]. In [216, 326], interacting amino acids across interfaces are studied and compared with regard to conservation and hot spots.

Protein–protein interactions can be classified in different ways, e.g., by how transient they are, and studies have been done to examine differences in size of interaction zones and sidechain propensities [366, 367].

Identification of individual sidechains that may play the role of "anchors" in protein–ligand recognition is studied in [409] via molecular dynamics simulations. Individual residues are identified that appear to fit into geometric features on paired protein surfaces both in crystal structures and in the dynamic simulations.

It is possible to refine the concept of sidechain interactions to one involving the interactions of individual atoms in structures. This approach has been suggested [96] as a way to discriminate between correct structures and incorrect ones. In [96], this concept was proposed as a way to critique structures being determined based on experimental imaging techniques, but the same concept could be applied to discriminate between native and **decoy structures** that are proposed via computational techniques.

7.6 Hot Spots

In [55], an attempt is made to identify so-called **hot spots** on protein surfaces. They report on the results of an experimental technique called **alanine scanning** in which residues are replaced by alanine and compared with the original protein by some activity assay. A hot spot is defined to be the set of residues X for which the change in binding affinity due to the mutation X→Ala is large.

What they discover is that the most common sidechains at hot spots are the ones that are bulkiest: Trp, Tyr, and Arg. This is not surprising since the replacement by Ala has the greatest geometric, electrostatic, and hydrophobic effects for these residues. However, such substitutions might be extremely rare in nature.

What might be a better test of importance would be other mutations, e.g., ones which do not change the volume or geometry of the sidechain. Systematic replacement of all amino acids by all other amino acids is clearly an order of magnitude more work than just replacing by a fixed sidechain. Having a better model of what governs protein–protein interactions could lead to a more directed study of sidechain mutation effects. A correlation has been found between experiments and the level of residue conservation among related interfaces [326].

Hot spots are known to be closely related to dehydrons [1, 152].

7.7 Conclusions About Interfaces

Two main conclusions were obtained. The first is that residue hydrophobicity is not the primary attribute that determines proximity of a residue to interaction sites. Instead, there is a different "interactivity" order that governs the likelihood of an amino acid residue being in an active zone. This interactivity scale is related strongly to the number of nonpolar constituents of sidechains, which governs the local dielectric environment. Thus, the likelihood of a residue being at an interface is to some extent *inversely* proportional to its hydrophobicity.

On the other hand, pairwise interactions with hydrophobic residues may play a secondary role in protein–protein interactions, especially in the interior, or core, regions of interaction domains. Moreover, their interactions tend to be less specific than might be the case in other pairings, such as in alpha helices and beta sheets. The role of hydrophobic sidechains in such interactions is not revealed by such an analysis. In particular, the definition of "interaction" has been taken to be simple proximity, so it is misleading to infer that there is any identified form of interaction.

The surprising conclusions of this chapter led us to look for additional mechanisms to explain protein–protein interactions. The subsequent chapters will explain a remarkable three-body interaction, namely wrapping of dehydrons, that proves to be essential in the description of such protein associations.

7.8 Exercises

Exercise 7.1. Compare the data for the surfaces in [28, 29, 81] by constructing a table analogous to Table 7.5.

Exercise 7.2. The amino acid frequencies for different data sets constitute probability distributions on the set of amino acids. Different data sets have different distributions. In [29], the distributions for nonspecific interaction surfaces are compared with the distributions for other surfaces [28, 81]. The comparison metric is the L^2 norm. Consider the effect of using the KL-divergence, Jensen–Shannon metric, and the earth-moving metric.

Exercise 7.3. The frequency of location at interfaces provides a linear ranking (Table 7.2) of residues that can be useful in making predictions based on techniques from learning theory. As an example, consider using this to identify under-wrapped hydrogen bonds in α-helices directly from sequence data. For an α-helix, there will be hydrogen bonds formed between residues at a distance of 3, 4, or 5 residues. Generate data from a protein sequence by computing the product of the product of interface ranks of two neighbors. That is, for a sequence abcd define x =rank(a)rank(b) and y =rank(c)rank(d). Thus for every four letter sequence, we assign a pair of numbers (x, y) in the unit square. If there is a dehydron associated with abcd then we expect (x, y) near zero. Using data from the PDB, construct a support-vector machine to separate dehydrons from wrapped hydrogen bonds. Then use this machine to predict dehydrons in sequences for which the sequence is not known.

Exercise 7.4. Compute the distribution of distances from water molecules (HOH) in PDB files to the nearest (a) CH_n, (b) NH, (c) OH, and (d) CO sidechain groups. Give the distances from the HOH coordinates to the nearest heavy atom (oxygen in the case of CO). Omit distances greater than 6 Å.

Exercise 7.5. Compute the distribution of distances from water molecules (HOH) in PDB files to the nearest (a) NH and (b) CO mainchain groups. Give the distances from the HOH coordinates to the nearest heavy atom (oxygen in the case of CO). Omit distances greater than 6 Å.

Exercise 7.6. Redo the analysis in the chapter using the abundance data in Table 4.3 instead of [62].

Exercise 7.7. How frequently does Cys appear in the PDB bonded to another Cys versus nonbonded to a Cys? Use a simple criterion of proximity of the terminal heavy atoms as a proxy for the formation of a disulfide bridge. Scan a large data set of PDB files and determine the relative frequencies of the two cases. Plot the distribution of distances between the terminal heavy atoms of the nearest Cys residues. Also plot the joint distribution of distances and the CSSC dihedral angle.

Exercise 7.8. Define a distance function for sidechains in some way, e.g., by computing the distance between the centroids [187]. Let $\Delta r > 0$ be fixed and define frequencies as follows for a given protein database. Let $f_{i,j,k}$ denote the frequency of occurrences of sidechain pairs i and j whose distance r is in the interval $(k-1)\Delta r \leq r < k\Delta r$. Let

$$f_k = \sum_{i,j} f_{i,j,k}$$

be the frequency of occurrence of any sidechain pair with this distance, and let

$$f_{i,j} = \sum_k f_{i,j,k}$$

be the frequency of occurrence of the pair i, j at any distance. Then the quotient

$$r_{i,j,k} := \frac{f_{i,j,k}}{f_k f_{i,j}}$$

measures the likelihood of finding a pair i, j at distance $k\Delta r$. Plotting $r_{i,j,k}$ as a function of $r = k\Delta r$ suggests the distances at which the two residues i and j are most likely to be located. This can be given an energetic interpretation [187].

Chapter 8
Wrapping Electrostatic Bonds

Francis Harry Compton Crick (1916–2004) was a co-discoverer of the structure of the DNA molecule in 1953 and was awarded the Nobel Prize in 1962. He was also a major contributor to the discovery of the genetic code. He is commemorated by the Francis Crick Institute in London.

We begin by reviewing the history of research regarding the hydrophobic effect and its impact on protein structure and stability. A key concept is the identification of a minimal unit of hydrophobicity: a single nonpolar carbonaceous group CH_n, $n = 0, 1, 2, 3$, a common constituent of protein sidechains. We then review the concepts leading to characterization of nonpolar versus polar groups in molecules, the key idea being electronegativity of atoms. Following these preliminaries, we explain how the hydrophobic effect may be put in a nonstandard context by examining the wrapping of hydrogen bonds by nonpolar carbonaceous groups. We then examine the variability of wrapping as dictated by the genetic code. A single-letter genetic mutation in a codon yields on average a minimal change in polarity and charge of the expressed residue. On the other hand, such a single-letter codon mutation can dramatically change the amount of wrapping of the expressed residue.

8.1 Role of Nonpolar Groups

By 1959, the role of hydrophobicity in protein chemistry was firmly established [263]. Soon afterward [192, 273], the role of hydrophobicity in enhancing the stability and strength of hydrogen bonds in proteins was demonstrated. However, the story developed slowly, and a careful interpretation is required. The paper [273] studied a model molecule, N-methylacetamide, that is, similar to the peptide backbone in structure and forms the same kind of amide–carbonyl (NH–OC) hydrogen bond formed by the backbone of proteins. Infrared absorption measurements were performed to assess the strength and stability of the hydrogen bonds formed by N-methylacetamide in various solvents (including water) with different degrees of polarity. The paper's main conclusion might be misinterpreted as saying that hydrogen bonds are not significant for proteins in water [273]: "It seems unlikely, therefore, that interpeptide hydrogen bonds contribute significantly to the stabilization of macromolecular configuration in aqueous solution." However, the authors did confirm the opposite view in

© Springer International Publishing AG 2017
L.R. Scott and A. Fernández, *A Mathematical Approach to Protein Biophysics*,
Biological and Medical Physics, Biomedical Engineering,
DOI 10.1007/978-3-319-66032-5_8

less polar solvents, so we would now say that their study indicated the value of hydrophobic protection of hydrogen bonds in proteins.

The subsequent paper [192] also studied model molecules, including N-methylacetamide, in solvents based on varying ratios of trans-dichloroethylene and cis-dichloroethylene, via infrared spectroscopy. They established that "the free energy and enthalpy of association of the amides can be expressed as a function of the reciprocal of the dielectric constant." Although the variation in dielectric constants achieved with these solvents reached only a level of one-tenth that of water, this paper quantified the effect of dielectric modulation on the strength and stability of hydrogen bonds in systems similar to proteins. Thus, it remained only to connect the variation in the dielectric constant to quantifiable variations in protein composition.

Although the energetic role of peptide hydrogen bonds remains a subject of significant interest [34, 35], it now seems clear that the variation in hydrophobicity in proteins has a significant and quantifiable effect on the behavior of proteins [116]. According to [535], "The prevailing view holds that the hydrophobic effect has a dominant role in stabilizing protein structures." The quantitative use of hydrophobicity as a marker for hot spots (Section 7.6) in proteins is reported by diverse groups [66, 171].

Attempts to quantify the hydrophobicity of different sidechains have a long history [307]. The role of hydrophobic residues in strengthening hydrogen bonds has been studied by many techniques. The concept we call wrapping here is very similar to what has been termed **blocking** [30] and **shielding** [202, 325]. We prefer the term wrapping since it evokes the image of providing a protective layer around a charged environment. The term "shielding" has a related meaning in electronics, but it is also easy to confuse with "screening" which for us is what the water dielectric performs. The material used for shielding in a coaxial cable is a type of cylindrical screen, and it is a conductor, not an insulator.

In an experimental study [30] of hydrogen exchange [31], the authors stated that (hydrophobic) "amino acid sidechains can enhance peptide group hydrogen bond strength in protein structures by obstructing the competing hydrogen bond to solvent in the unfolded state. Available data indicate that the steric blocking effect contributes an average of 0.5 kJ per residue to protein hydrogen bond strength and accounts for the intrinsic beta-sheet propensities of the amino acids." Although this result is clearly quantitative, it should be understood that the experimental technique is indirect. Hydrogen exchange [31] refers to the exchange of hydrogen for deuterium in a highly deuterated environment, and it most directly measures the lack of hydrogen bonds.

Numerical simulations of peptides also contributed to the growth in understanding of the quantitative effect of hydrophobic groups on hydrogen bonds. Based on computational simulations [503], the authors stated that their results provided "a sound basis with which to discuss the nature of the interactions, such as hydrophobicity, charge–charge interaction, and solvent polarization effects, that stabilize right-handed alpha-helical conformations."

One might ask what minimal quantum of wrapping might be identifiable as affecting the strength or stability of a hydrogen bond. The work on hydrogen exchange [30, 31] shows differences in the effect on hydrogen bonds for various hydrophobic sidechains (Ala, Val, Leu, Ile) which differ only in the number of carbonaceous groups. More recent experiments [325] have looked directly at the propensity to form alpha-helical structures of polypeptides (13 residues) which consisted of X=Gly, Ala, Val, Leu, or Ile flanked on either side by four alanine residues with additional terminal residues (Ac-KAAAAXAAAAKGY-NH2). These experiments directly measured the strength and stability of hydrogen bonds in these small proteins. The experimental evidence [325] again shows differences between the different sidechains X in terms of their ability to increase helix propensity, and hence their effect on the hydrogen bonds supporting helix formation. This observation was further developed in a series of papers [18–21]. More recent, and more complex, experiments [201] confirm that hydrogen bond strength is enhanced by a nonpolar environment.

8.1.1 Unit of Hydrophobicity

Based on the accumulated evidence described previously, we take a *single carbonaceous group to be an identifiable unit of hydrophobicity*. There is perhaps a smaller, or another, unit of interest, but at least this gives us a basis for quantification of the modulation of the dielectric effect. It is perhaps surprising that such a small unit could have a measurable effect on hydrophobicity, but we already remarked in Chapter 1 on comparable effects of a single carbonaceous group regarding the toxicity of alcohols and antifreezes.

It is possible that removal of water can be promoted by components of sidechains other than purely carbonaceous ones. For example, we noted that the arginine residue does not solvate well [336], in addition to the fact that it contains significant carbonaceous groups. A computational study [202] of a 21-residue peptide including a triple (tandem) repeat of the sidechains AAARA concluded that "the Arg sidechain partially shields the carbonyl oxygen of the fourth amino acid upstream from the Arg. The favorable positively charged guanidinium ion interaction with the carbonyl oxygen atom also stabilizes the shielded conformation." Note that the second sentence indicates a possible sidechain–mainchain hydrogen bond.

Since wrapping is of interest because of its implications for hydrophobicity, one could attempt to model hydrophobicity directly as a scalar quantity. Such an approach using a sidechain-based (cf. Section 8.3) evaluation has been taken [65, 66] based on estimates of hydrophobicity provided earlier [307] (see the values in Table 7.2). We have defined wrapping as an integer quantity defined for each bond, but this could (by interpolation) be extended as a function defined everywhere, and the use of a cutoff function [65, 66, 307] essentially does that. But the scalar quantity of real interest with regard to electrostatic bonds is the dielectric, which is described in Section 8.5.

8.1.2 Defining Hydrophobicity

There have been several attempts to define a hydrophobicity scale for protein sidechains as a guide to protein–ligand binding [130, 292, 344, 402]. The numbers for two of these are listed in Table 7.2. Also included in Table 7.2 is a scale for sidechains based on data mining protein interfaces that turns out to correlate closely with the amount of wrapping [181].

Although the scales are designated as hydrophobicity measures, they are really intended to be proxies for the local dielectric environment [344]. One characteristic feature of most hydrophobicity measures is that the scale attempts to balance hydrophobicity with hydrophilicity, in such a way that amphiphilic residues tend to be in the middle of the scale. However, a hydrophilic residue does not cancel the hydrophobic effect in a simple way, at least regarding its impact on the local dielectric. A hydrophilic residue surrounded by hydrophobic groups will not have a strong effect on the dielectric environment. It is both the abundance and the mobility of water molecules that contribute to the dielectric effect. A small number of (confined) water molecules hydrogen bonded to a singular polar group on a protein sidechain will not cause a significant increase in the local dielectric.

For a protein structure to persist in water, its electrostatic bonds must be shielded from water attack [30, 172, 182, 325, 392, 503]. This can be achieved through wrapping by nonpolar groups (such as CH_n, $n = 0, 1, 2, 3$) in the vicinity of electrostatic bonds to exclude surrounding water [172]. Such desolvation enhances the electrostatic energy contribution and stabilizes backbone hydrogen bonds [36]. Any amide and carbonyl partners in backbone hydrogen bonds can become separated temporarily due to thermal fluctuations or other movements of a protein. If such groups remain well wrapped, they are protected from being hydrated and more easily return to the bonded state [107], as depicted in Figure 2.3.

The thermodynamic benefit associated with water removal from preformed structure makes under-wrapped proteins adhesive [160, 173, 175]. As shown in [172], under-wrapped hydrogen bonds (UWHBs) are determinants of protein associations. In Section 9.1, we describe the average adhesive force exerted by an under-wrapped hydrogen bond on a test hydrophobe.

The dielectric environment of a chemical bond can be modified in different ways, but wrapping is a common factor. There are different ways to quantify wrapping. Here, we explore two that involve simple counting. One way of assessing a local environment around a hydrogen bond involves just counting the number of "hydrophobic" residues in the vicinity of a hydrogen bond. This approach is limited for two reasons.

The first difficulty of a "residue-based" approach relates to the taxonomy of residues being used. The concept of "hydrophobic residue" appears to be ambiguous for several residues. In some taxonomies, Arg, Lys, Gln, and Glu are listed as hydrophilic. However, we will see that they contribute substantially to a hydrophobic environment. On the other hand, Gly, Ala, Ser, Thr, Cys, and others are often listed variously as hydrophobic or hydrophilic or amphiphilic. We have identified these five residues in Chapter 4 as among the most likely to be neighbors of underwrapped hydrogen bonds, as discussed at more length in Chapter 7. As noted in Section 4.4.1, glycine, and to a lesser extent alanine, can be viewed as polar, and hence hydrophilic, but alanine has only a nonpolar group in its sidechain and thus would often be viewed as hydrophobic.

A second weakness of the residue-counting method is that it is based solely on the residue level and does not account for more subtle, "subresidue" features. We will see that these limitations can be overcome to a certain extent with the right taxonomy of residues. However, we will also consider (Section 8.4) a measure of wrapping that looks into the subresidue structure by counting all neighboring nonpolar groups.

The residue-counting method is included both for historical and for pedagogical reasons, although we would not recommend using it in general. It provides an example of how models are developed over time, with refinements made once better understanding is available. If an effect is important, then simple models ought to be able to represent it, even if later more refined models are preferred. Thus, we consider in detail some of the predictions from the residue-counting method to see how well they are holding up.

In the residue-counting measure of wrapping, we define precisely two classes of residues relevant to wrapping. This avoids potential confusion caused by using taxonomies of residues based on standard concepts. One could think of this dichotomy as defining hydrophilic versus hydrophobic residues, but that is not intended. In Section 8.3.2, we show that this definition is sufficient to give some insight into protein aggregation and to make predictions about protein behaviors.

However, it is also possible to provide a more refined measure that looks below the level of the residue abstraction and instead counts all nonpolar groups, independent of what type of sidechain they inhabit. We present this more detailed approach in Section 8.4. We will show in Section 9.1 that there is a measurable force associated with an UWHB that can be identified by the second definition. Later, we will define this force rigorously and use that as part of the definition of dehydron in Section 8.5. In Section 8.5, we will review a more sophisticated technique that incorporates the geometry of nonpolar groups as well as their number to assess the extent of protection via dielectric modulation.

8.2 Assessing Polarity

The key to understanding hydrophobicity is polarity. Nonpolar groups repel water molecules (or at least do not attract them strongly), and polar groups attract them. We have already discussed the concept of polarity, e.g., in the case of dipoles (Section 3.3). Similarly, we have noted that certain sidechains, such as glutamine, are polar, even though there the net charge on the sidechain is zero. Here, we explain how such polarity can arise due to more subtle differences in charge distribution.

8.2.1 Electronegativity Scale

atomic symbol	H	C	N	O	F	Na	Mg	P	S	Cl	Se
electronegativity	2.59	2.75	3.19	3.66	4.0	0.56	1.32	2.52	2.96	3.48	2.55
nuclear charge	1	6	7	8	9	11	12	15	16	17	34
outer electrons	1	4	5	6	7	1	2	5	6	7	6
missing electrons	1	4	3	2	1	7	6	5	2	1	2

Table 8.1 Electronegativity scale [297, 383, 432] of principal atoms in biology. The "outer electrons" row recapitulates the data in column 5 in Table 2.1, and the "missing electrons" row lists the number of electrons needed to complete the outer shell.

The key to understanding the polarity of certain molecules is the **electronegativity scale** [297, 383, 432], part of which is reproduced in Table 8.1. Atoms with similar electronegativity tend to form nonpolar groups, such as CH_n and to a lesser extent C-S. Atomic pairs with significant differences in electronegativity tend to form polar groups, such as >C=O and >N−H. The scaling of the electronegativity values is arbitrary, and the value for fluorine has been taken to be exactly four by convention [383].

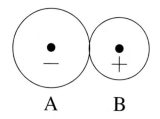

Fig. 8.1 Caricature showing how an atom A with greater electronegativity draws electrons from atom B to which it is covalently bonded: $\mathcal{E}(A) > \mathcal{E}(B)$. The circles indicate the extent of the electron distribution, and the dots indicate the atom nuclei. The $\pm$ signs indicate the resulting polarity of the AB pair.

Let us show how the electronegativity scale can be used to predict polarity. In a >C=O group, the O is more electronegative, so it will pull charge (electrons) from C, yielding a pair with a negative charge associated with the O side of the group, and a positive charge associated with the C side of the pair. Similarly, in an >N−H group, the N is more electronegative, so it pulls charge from the H, leaving a net negative charge near the N and a net positive charge near the H. This is depicted in Figure 8.1. In Section 8.2.2, we will see that molecular dynamics codes assign such partial charges.

Only the differences in electronegativity have any chemical significance. But these differences can be used to predict the polarity of atomic groups, as we now illustrate for the carbonyl and amide groups. For any atom X, let $\mathcal{E}(X)$ denote the electronegativity of X as defined in Table 8.1. Since $\mathcal{E}(O) > \mathcal{E}(C)$, we conclude that the dipole of the carbonyl group >C=O can be represented by a positive charge on the carbon and a negative charge on the oxygen. Similarly, because $\mathcal{E}(N) > \mathcal{E}(H)$, the dipole of the amide group >N−H can be represented by a positive charge on the hydrogen and a negative charge on the nitrogen. A more detailed comparison of the electronegativities of C, O, N, and H gives

$$\mathcal{E}(O) - \mathcal{E}(C) = 3.66 - 2.75 = 0.91 > 0.60 = 3.19 - 2.59 = \mathcal{E}(N) - \mathcal{E}(H). \qquad (8.1)$$

Thus, we conclude that the charge difference in the dipole representation of the carbonyl group ($>$C$=$O) should be larger than the charge difference in the dipole representation of the amide ($>$N$-$H) group. Thus, it would be expected to find larger partial charges for $>$C$=$O than for $>$N$-$H, as we will see. Of course, the net charge for both $>$C$=$O and $>$N$-$H must be zero.

It is beyond our scope to explain electronegativity here, but there is a simple way to comprehend the data. Electronegativity represents the power of an atom to attract electrons in a covalent bond [383]. Thus, a stronger positive charge in the nucleus would lead to a stronger attraction of electrons, which is reflected in the correlation between nuclear charge and electronegativity shown in Table 8.1. More precisely, there is a nearly linear relationship between the electronegativity scale and the number of electrons in the outer shell. The value for hydrogen can be explained by realizing that the outer shell is half full, as it is for carbon.

The noble gasses (helium, neon, argon, etc.) have a complete outer shell and thus have no electronegativity since they fulfill the octet rule. Similarly, atoms with just a few electrons in the outer shell seem to be more likely to donate electrons than to acquire them, so their electronegativity is quite small, such as sodium and magnesium. Hydrogen and carbon are in the middle of the scale, not surprisingly, since they are exactly in the middle between being empty and full of electrons.

Full name of amino acid	three letter	single letter	The various PDB codes for the nonpolar carbonaceous groups
Alanine	Ala	A	CB
Arginine	Arg	R	CB, CG
Asparagine	Asn	N	CB
Aspartate	Asp	D	CB
Cysteine	Cys	C	(CB)
Glutamine	Gln	Q	CB, CG
Glutamate	Glu	E	CB, CG
Glycine	Gly	G	na
Histidine	His	H	CB
Isoleucine	Ile	I	CB, CG1, CG2, CD1
Leucine	Leu	L	CB, CG, CD1, CD2
Lysine	Lys	K	CB, CG, CD
Methionine	Met	M	CB (CG, CE)
Phenylalanine	Phe	F	CB, CG, CD1, CD2, CE1, CE2, CZ
Proline	Pro	P	CB, CG
Serine	Ser	S	na
Threonine	Thr	T	CG2
Tryptophan	Trp	W	CB, CG, CD2, CE3, CZ2, CZ3, CH2
Tyrosine	Tyr	Y	CB, CG, CD1, CD2, CE1, CE2
Valine	Val	V	CB, CG1, CG2

Table 8.2 PDB codes for nonpolar carbonaceous groups. The carbonaceous groups (CG, CE) surrounding the sulfur in Met and (CB) adjacent to sulfur in Cys may be considered polar. The notation "na" indicates there are no nonpolar carbonaceous groups.

Full name of compound	PDB code	The various PDB codes for the nonpolar carbonaceous groups
pyroglutamic acid	PCA	CB, CG
phosphorylated tyrosine	PTR	CB, CG, CD1, CD2, CE1, CE2
staurosporine	STU	Ci, $i = 1, \ldots, 7$; $i = 11, \ldots, 16$; C24, C26

Table 8.3 Sample PDB codes and nonpolar carbonaceous groups for some nonstandard amino acids and other compounds.

8.2.2 Polarity of Groups

Using the electronegativity scale, we can now estimate the polarity of groups of atoms. For example, the near match of electronegativity of carbon and hydrogen leads to the correct conclusion that the carbonaceous groups CH_n, $n = 0, 1, 2, 3$ are not polar, at least in appropriate contexts. The typically symmetric arrangement of hydrogens also decreases the polarity of a carbonaceous group, at least when the remaining $4 - n$ atoms bonded to it are other carbons or atoms of similar electronegativity.

If a carbon is not covalently attached exclusively to carbon or hydrogen, then it is likely polarized and carries a partial charge. Thus, C_α carbons and the carbons in the carbonyl ($>C=O$) group in the peptide bonds of all residues are polar. Sidechain carbons are polar if they are covalently attached to heteroatoms such as N or O. Sulfur (S) is a closer electronegative match with carbon and polarizes carbon to a lesser extent. The case CH_n with $n = 0$ occurs in the aromatic sidechains in the C_γ position, and there are molecules (e.g., beta-Carotene) in which carbons are bonded only to other carbons. The number of carbon neighbors can be either three or four (e.g., in Fucoxanthin).

To illustrate the polarity of the atoms not listed in Table 8.2, we present the partial charges of the remaining atoms as utilized in the Gromos code in Table 8.4 and Table 8.5. In Table 12.1, partial charges for aromatic sidechains are listed. See [125] for partial charges in the Amber code.

In addition to the charges shown for the individual sidechain atoms, the backbone is assigned partial charges as follows: the charges of the amide group are ± 0.28 and the carbonyl group are ± 0.38 [17]. That is, in the amide (N–H) group, the N is given a partial charge of -0.28 and the H is given a partial charge of $+0.28$. Similarly, in the carbonyl (C–O) group, the O is given a partial charge of -0.38 and the C is given a partial charge of $+0.38$. Note that the partial charges for C–O are larger than the partial charges for N–H, in accord with our prediction using the electronegativity scale in (8.1).

The N-terminal and C-terminal groups also have appropriate modifications. The C-terminal oxygens have a charge of -0.635, and the attached carbon has a charge of 0.27. The N-terminal nitrogen has a charge of 0.129, and the attached three hydrogens have a charge of 0.248. All of the groups listed in Table 8.2 have zero partial charge.

8.3 Counting Residues

In [167], the definition of "well-wrapped" was based on the proximity of certain residues and defined in relation to the observed distribution of wrapping among a large sample set of proteins. The extent of hydrogen bond desolvation was defined by the number of residues ρ_R with at least two *nonpolar* carbonaceous groups (CH_n, $n = 0, 1, 2, 3$) whose β-carbon is contained in a specific desolvation domain, as depicted in Figure 8.2. In Section 8.2.2,

Residues	atom type	PDB codes	charge
ASP (GLU)	C	CG (CD)	0.27
	OM	ODi (OEi)$i = 1, 2$	-0.635
ASN (GLN)	NT	ND2 (NE2)	-0.83
	H	HD2i (HE2i), $i = 1, 2$	0.415
	C	CG (CD)	0.38
	O	OD1 (OE1)	-0.38
CYS	S	SG	-0.064
	H	HG	0.064
THR	CH1	CB	0.15
	OA	OG1	-0.548
	H	HG1	0.398
SER	CH2	CB	0.15
	OA	OG	-0.548
	H	HG	0.398

Table 8.4 Partial charges from the Gromos force field for polar and negatively charged amino acids.

Residue	atom type	PDB codes	charge
ARG	CH2	CD	0.09
	NE	NE	-0.11
	C	CZ	0.34
	NZ	NHi, $i = 1, 2$	-0.26
	H	HE, HHij, $i, j = 1, 2$	0.24
LYS	CH2	CE	0.127
	NL	NZ	0.129
	H	HZi, $i = 1, 3$	0.248
HIS (A/B)	C	CD2/CG	0.13
	NR	NE2/ND1	-0.58
	CR1	CE1	0.26
	H	HD1/HE2	0.19

Table 8.5 Partial charges from the Gromos force field for positively charged amino acids. The partial charges for His represent two possible ionized states which carry neutral charge.

we explained how to determine the polarity of groups using the electronegativity scale. The nonpolar carbonaceous groups are listed in Table 8.2.

The C_α carbons in all residues are covalently bonded to a nitrogen atom. The mismatch in electronegativity between carbon and nitrogen (Table 8.1) implies that the C_α carbons are polar and thus do not contribute to repelling water. Sidechain carbons are counted only if they are not covalently attached to heteroatoms such as N or O. The CH groups in serine and threonine are attached to an oxygen, which renders them polar. Similarly, a lone carbon that is attached to oxygens is also polar. Thus, the seven residues listed in Figure 4.3(a) are eliminated from the group of wrappers, as well as Met and His, in the residue-counting method.

8.3.1 Desolvation Domain

The desolvation domain was chosen in [167] to be the union of two (intersecting) 7 Å-radius spheres centered at the C_α-carbons of the residues paired by the hydrogen bond, as shown in

Figure 8.2. The desolvation circles in Figure 8.2 are drawn artificially large (corresponding to roughly 9 Å) in this two-dimensional depiction to show various possibilities.

The choice of the C_α-carbons as the centers of the desolvation spheres is justified in Figure 8.3. These figures show that the center of the line joining the centers of the desolvation spheres is often the center of the hydrogen bonds in typical secondary structures. In the case of a parallel β-sheet, the desolvation domain is the same for two parallel hydrogen bonds. The radius represents a typical cutoff distance to evaluate interactions between nearby residues. C_α-carbons which are neighboring in protein sequence are about 3.8 Å apart (cf. Exercise 2.3). The distance between other C_α-carbons is easily determined by data mining in the PDB (cf. Exercise 2.4).

An amide-carbonyl hydrogen bond was defined in [167] by an N–O (heavy-atom) distance within the range 2.6–3.4 Å (typical extreme bond lengths) and a 60-degree latitude in the N–H–O angle (cf. Section 6.4). As a scale of reference, at maximum density, water occupies a volume that corresponds to a cube of dimension just over 3.1 Å on a side (cf. Section 18.7.3).

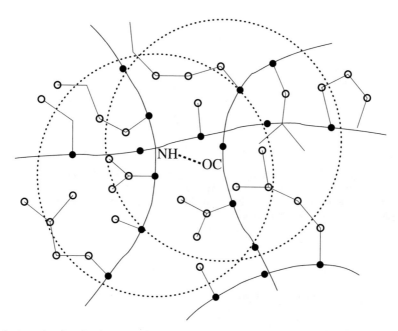

Fig. 8.2 Caricature showing desolvation spheres with various sidechains. The open circles denote the nonpolar carbonaceous groups, and the solid circles represent the C_α-carbons. The hydrogen bond between the amide (>N–H) and carbonyl (>C=O) groups is shown with a dashed line. Glycines appear without anything attached to the C_α-carbon. There are 22 nonpolar carbonaceous groups in the union of the desolvation spheres and six sidechains with two or more carbonaceous groups whose C_β carbon lies in the spheres.

The average extent of desolvation, ρ_R, over all backbone hydrogen bonds of a monomeric structure can be computed from any set of structures. In [167], a nonredundant sample of 2811 PDB-structures was examined. The average ρ_R over the entire sample set was found to be 6.6 [167]. For any given structure, the dispersion (standard deviation) σ from the mean value of ρ_R for that structure can be computed. The dispersion averaged over all sampled structures was found to be $\sigma = 1.46$ [167]. These statistics suggested a way to identify the extreme of the wrapping distribution as containing three or fewer wrapping residues in their desolvation domains. This can be interpreted as defining underwrapped as ρ_R values that are more than two standard deviations from the mean.

The distribution of the selected proteins as a function of their average wrapping as measured by ρ_R is shown in [167, Figure 5]. The probability distribution has a distinct inflection point at $\rho = 6.2$. Over 90% of the proteins studied have $\rho_R > 6.2$, and none of these are yet known to yield amyloid aggregation under physiological conditions. In addition, individual sites with low wrapping on selected proteins were examined and found to correlate with known binding sites.

In Section 8.3.2, we will see that the known disease-related amyloidogenic proteins are found in the relatively underpopulated $3.5 < \rho_R < 6.2$ range of the distribution, with the cellular prion proteins located at the extreme of the spectrum ($3.5 < \rho_R < 3.75$). We discuss there the implications regarding a propensity for organized aggregation. Approximately 60% of the proteins in the critical region $3.5 < \rho_R < 6.2$ which are not known to be amyloidogenic are toxins whose structures are stabilized mostly by disulfide bonds.

To further assess the effectiveness of the residue-based assessment of wrapping, we review additional results and predictions of [167].

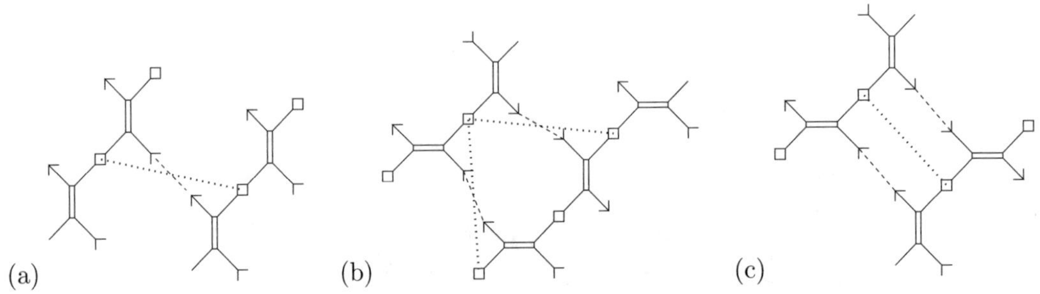

Fig. 8.3 The hydrogen bond (dashed line) configuration in (a) α-helix, (b) antiparallel β-sheet, and (c) parallel β-sheet. A dotted line connects the C_α-carbons (squares) that provide the centers of the spheres forming the desolvation domains in Figure 8.2. The amide (>N–H) groups are depicted by arrow heads, and the carbonyl (O–C) groups are depicted by arrow tails.

8.3.2 Predicting Aggregation

Prediction of protein aggregation can be based on locating regions of the protein surface with high density of defects which may act as aggregation sites [235, 284, 341]. [167, Figure 3a] depicts the (many) UWHBs for the human cellular prion protein (PDB file 1QM0) [406, 416, 541]. Over half of the hydrogen bonds are UWHBs, indicating that many parts of the structure must be open to water attack. For example, α-helix 1 has the highest concentration of UWHBs and therefore may be prone to structural rearrangement.

In helix 1 (residues 143 to 156), all of the hydrogen bonds are UWHBs, and this helix has been identified as undergoing an α-helix to β-strand transition [406, 416, 541]. Furthermore, helix 3 (residues 199 to 228) contains a significant concentration of UWHBs at the C-terminus, a region assumed to define the epitope for protein-X binding [406]. The remaining UWHBs occur at the helix-loop junctures and may contribute to flexibility required for rearrangement.

The average underwrapping of hydrogen bonds in an isolated protein may be a significant indicator of aggregation, but it is not likely to be sufficient to determine amyloidogenic propensity. For instance, protein L (PDB file 2PTL) is not known to aggregate even though its $\rho_R = 5.06$ value is outside the standard range of sufficient wrapping. Similarly, trp-repressor

(PDB file 2WRP) has $\rho_R = 5.29$, and the factor for inversion stimulation (PDB file 3FIS) has $\rho_R = 4.96$. Many neurotoxins (e.g., PDB file 1CXO with $\rho_R = 3.96$) are in this range as well.

The existence of short fragments endowed with fibrillogenic potential [23, 115, 137, 214, 235, 284, 341] suggests a localization or concentration of amyloid-related structural defects. In view of this, a local wrapping parameter, the maximum density δ_{max} of UWHBs on the protein surface, was introduced [167]. A statistical analysis involving δ_{max} [167] established that a threshold $\delta_{max} > 0.38/\text{nm}^2$ distinguishes known disease-related amyloidogenic proteins from other proteins with a low extent of hydrogen bond wrapping. On the basis of a combined assessment, identifying both low average wrapping and high maximum density of underwrapping, it was predicted [167] that six proteins might possess amyloidogenic propensity. Four of them,

- angiogenin (cf. PDB files 1B1E and 2ANG),
- meizothrombin (cf. PDB file 1A0H),
- plasminogen (cf. PDB file 1B2I), and
- anti-oncogene A (cf. PDB file 1A1U),

are involved in some form of blood clotting, wound healing, or cell adhesion.

Not all protein aggregation is related to disease. Angiogenesis refers to the growth of new capillaries from an existing capillary network, and many processes involve this, including wound healing. Angiogenin is only one of many proteins involved in the angiogenesis process, but it appears to have certain unique properties [306]. Meizothrombin is formed during prothrombin activation, is known to be involved in blood clotting [262], and is able to bind to procoagulant phospholipid membranes [388]. Plasminogen has been identified as being a significant factor in wound healing [422].

An extreme in dehydron concentration could be a contributor to cell adhesion. Surprisingly, "cell-cell adhesiveness is generally reduced in human cancers" [231]. Thus, anti-oncogene A could use its lack of wrapping as a mechanism to foster cell adhesion.

8.4 Counting Nonpolar Groups

A more refined measure of hydrogen bond protection has been proposed based on the number of vicinal nonpolar groups [160, 172]. The desolvation domain for a backbone hydrogen bond is defined again as the union of two intersecting spheres centered at the α-carbons of the residues paired by the hydrogen bond, as depicted in Figure 8.2. In this case, all of the open circles are counted, whether or not the base of the sidechain lies within the desolvation domain. The extent of intramolecular desolvation of a hydrogen bond, ρ_G, is defined by the number of sidechain nonpolar groups (CH_n, $n = 0, 1, 2, 3$) in the desolvation domain (see Table 8.2).

The distribution of wrapping for a large sample of nonredundant proteins is given in [194] for a different radii for the definition of the desolvation domain. In [173], an UWHB was defined by the inequality $\rho_G < 12$ for this value of the radius. Statistical inferences involving this definition of ρ_G were found to be robust to variations in the range 6.4 ± 0.6 Å for the choice of desolvation radius [172, 182]. In Figure 8.4, the distribution of wrapping is presented for a particular PDB file.

The "group" definition of wrapping is superficially similar to the definition of **buried** groups [311]. The latter provides a way of defining the difference between entities at the "surface" of a protein versus the "core" of the protein. The definition of "buried" utilized a sphere of radius 15.5 Å around each atom. If this sphere contains more than 400 heavy (nonhydrogen) atoms, then the atom is declared to be buried. We can think of this in terms of heavy-atom density, which allows us to compare with the known sizes (Section 5.6). Roughly speaking, when the local density of heavy atoms is greater than one per $39\,\text{Å}^3$ (corresponding to a box of side about $3.4\,\text{Å}$), that region is considered to be buried. For comparison, the average density of water is about one water per $30\,\text{Å}^3$ (corresponding to a box of side about $3.1\,\text{Å}$) (cf. Section 18.7.3), whereas we see in Tables 5.4–5.6 that most protein atom groups have a volume less than $39\,\text{Å}^3$ and thus a density of greater than one per $39\,\text{Å}^3$.

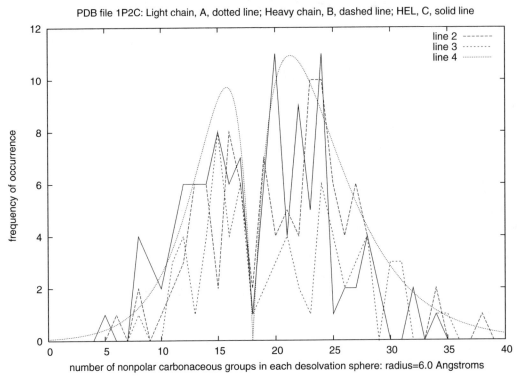

Fig. 8.4 Distribution of wrapping for PDB file 1P2C. The horizontal axis is $\rho = \rho_G$, the number of nonpolar carbonaceous groups wrapping a hydrogen bond. There are three chains: light, heavy chains of the antibody, and the antigen (HEL) chain. The desolvation radius is $6.0\,\text{Å}$. Smooth curves (8.2) are added as a guide to the eye.

8.4.1 Distribution of Wrapping for an Antibody Complex

It is instructive to consider wrapping of hydrogen bonds from a more detailed statistical point of view. In Figure 8.4, the distribution of wrapping is presented for the antibody complex whose structure is recorded PDB file 1P2C. There are three chains: two in the antibody (the light and heavy chains) and one in the antigen, hen egg-white lysozyme (HEL).

What is striking about the distributions is that they are bimodal and roughly comparable for all three chains. We have added a smooth curve representing the distributions

$$d_i(\rho) = a_i|\rho - \rho_0|e^{-|\rho-\rho_0|/w_i} \tag{8.2}$$

to interpolate the actual distributions. More precisely, d_1 represents the distribution for $\rho < \rho_0$, and d_2 represents the distribution for $\rho > \rho_0$. The coefficients chosen were $w_1 = 2.2$ and $w_2 = 3.3$. The amplitude coefficients were $a_1 = 12$ and $a_2 = 9$, and the offset $\rho_0 = 18$ for both distributions. In this example, there seems to be a line of demarcation at $\rho = 18$ between hydrogen bonds that are well wrapped and those that are underwrapped.

The distributions in Figure 8.4 were computed with a desolvation radius of $6.0\,\text{Å}$. Larger desolvation radii were also used, and the distributions are qualitatively similar. However, the sharp gap at $\rho = 18$ becomes blurred for larger values of the desolvation radius.

8.4.2 Residues Versus Polar Groups

The two measures considered here for determining UWHBs share some important key features. Both count sidechain indicators which fall inside of desolvation domains that are centered at the C_α backbone carbons. The residue-based method counts the number of residues (of a restricted type) whose C_β carbons fall inside the desolvation domain. The group-based method counts all of the carbonaceous groups that are found inside the desolvation domain, independent of which residue that they come from.

We observed that the average measure of wrapping based on counting residues was $\rho_R = 6.6$, whereas the average measure of wrapping based on counting nonpolar groups is $\rho_G = 15.9$. The residues in the former count represent at least two nonpolar groups, so we would expect that $\rho_G > 2\rho_R$. We see that this holds and that the excess corresponds to the fact that some residues have three or more nonpolar groups. Note that these averages were obtained with different desolvation radii, $6.0\,\text{Å}$ for ρ_G and $7.0\,\text{Å}$ for ρ_R. Adjusting for this difference would make ρ_G even larger, indicating an even greater discrepancy between the two measures. This implies that ρ_G provides a much finer estimation of local hydrophobicity.

The structural analysis in [167] identified site mutations which might stabilize the part of the cellular prion protein (PDB file 1QM0) believed to nucleate the cellular-to-scrapie transition. The (Met134, Asn159)-hydrogen bond has a residue wrapping factor of only $\rho_R = 3$ and is protected only by Val161 and Arg136 locally, which contribute only a minimal number (five) of nonpolar carbonaceous groups. Therefore, it is very sensitive to mutations that alter the large-scale context preventing water attack. It was postulated in [167] that a factor that triggers the prion disease is the stabilization of the (Met134, Asn159) β-sheet hydrogen bond by mutations that foster its desolvation beyond wild-type levels.

In the wild type, the only nonadjacent residue in the desolvation domain of hydrogen bond (Met134, Asn159) is Val210, thus conferring marginal stability with $\rho_R = 3$. Two of the three known pathogenic mutations (Val210Ile and Gln217Val) would increase the number of nonpolar carbonaceous groups wrapping the hydrogen bond (Met134, Asn159), even though the number of wrapping residues would not change. Thus, we see a clearer distinction in the wrapping environment based on counting nonpolar carbonaceous groups instead of just residues.

The third known pathogenic mutation, Thr183Ala, may also improve the wrapping of the hydrogen bond (Met134, Asn159) even though our simple counting method will not show this, as both Thr and Ala contribute only one nonpolar carbonaceous group for desolvation. However, Ala is four positions below Thr in Table 7.2 and is less polar than Thr. Table 7.2 reflects a more refined notion of wrapping for different sidechains, but we do not pursue this here.

8.5 Defining Dehydrons via Geometric Requirements

The enhancement of backbone hydrogen bond strength and stability depends on the partial structuring, immobilization, or removal of surrounding water. In this section, we review an attempt [174] to quantify this effect using a continuous representation of the local solvent environment surrounding backbone hydrogen bonds [64, 160, 172, 182, 233, 372, 510]. The aim is to estimate the changes in the permittivity (or dielectric coefficient) of such environments and the sensitivity of the Coulomb energy to local environmental perturbations caused by protein interactions [160, 182]. However, induced-fit distortions of monomeric structures are beyond the scope of these techniques.

The new ingredient is a sensitivity parameter M_k assessing the net decrease in the Coulomb energy contribution of the kth hydrogen bond which would result from an exogenous immobilization, structuring, or removal of water due to the approach by a hydrophobic group. This perturbation causes a net decrease in the permittivity of the surrounding environment which becomes more or less pronounced, depending on the preexisting configuration of surrounding hydrophobes in the monomeric state of the protein. In general, nearby hydrophobic groups induce a structuring of the solvent needed to create a cavity around them, and the net effect of this structuring is a localized reduction in the solvent polarizability with respect to reference bulk levels. This structuring of the solvent environment should be reflected in a decrease of the local dielectric coefficient ϵ. This effect has been quantified in recent work which delineated the role of hydrophobic clustering in the enhancement of dielectric-dependent intramolecular interactions [160, 182].

We now describe an attempt to estimate ϵ as a function of the fixed positions $\{r_j \mid j = 1, \ldots, n_k\}$ of surrounding nonpolar hydrophobic groups (CH$_n$, with $n = 1, 2, 3$, listed in Table 8.2). The simpler estimates of wrapping considered so far could fail to predict an adhesive site when it is produced by an uneven distribution of desolvators around a hydrogen bond, rather than an insufficient number of such desolvators. Based on the fixed atomic framework for the monomeric structure, we now identify Coulomb energy contributions from intramolecular hydrogen bonds that are most sensitive to local environmental perturbations by subsuming the effect of the perturbations as changes in ϵ.

Suppose that the carbonyl oxygen atom is at $\mathbf{r}_O$ and that the partner hydrogen net charge is at $\mathbf{r}_H$. The electrostatic energy contribution E_{COUL} for this hydrogen bond in a dielectric medium with dielectric permittivity ϵ (at the center of the hydrogen bond) is approximated (see Chapter 15) by

$$E_{\text{COUL}} = \frac{-1}{4\pi\epsilon} \frac{qq'}{|\mathbf{r}_O - \mathbf{r}_H|}, \tag{8.3}$$

where q, q' are the net charges involved and $|\cdot|$ denotes the Euclidean norm.

Now suppose that some agent enters in a way to alter the dielectric field, e.g., a hydrophobe that moves toward the hydrogen bond and disrupts the water that forms the dielectric material. This movement will alter the Coulombic energy as it modifies ϵ, and we can use equation (8.3) to determine an equation for the change in ϵ in terms of the change in E_{COUL}. Such a change in E_{COUL} can be interpreted as a force (cf. Chapter 3). We can compute the resulting force as a derivative with respect to the position R of the hydrophobe relative to the center of the hydrogen bond:

$$\nabla_R(1/\epsilon(R)) = \frac{4\pi|\mathbf{r}_O - \mathbf{r}_H|}{qq'}(-\nabla_R E_{\text{COUL}}(R)) = \frac{4\pi|\mathbf{r}_O - \mathbf{r}_H|}{qq'}F(R), \tag{8.4}$$

where $F(R) = -\nabla_R E_{\text{COUL}}(R)$ is a net force exerted on the hydrophobe by the fixed pre-formed hydrogen bond. This force represents a net 3-body effect [160], involving the bond, the dielectric material (water), and the hydrophobe. If E_{COUL} is decreased in this process, the hydrophobe is attracted to the hydrogen bond because, in so doing, it decreases the value of $E_{\text{COUL}}(R)$.

To identify the "opportune spots" for water exclusion on the surface of native structures, we need to first cast the problem within the continuous approach, taking into account that $1/\epsilon$ is the factor in the electrostatic energy that subsumes the influence of the environment. Thus to identify the dehydrons, we need to determine for which Coulombic contributions the exclusion or structuring of surrounding water due to the proximity of a hydrophobic "test" group produces the most dramatic increase in $1/\epsilon$. The quantity M_k was introduced [174] to quantify the sensitivity of the Coulombic energy for the kth backbone hydrogen bond to variations in the dielectric. For the kth backbone hydrogen bond, this sensitivity is quantified as follows.

Define a desolvation domain D_k with border ∂D_k circumscribing the local environment around the kth backbone hydrogen bond, as depicted in Figure 8.2. In [174], a radius of 7 Å was used. The set of vector positions of the n_k hydrophobic groups surrounding the hydrogen bond is extended from $\{R_j \mid j = 1, 2, \ldots, n_k\}$ to $\{R_j \mid j = 1, 2, \ldots, n_k; R\}$ by adding the test hydrophobe at position R. Now, compute the gradient $\nabla_R(1/\epsilon)|_{R=R_o}$, taken with respect to a perpendicular approach by the test hydrophobe to the center of the hydrogen bond at the point $R = R_o$ located on the circle consisting of the intersection C of the plane perpendicular to the hydrogen bond with the boundary ∂D_k of the desolvation domain. Finally, determine the number

$$M_k = \max\left\{ |\nabla_R(1/\epsilon(\{R_j\}, R_o, r_H - r_O))| \mid R_o \in C \right\}. \tag{8.5}$$

The number M_k quantifies the maximum alteration in the local permittivity due to the approach of the test hydrophobe in the plane perpendicular to the hydrogen bond, centered in the middle of the bond, at the surface of the desolvation domain.

The quantity M_k may be interpreted in physical terms as a measure of the maximum possible attractive force exerted on the test hydrophobic group by the preformed hydrogen bond. The only difficulty in estimating M_k is that it requires a suitable model of the dielectric permittivity ϵ as a function of the geometry of surrounding hydrophobic groups. We will consider the behavior of the dielectric permittivity more carefully in Chapter 15, but for now we consider a heuristic model used in [174].

The model in [174] for the dielectric may be written

$$\epsilon^{-1} = (\epsilon_o^{-1} - \epsilon_w^{-1})\Omega(\{R_j\})\Phi(\mathbf{r}_H - \mathbf{r}_O) + \epsilon_w^{-1}, \tag{8.6}$$

where ϵ_w and ϵ_o are the permittivity coefficients of bulk water and vacuum, respectively, and

$$\Omega(\{R_j\}) = \prod_{j=1,\ldots,n_k} \left(1 + e^{-|\mathbf{r}_O - R_j|/\Lambda}\right)\left(1 + e^{-|\mathbf{r}_H - R_j|/\Lambda}\right) \tag{8.7}$$

provides an estimate of the change in permittivity due to the hydrophobic effects of the carbonaceous groups. In [174], a value of $\Lambda = 1.8$ Å was chosen to represent the characteristic length associated with the water-structuring effect induced by the solvent organization around the hydrophobic groups. Further, a cutoff function

$$\Phi(\mathbf{r}) = (1 + |\mathbf{r}|/\xi)\, e^{-|\mathbf{r}|/\xi}, \tag{8.8}$$

where $\xi = 5$ Å is a water dipole–dipole correlation length, approximates the effect of hydrogen bond length on its strength [174].

We can write the key expression Ω in (8.7) as

$$\Omega(\{R_j\}) = \prod_{j=1,\dots,n_k} \omega(R_j),\tag{8.9}$$

where the function ω is defined by

$$\omega(R) = \left(1 + e^{-|\mathbf{r}_O - R|/\Lambda}\right)\left(1 + e^{-|\mathbf{r}_H - R|/\Lambda}\right).\tag{8.10}$$

The function ω is never smaller than one, and it is maximal in the plane perpendicular to the line connecting r_H and r_O. Moreover, it is cylindrically symmetric around this axis. The values of ω are plotted in Figure 8.5 as a function of the distance from the perpendicular bisector of the axis connecting $\mathbf{r}_H$ and $\mathbf{r}_O$, for three different values of the distance y from the line connecting $\mathbf{r}_H$ and $\mathbf{r}_O$.

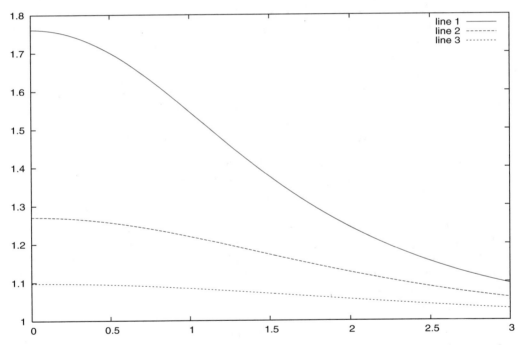

Fig. 8.5 The function $\omega(x, y)$ plotted as a function of the distance along the x-axis connecting r_H and r_O, for three different values of the distance y from that axis: $y = 1$ (solid line), $y = 2$ (dashed line), $y = 3$ (dotted line). The coordinates have been scaled by Λ and the value of $|\mathbf{r}_O - \mathbf{r}_H| = 1$ was assumed.

We see that the deviation in ω provides a strong spatial dependence on the dielectric coefficient in this model. Thus, hydrophobes close to the plane bisecting the line connecting r_H and r_O are counted more strongly than those away from that plane, for a given distance from the axis, and those closer to the line connecting r_H and r_O are counted more strongly than those further away. When the product $\Phi\Omega = 1$, we get $\epsilon = \epsilon_o$ reflecting the maximal amount of water exclusion possible. Correspondingly, if $\Phi\Omega$ tends to zero, then ϵ tends to ϵ_w, yielding a dielectric similar to bulk water. Thus, bigger values of Ω correspond to the effect of wrapping.

The definition (8.6) of the dielectric has not been scaled in a way that assures a limiting value of $\epsilon = \epsilon_o$. However, since we are interested only in comparing relative dielectric strength, this scaling is inessential. What matters is that larger values of Ω correspond to a lower dielectric and thus stronger bonds.

The computation of M_k involves computing the gradient of

$$\Omega(\{R_1, \ldots, R_{n_k}, R\}) = \omega(R)\Omega(\{R_1, \ldots, R_{n_k}\}) \tag{8.11}$$

with respect to R. Due to the cylindrical symmetry of ω, $|\nabla_R \omega|_{R=R_o}$ is a constant depending only on the desolvation radius $|R_o|$ and the hydrogen bond length $|\mathbf{r}_O - \mathbf{r}_H|$ for all $R_o \in C$. Thus, for a fixed desolvation radius $|R_o|$, M_k may be written as a function of $|\mathbf{r}_O - \mathbf{r}_H|$ times $\Omega(\{R_1, \ldots, R_{n_k}\})$ when using the model (8.6).

A sensitivity threshold for hydrogen bonds was established in [174] by statistical analysis on a sample of native structures for soluble proteins. Only 8% of backbone hydrogen bonds from a sample of 702 proteins, of moderate sizes ($52 < N < 110$) and free from sequence redundancies [232], were found to be highly sensitive in the sense that

$$M_k > \lambda/10, \tag{8.12}$$

where λ was defined to be

$$\lambda = \frac{\epsilon_o^{-1} - \epsilon_w^{-1}}{2\,\text{Å}}. \tag{8.13}$$

On the other hand, 91.6% of backbone hydrogen bonds were found to be relatively insensitive to water removal, namely

$$0 < M_k < \lambda/100 \tag{8.14}$$

This remarkable separation in the (nearly bimodal) distribution of sensitivities led [174] to the definition of a dehydron as a backbone hydrogen bond satisfying (8.12).

8.6 Dynamic Models

A dehydron is, by definition, a hydrogen bond that becomes strengthened and stabilized by dehydration, e.g., upon binding to a ligand. This is an inherently dynamic description. In Section 8.5, an attempt [174] was described to approximate this dynamic picture. Instead of simulating the approach of a dehydrating group, a derivative was defined to estimate the force associated with the change in dielectric due to the approaching hydrophobic group. A dynamic assessment for a model problem was carried out in [166]. This provides a prototype for a potentially improved prediction of dehydrons.

An alternate approach to quantifying hydrophobicity is to consider the behavior of water directly. When doing molecular dynamics simulations, it is possible to measure the residence times for water molecules near a hydrogen bond [107–109, 165, 170, 345] or other sites of interest. The residence times are distinctly different for well-wrapped and under-wrapped hydrogen bonds [107]. Thus, such residence times may be used as predictors of sites of interest [170, 345]. It has similarly been found that water behavior near "wet" residues, namely those involved in intermolecular interactions mediated by a water molecule, is also distinct [428].

8.7 Genetic Code

The genetic code, depicted in Figure 8.6, describes the mapping from DNA (or RNA) to protein residues. DNA is first translated (in a one-to-one fashion) to RNA, and so we use the RNA code letters A, C, G, U to describe the code. A DNA sequence or an RNA sequence consists of sequences of triplets, called **codons**, and each codon specifies a particular residue, with the exception of three that indicate that translation of DNA/RNA to proteins is to stop. There are 64 such codon values, so there is significant redundancy, but two-letter codons (with only 16 values) would not be sufficient to describe 20 residues. To read a gene sequence, one must first determine the **reading frame**, that is, where the codons begin, and there are three possible reading frames, each shifted by one letter.

The initial discovery of the code was not easy, and there were many false steps in the process [221]. One proposed code by Crick and co-workers was a **comma free code** with the property that there was only one possible reading frame for the code [221]. Such codes were mathematically appealing but would be too brittle: most single-letter mutations would produce codons that did not code for a protein residue. Evolution requires mutation of DNA, and resulting mutation of residues, and the simplest mutation would be to a single letter in the DNA (and resulting RNA) sequence. The genetic code insures that most such mutations lead from one residue to another, although some would lead to a stop code, which could be fatal.

Second Position

	u	c	a	g	
u	uuu ⎤ Phe 7 ǀǀ uuc ⎦ uua ⎤ Leu 4 ǀǀ uug ⎦	ucu ⎤ ucc ⎟ Ser 0 + − uca ⎟ ucg ⎦	uau ⎤ Tyr 6 + − uac ⎦ uaa stop uag stop	ugu ⎤ Cys 0 + − ugc ⎦ uga stop ugg Trp 7 + −	u c a g
c	cuu ⎤ cuc ⎟ Leu 4 ǀǀ cua ⎟ cug ⎦	ccu ⎤ ccc ⎟ Pro 2 ǀǀ cca ⎟ ccg ⎦	cau ⎤ His 1 + − cac ⎦ caa ⎤ Gln 2 + − cag ⎦	cgu ⎤ cgc ⎟ Arg 2 + + cga ⎟ cgg ⎦	u c a g
a	auu ⎤ auc ⎟ Ile 4 ǀǀ aua ⎦ aug Met 1 + −	acu ⎤ acc ⎟ Thr 1 + − aca ⎟ acg ⎦	aau ⎤ Asn 1 + − aac ⎦ aaa ⎤ Lys 3 + + aag ⎦	agu ⎤ Ser 0 + − agc ⎦ aga ⎤ Arg 2 + + agg ⎦	u c a g
g	guu ⎤ guc ⎟ Val 3 ǀǀ gua ⎟ gug ⎦	gcu ⎤ gcc ⎟ Ala 1 ǀǀ gca ⎟ gcg ⎦	gau ⎤ Asp 1 − − gac ⎦ gaa ⎤ Glu 2 − − gag ⎦	ggu ⎤ ggc ⎟ Gly 0 + − gga ⎟ ggg ⎦	u c a g

First Position (left side)   Third Position (right side)

Fig. 8.6 The genetic code. The first digit after the residue name is the amount of wrapping, and the second indicator is polarity: ǀǀ is nonpolar, + − is polar, − − is negatively charged, and + + is positively charged. The code segregates residues by polarity, nonpolars on the left, strongly polar residues on the right. Wrapping is more diversely distributed.

The topology of the genetic code is fairly simple using the standard representation used in Figure 8.6. As a first approximation, it is a two-dimensional toroidal grid, and single-letter mutations in codons represent a shift in the horizontal or vertical directions (with wraparound). Thus, Leu to Pro is a mutation that can be done with a single-letter mutation, as well as Pro to Thr and Thr to Ala. But there are some grid boxes divided into two sub-boxes, and these should be visualized as extending vertically in a third coordinate perpendicular to the page. Some single-letter mutations will thus be vertical in direction, and mutations between vertical boxes require appropriate interpretation. A movement in the coordinate directions of the plane of the page is a single-point mutation if it goes between two upper boxes, or between two lower boxes, but not between an upper and a lower box in neighboring squares. Thus, His to Lys requires a two-letter mutation. Boxes without dashed lines separating them (like the one for Pro) should be visualized as extending vertically as well, so that a horizontal step represents single-letter mutations between Pro and His as well as between Pro and Gln.

The individual residue cells are all connected with the exception of Ser. The Arg cell may appear at first to be disconnected, but it is an L-shaped domain. Not surprisingly, the cell for Trp is one of the smallest, but so is that of Met, whose codon plays the dual role of a start codon to signal the initiation of DNA/RNA translation.

8.7.1 Interpretation of the Genetic Code

Once the actual code was determined, there was extensive effort to explain why it might have evolved into an "optimal" code according to some criteria [11, 119, 206, 526]. Certain properties of the genetic code have been long recognized, such as the segregation of residues into polar and hydrophobic groups [304]. In Figure 8.6, we see that most of the residues on the left half of the code are hydrophobic, and all of the residues on the right half have some polar component. This property of the genetic code implies that most single-codon mutations will not change the polarity of the

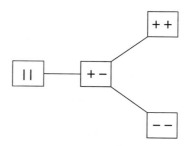

Fig. 8.7 A polarity scale for sidechains: | | is nonpolar, $+-$ is polar, $--$ is negatively charged, and $++$ is positively charged.

residue dramatically. To measure polarity, we propose a simple system consisting of four states: nonpolar, polar, and charged, where there are two signs for charged residues. This uses a simple interpretation of the polar scale as indicated in Figure 8.7.

The genetic code has only a few places where a single-letter codon mutation can change the polarity more than one step in the polarity scale in Figure 8.7. Mutations between Lys and Glu occur when the first codon changes between a and g. Similarly, Ala can mutate to Asp and Glu, and Arg can mutate to Ile and Leu. Such mutations might be catastrophic, but they represent only a small fraction of the possible mutations. Otherwise, all single-letter codon mutations either change between nonpolar and polar $(+-)$ residues, or between polar and charged residues. Moreover, mutations occurring on the left-hand side of the code involve mostly nonpolar residues.

Although the genetic code has the property that single-letter codon mutations cause minimal change in polarity, they can cause a significant change in wrapping. For example, the three residues with maximal wrapping (Phe, Tyr, Trp) are isolated in the genetic code (requiring more than a single-letter mutation to move between them). By contrast, Arg and Lys are neighbors, as are Asp and Glu. Thus, the genetic code allows radical changes in wrapping to be explored while keeping changes in polarity to a minimum.

There are only 50 possible mutations from one residue to another (or 100 if you count both directions) based on single-letter mutations in DNA/RNA. These mutations are listed in Table 8.6. The rest of the 190 pairs of possible transitions from one residue to another (or 380 if you count both directions) require at least two-letter mutations and thus would require a more coordinated mutation strategy, or a sequence of changes. The sparsity of the genetic code in terms of possible residue mutations is a significant feature.

The number of possible single-letter DNA/RNA mutations is of course much larger, on the order of 64 squared. Thus, most single-letter DNA/RNA mutations are **synonymous mutations** in that they leave the residue unchanged. This is another important property of the genetic code. Since the genetic code is a three-letter code, any residue can be mutated to any other by a three-letter mutation. However, not all pairs of residues can be linked by a two-letter mutation (Exercise 8.9).

Of the 50 possible mutations from one residue to another based on single-letter mutations in DNA/RNA, 23 of them involve no change in polarity. We might call them **polarity-synonymous mutations**. Thus, only 27 mutations based on single-letter mutations in DNA/RNA, of the possible 190 pairs of possible transitions from one residue to another, have any polarity change, and 22 of these involve just one step in our simple polarity scale depicted in Figure 8.7. On the other hand, only 12 of the mutations listed in Table 8.6 are neutral with respect to wrapping change. The overwhelming majority of mutations involve a change in wrapping, and over half of the total involve a change of two or more units.

Res	Mutations
Ala	Asp(0,↑↑), Glu(+1,↑↑), Ser(−1,↑), Thr(0,↑), Val(+2,◯)
Arg	Gln(0,↓), Gly(−2,↓), His(−1,↓), Ile(+2,↓↓), Leu(+2,↓↓), Lys(+1,◯), Met(−1,↓), Ser(−2,↓)
Asn	Asp(0,↑), His(0,◯), Ser(−1,◯), Thr(0,◯)
Asp	Glu(+1,◯), Gly(−2,↓), Tyr(+6,◯)
Cys	Gly(0,◯), Phe(+7,◯), Trp(+7,◯), Tyr(+6,◯)
Gln	His(−1,◯), Lys(+1,↑), Pro(0,↓)
Glu	Gly(−2,↓), Lys(+1,↑↑)
Gly	Ser(0,◯), Val(+3,↓), Trp(+7,◯)
His	Pro(0,↓), Tyr(+5,◯)
Ile	Leu(0,◯), Met(−3,↑), Ser(−2,↓), Thr(−3,↑)
Leu	Phe(+3,◯), Pro(−2,◯), Trp(+3,◯), Val(-1,◯)
Lys	Thr(−2,↓)
Met	Thr(0,◯), Val(+2,↓)
Phe	Ser(−7,↑), Val(−4,◯)
Pro	Ser(−2,↑), Thr(−1,↑)
Ser	Tyr(+6,◯)

Table 8.6 Amino acid mutations caused by single-letter DNA/RNA mutations. Only mutations to residues later in the alphabet are listed. The table can be extended antisymmetrically (Exercise 8.8). The symbol ↑ indicates a move to the right in the polarity scale in Figure 8.7, and ↓ indicates a move to the left. The symbol ◯ indicates no polarity change.

Thus, we can describe the genetic code according to three attributes. First, it minimizes the effect of most single-letter codon mutations by making them synonymous, not changing the residue at all. This provides basic protection against any kind of DNA/RNA mutation. Second, the code minimizes polarity changes due to single-letter codon mutations, limiting substantial chemical changes due to single-letter codon mutations. Finally, the genetic code enables wrapping changes via single-letter codon mutations, so that the inevitable changes due to natural mutations in DNA/RNA are directed toward a useful but less dangerous path.

These properties of the genetic code are easy to understand. The first property relaxes constraints on the system that translates DNA/RNA to protein residues, and it allows simpler systems to function adequately and with less energetic cost. Changes in polarity have a strong impact on chemical properties, and these can likely cause an unwanted change in function. Thus, the genetic code makes it hard to make radical changes in polarity. On the other hand, the genetic code does not restrict radical changes in wrapping. As we know, these do cause important changes in function, but they do so in a much more subtle way.

8.7.2 Size Matters

Another variable that is of interest is the size of sidechains. A mutation that changes the size of a sidechain could have a significant effect. Fortunately, the number of wrappers provides a very simple proxy for the size of the sidechain. For nonpolar residues, including Ala and Gly, the number of wrappers is the number of heavy atoms in the sidechain. For polar and charged residues, there is a difference, but we see in Table 8.7 that the differences are largely predictable. That is, the difference between the number heavy atoms and the number of wrappers, which is the number of polar atoms or charged atoms, is for the most part either two or three. The exceptions are Arg and His. Therefore, we can take the change in the number of wrappers as a good proxy for the size variation upon mutation among polar and charged residues.

The genetic code is not a simple translation table [50]. There is an amino acid residue called selenocysteine (with three-letter code Sec and single-letter code U) which is a cysteine with the sulfur atom replaced by selenium. The genetic code for this is UGA, which is typically a stop code, and the interpretation of UGA is then context dependent. Similarly, the amino acid residue pyrrolysine (abbreviated by Pyl and O) uses the the (normally stop) codon UAG, and the translation is again context dependent. Pyl is related to Lys in structure, but has a significant modification, including a pentagonal ring, at the terminal end of the sidechain. Finally, the codon CUG can be translated ambiguously as Ser or Lys depending on context in some organisms [50]. Further complexities in code translation, including the time dependence of different synonymous codes (codons that code for the same residue), are discussed in [50].

3-letter code	1-letter code	Nonpolar Carbons	Heavy Atoms	Polar Atoms
Asp	D	1	4	3
Asn	N	1	4	3
Cys	C	0	2	2
Glu	E	2	5	3
Gln	Q	2	5	3
Lys	K	3	5	2
Met	M	1	4	3
Ser	S	0	2	2
Thr	T	1	3	2
Tyr	Y	6	8	2
Arg	R	2	7	5
His	H	1	6	5

Table 8.7 The size of residues (the number of heavy atoms) compared with the number of nonpolar carbons. The difference is the number of polar and charged amino acids residues and is typically only two or three.

8.8 Exercises

Exercise 8.1. It was observed [167] that the two proteins

- RADR zinc finger peptide (PDB file 1A1K) and
- rubredoxin (PDB file 1B20).

have extreme concentrations of dehydrons. Investigate these two proteins to see why this might influence their function.

Exercise 8.2. In Section 5.3.4, we noted that sidechains have different conformations. Determine and list the number of different rotameric states possible for each sidechain (hint: read [323]). Explain your reasoning. Compare the number of rotameric degrees of freedom (number of states) for the seven residues listed in Figure 4.3(a) with the remaining group of thirteen sidechains. (Hint: review Exercise 5.6 and 5.7.)

Exercise 8.3. In Figure 8.3(a), it appears that the dotted line joining the two C_α-carbons intersects the dashed line joining the amide and carbonyl groups. By searching the PDB, determine the distribution of distances between the midpoints of these two lines for α-helices.

Exercise 8.4. The residues S, T, and Y can become phoshorylated. Various Web sites record the locations where these residues can be expected to become phoshorylated. Investigate any possible correlation between wrapping and phosphorylation sites.

Exercise 8.5. Explain whether you would expect methyl fluoride to have a polar carbon, based on the electronegativity scale.

Exercise 8.6. Determine the extent to which the wrapping in a given protein is bimodal, as depicted in Figure 8.4. Determine a way to measure consistently the degree to which a distribution is bimodal, and survey a large set of PDB structures with this measure.

Exercise 8.7. Count the number of synonymous mutations of DNA/RNA, that is, single-letter mutation in DNA/RNA, that do not change the residue.

Exercise 8.8. Make a table of all residue mutations that can arise from a single-letter mutation using the genetic code topology. (Hint: complete Table 8.6 by including the mutations to residues lower in the alphabet; the wrapping number should switch sign, and the direction of the polarity change should invert.)

Exercise 8.9. Make a list of all residue pairs that can be linked only by three-letter codon mutations, such as Met and His, Ile and Trp, Lys and Phe. That is, determine the complement of pairs that can be reached by either one-letter or two-letter codon mutations. To be precise, we say that two residues can be linked by an n-letter mutation if, for any three-letter code representing one residue, there is a three-letter code representing the other residue that differs by n letters. Thus, any residue can be linked to any other by an n-letter mutation for some $n \leq 3$. We are asking for pairs where there is no link for $n \leq 2$. (Hint: look at diagonal moves in the genetic code that are two squares away, and the moves that a knight can make on a chessboard, e.g., two over and one up/down. Examine whether these can be reached by a two-letter codon mutation.)

Exercise 8.10. Make a 20×20 matrix indexed by the 20 standard amino acids whose entries are the wrapping change arising from a single-letter mutation using the genetic code topology. If no such mutation exists, let the matrix entry be NaN. Make sure that matrix is antisymmetric. Leave the diagonal blank. (Hint: capture the wrapping data from Table 8.6 to generate the upper-triangular part of the matrix; using octave, set the nonexisting entries to be 1/0. Extend antisymmetrically.)

Chapter 9
Stickiness of Dehydrons

Heraclitus (c. 535–c. 475 BCE) advised us that $\phi\upsilon\sigma\iota\varsigma$ $\kappa\rho\upsilon\pi\tau\epsilon\sigma\theta\alpha\iota$ $\phi\iota\lambda\epsilon\iota$, which can be translated as "nature keeps its secrets."

We have explained why under-wrapped hydrogen bonds get strengthened and stabilized by the removal of surrounding water. This makes them promote interaction with ligands that can replace water molecules in the vicinity of the hydrogen bond upon binding. Conceptually, this implies that under-wrapped hydrogen bonds attract entities that can dehydrate them. Thus they must be sticky. If so, it must be possible to observe this experimentally. Here, we review several papers that substantiate this conclusion. One of them involves a mesoscopic measurement of the mechanical equivalent of the dehydration propensity of a dehydron [173], which is in effect an attractive force exerted by a dehydron on a test hydrophobe. A second presents data on the direct measurement of the dehydronic force using atomic force microscopy [147]. Another paper examines the effect of such a force on a deformable surface [175].

9.1 Surface Adherence Force

We defined the notion of an under-wrapped hydrogen bond by a simple counting method in Chapter 8 and have asserted that there is a force associated with dehydrons. Here, we describe measurements of the adhesion of an under-wrapped hydrogen bond by analyzing the flow-rate dependence of the adsorption uptake of soluble proteins onto a phospholipid bilayer.

9.1.1 Biological Surfaces

A biological surface of considerable interest is provided by the cell membrane. This is a complex system, but a key component is what is called a **phospholipid bilayer**. The term **lipid** refers to a type of molecule that is a long carbonaceous polymer with a polar (phospho) group at the "head." This is hydrophobic at one end and hydrophilic at the other end. These molecules align to form a complex that could be described as a bundle of pencils, with the hydrophilic head group (the eraser) at one side of the surface and the hydrophobic "tail" on the other side. These bundles can grow to form a surface when enough pencils are added. A second surface can form in the opposite orientation, with the two hydrophobic surfaces in

© Springer International Publishing AG 2017
L.R. Scott and A. Fernández, *A Mathematical Approach to Protein Biophysics*,
Biological and Medical Physics, Biomedical Engineering,
DOI 10.1007/978-3-319-66032-5_9

close proximity. This results in a membrane that is hydrophilic on both sides and thus can persist in an aqueous environment.

One might wonder what holds together a lipid bilayer. We have noted that there is a significant volume change when a hydrophobic molecule gets removed from water contact in Section 5.6. The volume change causes self-assembly of lipids and provides a substantial pressure that holds the surface together. The architecture of a lipid bilayer is extremely adaptive. For example, a curved surface can be formed simply by allocating more lipid to one side than the other. Moreover, it easily allows insertion of other molecules of complex shape but with other composition. Much of a cell membrane is lipid, but there are also proteins with various functions as well as other molecules such as cholesterol. However, a simple lipid bilayer provides a useful model biological surface.

9.1.2 Soluble Proteins on a Surface

One natural experiment to perform is to release soluble proteins in solution near a lipid bilayer and to see to what extent they attach to the bilayer. Such an experiment [161] indicated a significant correlation between the under-wrapping of backbone hydrogen bonds and bilayer attachment. The results were explained by assuming that the probability of successful landing on the liquid–solid interface is proportional to the ratio of dehydrons to all backbone hydrogen bonds on the protein surface. Here, the number of surface hydrogen bonds is taken simply as a measure of the surface area. The experiments in [161] indicated that more dehydrons lead to more attachments, strongly suggesting that dehydrons are sticky. However, such indications were only qualitative.

A more refined analysis of lipid bilayer experiments was able to quantify a force of attachment [173]. The average magnitude of the attractive force exerted by a dehydron on a surface was assessed based on measuring the dependence of the adsorption uptake on the flow rate of the ambient fluid above the surface. The adhesive force was measured via the decrease in attachment as the flow rate was increased.

Six proteins were investigated in [173], as shown in Table 9.1, together with their numbers of well-wrapped hydrogen bonds as well as dehydrons. The dehydrons for three of these are shown in [173, Fig. 1a-c]. The particular surface was a Langmuir–Blodgett bilayer made of the lipid DLPC (1,2 dilauroyl-sn-glycero-3 phosphatidylcholine) [418]. We now review the model used in [173] to interpret the data.

9.2 A Two-Zone Model

In [173], a two-zone model of surface adhesion was developed. The first zone deals with the experimental geometry and predicts the number of proteins that are likely to reach a fluid boundary layer close to the lipid bilayer. The probability Π of arrival is dependent on the particular experiment, so we only summarize the model results from [173]. The second zone is the fluid boundary layer close to the lipid bilayer, where binding can occur. In this layer, the probability P of binding is determined by the thermal oscillations of the molecules and the solvent as well as the energy of binding.

The number M of adsorbed molecules is given by

$$M = \Pi P(n_{UW}, n_W, T)N \tag{9.1}$$

where Π is the fraction of molecules that reach the (immobile) bottom layer of the fluid, $P(n_{UW}, n_W, T)$ is the conditional probability of a successful attachment at temperature T given that the bottom layer has been reached, and N is the average number of protein molecules in solution in the cell. The quantities n_{UW} and n_W are the numbers of underwrapped and well-wrapped hydrogen bonds on the surface of the protein, respectively. These will be used to estimate the relative amount of protein surface area related to dehydrons. The fraction Π depends on details of the experimental design, so we focus initially on the second term P.

9.2.1 Boundary Zone Model

Suppose that ΔU is the average decrease in Coulombic energy associated with the desolvation of a dehydron upon adhesion. It is the value of ΔU that we are seeking to determine. Let ΔV be the Coulombic energy decrease upon binding at any other site. Let f be the fraction of the surface covered by dehydrons. As a simplified approximation, we assume that

$$f \approx \frac{n_{UW}}{n_{UW} + n_W}. \tag{9.2}$$

Then, the probability of attachment at a dehydron is predicted by thermodynamics as

$$P(n_{UW}, n_W, T) = \frac{f e^{\Delta U/k_B T}}{(1-f)e^{\Delta V/k_B T} + f e^{\Delta U/k_B T}} \approx \frac{n_{UW} e^{\Delta U/k_B T}}{n_W e^{\Delta V/k_B T} + n_{UW} e^{\Delta U/k_B T}}, \tag{9.3}$$

with $k_B =$ Boltzmann's constant. In [173], ΔV was assumed to be zero. In this case, (9.3) simplifies to

$$P(n_{UW}, n_W, T) = \frac{f e^{\Delta U/k_B T}}{(1-f) + f e^{\Delta U/k_B T}} \approx \frac{n_{UW} e^{\Delta U/k_B T}}{n_W + n_{UW} e^{\Delta U/k_B T}} \tag{9.4}$$

(cf. equation (2) of [173]). Note that this probability is lower if $\Delta V > 0$.

9.2.2 Diffusion Zone Model

The probability Π in (9.1) of penetrating the bottom layer of the fluid is estimated in [173] by a model for diffusion via Brownian motion in the plane orthogonal to the flow direction. This depends on the solvent bulk viscosity μ, and the molecular mass m and the **hydrodynamic radius** [269] or **Stokes radius** [222] of the protein. This radius R associates with each protein an equivalent sphere that has approximately the same flow characteristics at low Reynolds numbers. This particular instance of a "spherical cow" approximation [124, 285] is very accurate, since the variation in flow characteristics due to shape variation is quite small

[269]. The drag on a sphere of radius R, at low Reynolds numbers, is $F = 6\pi R\mu v$ where v is the velocity. The drag is a force that acts on the sphere through a viscous interaction. The coefficient

$$\xi = 6\pi R\mu/m = F/mv \tag{9.5}$$

where m is the molecular mass is a temporal frequency (units: inverse time) that characterizes Brownian motion of a protein. The main nondimensional factor that appears in the model is

$$\alpha = \frac{m\xi^2 L^2}{2k_B T} = \frac{L^2(6\pi R\mu)^2/m}{2k_B T}, \tag{9.6}$$

which has units of energy in numerator and denominator. We have [5]

$$
\begin{aligned}
\Pi(v, R, m) &= \int_\Lambda \int_{\Omega\backslash\Lambda} \int_{[0,\tau]} \frac{\alpha L^{-2}}{\pi\Gamma(t)} e^{-\alpha L^{-2}|\mathbf{r}-\mathbf{r}_0|^2/\Gamma(t)} \, dt d\mathbf{r}_0 d\mathbf{r} \\
&= \int_{\tilde{\Lambda}} \int_{\tilde{\Omega}\backslash\tilde{\Lambda}} \int_{[0,L/v]} \frac{\alpha}{\pi\Gamma(t)} e^{-\alpha|\tilde{\mathbf{r}}-\tilde{\mathbf{r}}_0|^2/\Gamma(t)} \, dt d\tilde{\mathbf{r}}_0 d\tilde{\mathbf{r}}
\end{aligned}
\tag{9.7}
$$

where $\mathbf{r}$ is the two-dimensional position vector representing the cell cross section Ω, $|\mathbf{r}|$ denotes the Euclidean norm of $\mathbf{r}$, Λ is the $6\,\text{Å} \times 10^8\,\text{Å}$ cross section of the bottom layer, and $\Gamma(t) = 2\xi t - 3 + 4e^{-\xi t} - e^{-2\xi t}$. Note that Γ grows like $2\xi t$ for t large, but initially there is a different behavior that corresponds to a correction to account for the discrete nature of physical diffusion of particles of finite size [495].

The domains $\tilde{\Lambda}$ and $\tilde{\Omega}$ represent domains scaled by the length L, and thus the variables $\tilde{\mathbf{r}}$ and $\tilde{\mathbf{r}}_0$ are nondimensional. In particular, the length of $\tilde{\Lambda}$ and $\tilde{\Omega}$ is one in the horizontal coordinate. Note that $\Gamma(t) = \frac{2}{3}(\xi t)^3 + \mathcal{O}((\xi t)^4)$ for ξt small. Also, since the mass m of a protein tends to grow with the radius cubed, α actually decreases like $1/R$ as the Stokes radius increases.

9.2.3 Model Validity

The validity of the model represented by equations (9.1—9.7) was established by data fitting. The only parameter in the model, ΔU, was varied, and a value was found that consistently fits within the confidence band for the adsorption data for the six proteins (see Fig. 3 of [173]) across the entire range of flow velocities v. This value is

$$\Delta U = 3.91 \pm 0.67\,\text{kJ/mole} = \Delta U = 0.934 \pm 0.16\,\text{kcal/mole}. \tag{9.8}$$

This value is within the range of energies associated with typical hydrogen bonds. Thus, we can think of a dehydron as a hydrogen bond that gets turned "on" by the removal of water due to the binding of a ligand.

Using the estimate (9.8) of the binding energy for a dehydron, an estimate was made [173] of the force

$$|F| = 7.78 \pm 1.5\,\text{pN} \tag{9.9}$$

exerted by a single dehydron on a test hydrophobe at a $6\,\text{Å}$ distance from the dehydron.

9.3 Direct Force Measurement

protein name	PDB code	residues	WWHB	dehydrons
apolipoprotein A-I	1AV1	201	121	66
β lactoglobulin	1BEB	150	106	3
hen egg-white lysozyme	133L	130	34	13
human apomyoglobin	2HBC	146	34	3
monomeric human insulin	6INS	50	30	14
human β_2-microglobulin	1I4F	100	17	9

Table 9.1 Six proteins and their hydrogen bond distributions. WWHB=well-wrapped hydrogen bonds.

The experimental techniques reviewed in the previous section suggest that the density of dehydrons correlates with protein stickiness. However, the techniques are based on measuring the aggregate behavior of a large number of proteins. One might ask for more targeted experiments seeking to isolate the force of a dehydron or at least a small group of dehydrons. Such experiments were reported in [147] based on atomic force microscopy (AFM).

We will not give the details of the experimental setup, but just describe the main points. The main concept was to attach hydrophobic groups to the tip of an atomic force microscope. These were then lowered onto a surface capable of forming arrays of dehydrons. This surface was formed by a self-assembling monolayer of the molecules $SH-(CH_2)_{11}-OH$. The OH "head" groups are capable of making OH–OH hydrogen bonds, but these will be exposed to solvent and not well protected.

The data obtained by lowering a hydrophobic probe on such a monolayer are complex to interpret. However, they become easier when they are compared with a similar monolayer not containing dehydrons. In [147], the molecule $SH-(CH_2)_{11}-Cl$ was chosen.

The force–displacement curve provided by the AFM has similarities for both monolayers [147]. For large displacements, there is no force, and for very small displacements, the force grows substantially as the tip is driven into the monolayer. However, in between, the characteristics are quite different.

For the OH-headed monolayer, as the displacement is decreased to the point where the hydrophobic group on the tip begins to interact with the monolayer, the force on the tip decreases, indicating a force of attraction. Near the same point of displacement, the force on the tip increases for the chlorine-headed monolayer. Thus, we see the action of the dehydronic force in attracting the hydrophobes to the dehydron-rich OH-headed layer. On the other hand, there is a resistance at the similar displacement as the hydrophobic tip begins to dehydrate the chlorine-headed monolayer. Ultimately, the force of resistance reaches a maximum, and then, the force actually decreases to a slightly negative (attractive) value as the monolayer becomes fully dehydrated. It is significant that the displacement for the force minimum is approximately the same for both monolayers, indicating that they both correspond to a fully dehydrated state.

The force–displacement curves when the tip is removed from the surface also provide important data on the dehydronic force. The force is negative for rather large displacements, indicating the delay due to the requirements of rehydration. Breaking the hydrophobic bond formed by the hydrophobic groups on the tip and the monolayer requires enough force to be accumulated to completely rehydrate the monolayer. This effect is similar to the force that is required to remove sticky tape, in which one must reintroduce air between the tape and the surface to which it was attached. For the chlorine-headed monolayer, there is little change in force as the displacement is increased by four Ångstroms from the point where the force is minimal. Once the threshold is reached, then the force returns abruptly to zero, over a distance of about one Ångstrom. For the OH-headed monolayer, the threshold is delayed by another two Ångstroms, indicating the additional effect of the dehydronic force.

The estimation of the dehydronic force is complicated by the fact that one must estimate the number of dehydrons that will be dehydrated by the hydrophobic groups on the tip. But the geometry of AFM tips is well characterized, and the resulting estimate [147] of

$$|F| = 5.9 \pm 1.2 \text{pN} \tag{9.10}$$

at a distance of 5 Å is in close agreement with the estimate (9.9) of 7.78 ± 1.5pN at a distance of 6 Å in [173]. Part of the discrepancy could be explained by the fact that in [173] no energy of binding was attributed to the attachment to areas of a protein lacking dehydrons. If there were such an energy decrease, due, e.g., to the formation of intermolecular interactions, the estimate of the force obtained in [173] would be reduced.

9.4 Membrane Morphology

Since dehydrons have an attractive force that causes them to bind to a membrane, then the equal and opposite force must pull on the membrane. Since membranes are flexible, then this will cause the membrane to deform.

The possibility of significant morphological effect of dehydrons on membranes was suggested by the diversity of morphologies [453] of the inner membranes of cellular or subcellular compartments containing soluble proteins [175]. These vary from simple bag-like membranes [131] (e.g., erythrocytes, a.k.a. red blood cells) to highly invaginated membranes [505] (e.g., mitochondrial inner membranes). This raises the question of what might be causing the difference in membrane structure [279, 313, 359, 509].

Some evidence [175] suggests that dehydrons might play a role: hemoglobin subunits (which comprise the bulk of erythrocyte contents) are generally well wrapped, whereas two mitochondrial proteins, cytochrome c and pyruvate dehydrogenase, are less well wrapped. The correlation between the wrapping difference and the morphology difference provided motivation to measure the effect experimentally [175].

9.4.1 Protein Adsorption

Morphology induction was tested in fluid phospholipid (DLPC) bilayers (Section 9.1) coating an optical waveguide [175]. The density of bilayer invaginations was measured by a technology called evanescent field spectroscopy which allowed measurement of both the thickness and refractive index of the adlayer [413, 476]. DLPC was added as needed for membrane expansion, with the portion remaining attached to the waveguide serving as a nucleus for further bilayer formation. Stable invaginations in the lipid bilayer formed after 60-hour incubation at T = 318K.

9.4.2 Density of Invaginations

The density of invaginations correlates with the extent of wrapping, ρ, of the soluble protein structure (Fig. 1, 2a in [175]). Greater surface area increase corresponds with lack of wrapping of backbone hydrogen bonds. The density of invaginations as a function of concentration (Figure 2b in [175]) shows that protein aggregation is a competing effect in the protection of solvent-exposed hydrogen bonds ([144, 160, 161, 172, 182]): for each protein, there appears to be a concentration limit beyond which aggregation becomes more dominant.

9.5 Kinetic Model of Morphology

The kinetics of morphology development suggest a simple morphological instability similar to the development of moguls on a steep ski run. When proteins attach to the surface, there is a force that binds the protein to the surface. This force pulls upward on the surface (and downward on the protein) and will increase the curvature in proportion to the local density of proteins adsorbed on the surface [161]. The rate of change of curvature $\frac{dg}{dt}$ is an increasing function of the force f:

$$\frac{dg}{dt} = \phi(f) \tag{9.11}$$

for some increasing function ϕ. Note that

$$\phi(0) = 0; \tag{9.12}$$

If there is no force, there will be no change. The function ϕ represents a material property of the surface.

The probability p of further attachment increases as a function of the curvature at that point since there is more area for attachment where the curvature is higher. That is, $p(g)$ is also an increasing function.

Of course, attachment also reduces surface area, but we assume this effect is small initially. However, as attachment grows, this neglected term leads to a "saturation" effect. There is a point at which further reduction of surface area becomes the dominating effect, quenching further growth in curvature. But for the moment, we want to capture the initial growth of curvature in a simple model. We leave as Exercise 9.2 the development of a more complete model.

Assuming equilibrium is attained rapidly, we can assert that the force f is proportional to $p(g)$: $f = cp(g)$ at least up to some saturation limit, which we discuss subsequently. If we wish to be conservative, we can assert only that

$$f = \psi(p(g)) \tag{9.13}$$

with ψ increasing. In any case, we conclude that f may be regarded as an increasing function of the curvature g, say

$$f = F(g) := \phi(\psi(p(g))). \tag{9.14}$$

To normalize forces, we should have no force for a flat surface. That is, we should assume that $p(0) = 0$. This implies, together with the condition $\phi(0) = 0$, that

$$F(0) = 0. \tag{9.15}$$

The greater attachment that occurs locally causes the force to be higher there and thus the curvature to increase even more, creating an exponential runaway (Fig. 4 in [175]). The repeated interactions of these two reinforcing effects causes the curvature to increase in an autocatalytic manner until some other process forces it to stabilize.

The description above can be captured in a semi-empirical differential equation for the curvature g at a fixed point on the bilayer. It takes the form

$$\frac{dg}{dt} = F(g), \tag{9.16}$$

where F is the function in (9.14) that quantifies the relationships between curvature, probability of attachment, and local density of protein described in the previous paragraph. Abstractly, we know that F is increasing because it is the composition of increasing functions. Hence, F has a positive slope s at $g = 0$. Moreover, it is plausible that $F(0) = 0$ using our assumptions made previously.

Thus, the curvature should grow exponentially at first with rate s. In the initial stages of interface development, F may be linearly approximated by virtue of the mean value theorem, yielding the autocatalytic equation:

$$\frac{dg}{dt} = sg. \tag{9.17}$$

Figure 4 in [175] indicates that the number of invaginations appears to grow exponentially at first and then saturates.

We have observed that there is a maximum amount of protein that can be utilized to cause morphology (Figure 2b in [175]) beyond which aggregation becomes a significantly competitive process. Thus, a "crowding problem" at the surface causes the curvature to stop increasing once the number of adsorbed proteins gets too high at a location of high curvature.

9.6 Exercises

Exercise 9.1. Determine the minimal distance between a hydrophobe and a backbone hydrogen bond in protein structures. That is, determine the number of wrappers as a function of the desolvation radius, and determine when, on average, this tends to zero.

Exercise 9.2. Derive a more refined model of morphological instability accounting for the reduction in surface area upon binding. Give properties of a function F as in (9.14) that incorporate the effect of decreasing surface area, and show how it would lead to a model like (9.16) which would saturate (rather than grow exponentially forever), reflecting the crowding effect of the molecules on the lipid surface.

Exercise 9.3. The logistic equation

$$\frac{dg}{dt} = g(1 - g), \tag{9.18}$$

naturally includes both an exponential rise and a saturation effect. Show how this might be used to include a saturation effect in the model in Section 9.5.

Exercise 9.4. Determine the distribution of distances between waters in PDB files and (a) the nearest nonpolar carbonaceous group and (b) the nearest of all other heavy atoms. Also refine your data mining to compute different distributions in (a) for the different nonpolar carbonaceous groups CH_n for different values of $n = 0, 1, 2, 3$. Similarly, refine your data mining to compute different distributions in (b) for different atom classes (e.g., mainchain atoms versus ones in sidechains, and C versus N versus O).

Chapter 10
Electrostatic Force Details

Paul John Flory (1910–1985) was a Nobel laureate cited for his work
on macromolecules, and is known for "Flory's Rule" that says that the
orientation of sidechains of residues i and $i + 1$ should be independent.
He was born in Sterling, Illinois and got his Ph.D. from the Ohio State
University in 1934.

In Section 3.3, we introduced some basic electrostatic interactions. Here we look at electrostatic interactions in more detail. Our objective is to understand the expected configuration of interacting charged and polar sidechains with each other and with other entities such as ions and polar groups in drugs. We make the assumption that the minimum energy configuration will be informative. Since we cannot know what the global electrostatic environment will be in general, we use local electrostatic energy as the quantity to be minimized. The resulting configurations provide only a guide to what we might expect in practice, but we will see that there are some surprising results.

The basic electrostatic entities are groups of charges that are constrained to be together, such as dipoles. In Section 10.2.1 we study dipole–dipole interactions. In Section 10.3.1, we consider charge-dipole interactions such as arise in cation-pi pairs such as Arg-Tyr or Lys-Phe. We also consider like-charge repulsion such as occurs with Arg-His or Asp-Glu pairs in Section 10.3.2.

There is a natural hierarchy of charged groups. These can be ranked by the rate of decay of their potentials, and thus by how localized they are. At the highest (most global) level is the single charge, with a potential r^{-1}. A dipole is a combination of opposite charges at nearby locations, with a potential r^{-2}. A quadrapole is a collection of four or more charges arranged in appropriate positions with a potential r^{-3}. Some important entities, such as water, are often modeled as being four charges at positions with substantial symmetry, and it is important to know whether they constitute quadrapoles or just dipoles. This determines the global accumulation of charge and thus has significant implications as we now discuss. We subsequently return to the question of whether water is a dipole or quadrapole.

10.1 Global Accumulation of Electrostatic Force

The reason that we need to know the order of decay of the potential, or the associated force, for various types of charged groups is quite simple to explain. Suppose that we have a material made of an assembly of electrostatic entities, such as water. We would like to understand the

© Springer International Publishing AG 2017

L.R. Scott and A. Fernández, *A Mathematical Approach to Protein Biophysics*,
Biological and Medical Physics, Biomedical Engineering,
DOI 10.1007/978-3-319-66032-5_10

locality of forces exerted by the entities on each other. In particular, are they local, or do global contributions have a significant effect [459]

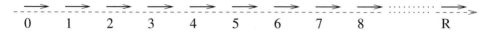

Fig. 10.1 A one dimensional array of dipoles.

To quantify this question, suppose we try to estimate the potential or force due to an organized collection molecular groups each of whose effects at a distance r is proportional to r^{-n} for some n. In Figure 10.1 we depict a simple one-dimensional array of dipoles. Summing over all space, we determine the total effect. In the one dimensional case, as depicted in Figure 10.1, the resulting force or potential at the origin is proportional to

$$\sum_{r=1}^{R} r^{-n} \leq C \quad \text{provided } n > 1. \tag{10.1}$$

The potential for dipoles has $n = 2$, and the result (10.1) remains bounded as $R \to \infty$. On the other hand, if instead of dipoles at the points $1, 2, \ldots, R$ we had point charges of all the same sign, then the resulting potential as in (10.1) would diverge logarithmically. By contrast, the electric field due to a point charge has $n = 2$, and the resulting electric field (10.1) due to a sequence of charges remains bounded as $R \to \infty$.

In Figure 10.2, we depict a two-dimensional arrangement of dipoles. One each circle of radius r, we place r dipoles pointing toward the origin, at a spacing of 2π units along the circles. We can also imagine that instead of dipoles we place point charges or other entities. In the two dimensional case, there are r such entities all at a distance r from the origin, so the resulting potential is

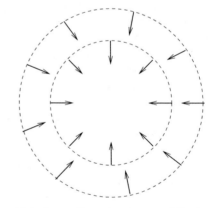

$$\sum_{r=1}^{R} r\, r^{-n} \leq C \quad \text{provided } n > 2. \tag{10.2}$$

Since the potential for dipoles has $n = 2$, the result becomes unbounded as $R \to \infty$, although the resulting electric field remains bounded independent of R.

Fig. 10.2 A two-dimensional array of dipoles.

Now let us imagine a three-dimensional arrangement of dipoles arrayed on concentric spherical shells of radius r, all pointing towards the origin in space, that is, the center of the concentric spheres. We leave the depiction of this to the imagination. Placing r^2 points on each sphere of radius r is not trivial, similar to the problem of putting dimples on a golf ball. But as r gets larger, it becomes possible to place them in an asymptotically symmetric arrangement.

We can estimate the resulting potential or force at the origin by computing sums over expanding spherical shell sets

$$\left\{ \mathbf{r} \in \mathbb{R}^3 \mid r - 1 \leq |\mathbf{r}| < r \right\} \text{ for } r = 1, 2, 3, \ldots.$$

In each spherical shell region, the sum of all contributions, ignoring possible cancellations, would be approximately cr^{2-n} since all values of $\mathbf{r}$ in the set would be comparable to r, and there would be approximately cr^2 of them (assuming as we do that they are uniformly distributed). Then the total result would be proportional to

$$\sum_{r=1}^{R} r^{2-n} \tag{10.3}$$

which is divergent (as R increases) for $n \leq 3$.

The issue of charge cancellations is quite complex [58, 414, 459]. However, the general thrust of the arguments here is born out [414]: entities with faster decay do not have global effects, but ones with longer range can accumulate to yield significant effects. For example, collections of dipoles can have net macroscopic charge, and the field induced can depend on the macroscopic geometry of the material. Similarly, assemblages of quadrapoles can have a potential that depend on the macroscopic geometry of the material [414].

Note that R is the size of a macroscopic system in microscopic units, so it is related to Avogadro's number, hence R should be viewed as nearly infinite. The borderline case $n = 3$, for which the divergence is only logarithmic, corresponds to the quadrapole potential, as we shall show later in this chapter. For the octapole potential, $n = 4$, the first exponent where the interaction potential can be said to be local, but the convergence rate is rather slow: $\mathcal{O}(1/R_{\mathrm{cut}})$ if we take R_{cut} to be a cut-off radius beyond which we ignore external effects. This explains to some extent why molecular dynamics simulations have to expend so much computational effort to compute electrostatic interactions in order to represent the forces accurately.

The electrostatic potential for a system of dipoles exhibits a possible divergence if no further organization (i.e., formation of octapoles by groups of dipoles) obtains. This would imply that dipoles *must* form such structures in larger aggregates, or otherwise infinite potentials would arise.

We list in Table 10.1 several different interaction potentials that we will analyze together with the different power laws associated with them and common names associated with them.

The divergence of sums of potentials or electrostatic fields is indicative of more complex physical behavior, but the details of the mathematics and physics associated with this is more complicated to describe [414]. For example, materials made of organized dipoles can have permanent electric fields. Such materials are called **ferro-electric** [138, 197, 248, 412] by analogy with ferro-magnetic materials that form permanent magnets. Water ice can be ferro-electric [301, 546].

charge groups	force name	power law	equation
charge-charge	salt bridge	r^{-1}	(3.1)
charge-dipole	Keesom force	$\cos\theta\ r^{-2}$	(10.33)
dipole–dipole	hydrogen bond	r^{-3}	(3.11), (10.28)
dipole-quadrapole	van der Waals	r^{-4}	(3.33)
induced dipole	London-Debye	r^{-6}	(3.30)

Table 10.1 Different power law behaviors for various interactions, together with either common names frequently used for them or examples of interactions where they occur. The 'equation' column indicates where they can be found in the text.

10.2 Dipole–Dipole Interactions

We have seen that certain bonds can be modeled by simple interactions between charge groups. For example, polar groups can be modeled simply by placing partial charges appropriately at atom centers, as described in Section 8.2.2. Typically, these groups can be represented as dipoles (Section 3.3). Here we investigate in detail the angular dependence of some of these models. In Section 3.5.3, we saw that dipoles can interact even with neutrally charged groups; this is one manifestation of the van der Waals force. However, we will see that this can really be rationalized as a quadrapole-dipole interaction.

In this section, we will restrict most of our attention to charge groups in two dimensions. In some cases, we will explicitly include the third coordinate $z = 0$ in the representations (see Figure 3.12), but in others we will simply omit it (see Figure 10.3(a)).

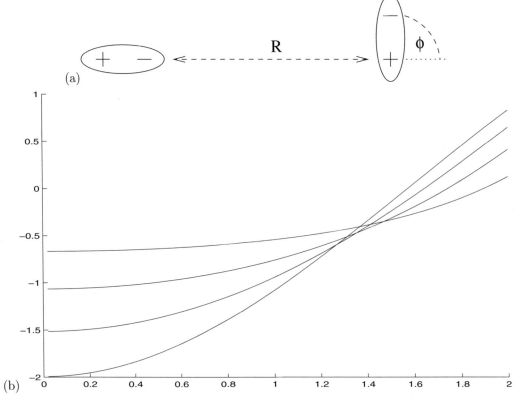

Fig. 10.3 Dipole–dipole interaction: rotated in-line configuration. (a) The positive (resp., negative) charge center of the dipole on the right is at $(0,0)$ (resp., at $(\cos\phi, \sin\phi)$). The positive (resp., negative) charge center of the dipole on the left is at $(-R-1, 0)$ (resp., at $(-R, 0)$). (b) Dipole–dipole (in-line) interaction energy (vertical axis), scaled by R^3, for $R = 2, 4, 10, 1000$. Horizontal ϕ-axis measured in radians. The flattest curve (the top one near ϕ=0) corresponds to $R = 2$.

10.2.1 Dipole–Dipole Configurations

Let us consider the effect of angular orientation on the strength of interaction of two dipoles. Since the possible set of configurations has a high dimension, we break down into special cases.

10.2.1.1 In-Line Interaction Configuration

Suppose we have two dipoles as indicated in Figure 10.3(a). The exact positions of the charges are as follows. The position of the positive charge on the right we take as the origin, and we assume the separation distance of the charges is one. The separation between the positive charge on the right and the negative charge on the left is R. Thus the charge centers of the dipole on the left are at $(-R - 1, 0)$ (positive charge) and $(-R, 0)$ (negative charge). The negative charge on the right is at $(\cos \phi, \sin \phi)$.

The distances between the various charges are easy to compute. The distance between the negative charge on the left and the positive charge on the right is R, and the distance between the two positive charges is $R + 1$. The distance between the two negative charges is

$$|(\cos \phi, \sin \phi) - (-R, 0)| = \sqrt{(R + \cos \phi)^2 + \sin^2 \phi}$$
$$= \sqrt{1 + R^2 + 2R \cos \phi}, \tag{10.4}$$

and the distance between the positive charge on the left and the negative charge on the right is

$$|(\cos \phi, \sin \phi) - (-R - 1, 0)| = \sqrt{(1 + R + \cos \phi)^2 + \sin^2 \phi}$$
$$= \sqrt{1 + (R + 1)^2 + 2(R + 1) \cos \phi}. \tag{10.5}$$

Thus the interaction energy for the dipole pair (assuming unit charges) is

$$\frac{1}{R + 1} - \frac{1}{R} + \frac{1}{\sqrt{1 + R^2 + 2R \cos \phi}} - \frac{1}{\sqrt{1 + (R + 1)^2 + 2(R + 1) \cos \phi}}. \tag{10.6}$$

A plot of the interaction energy (10.6) is given in Figure 10.3(b) as a function of ϕ for various values of R. Since we know (cf. (3.11)) that the interaction energy will decay like R^{-3}, we have scaled the energy in Figure 10.3(b) by R^3 to keep the plots on the same scale. The value of $R = 1000$ indicates the asymptotic behavior; see Exercise 10.1 for the analytical expression of the asymptotic limit. Indeed, there is little difference between $R = 100$ (not shown) and $R = 1000$. The flatter curve is the smallest value of R (=2) and shows only limited angular dependence. Thus we reach the following striking conclusion:

> modeling a hydrogen bond using a simple dipole–dipole
> interaction does not yield a very strong angular dependence.

10.2.1.2 Parallel Interaction Configuration

Let us consider the effect of a different angular orientation on the strength of interaction of two dipoles. Suppose we have two dipoles as indicated in Figure 10.4(a). Here the dipoles stay parallel, but the one on the right is displaced by an angle θ from the axis through the dipole on the left. The exact positions of the charges are as follows.

The position of the negative charge on the left we take as the origin, and we assume the separation distance of the charges is one. The separation between the positive charge on the right and the negative charge on the left is R. Thus the charge centers of the dipole on the right are at $R(\cos \theta, \sin \theta)$ (positive charge) and $(1 + R \cos \theta, R \sin \theta)$ (negative charge).

The distance between the positive charges is the same as the distance between the negative charges because the dipoles are parallel:

$$|(1 + R\cos\theta, R\sin\theta)| = \sqrt{1 + R^2 + 2R\cos\theta}. \tag{10.7}$$

Similarly, the distance between the positive charge on the left and the negative charge on the right is

$$|(2 + R\cos\theta, R\sin\theta)| = \sqrt{4 + R^2 + 4R\cos\theta}. \tag{10.8}$$

Thus the interaction energy for the dipole pair (assuming unit charges) is

$$-\frac{1}{R} + \frac{2}{\sqrt{1 + R^2 + 2R\cos\theta}} - \frac{1}{\sqrt{4 + R^2 + 4R\cos\theta}}. \tag{10.9}$$

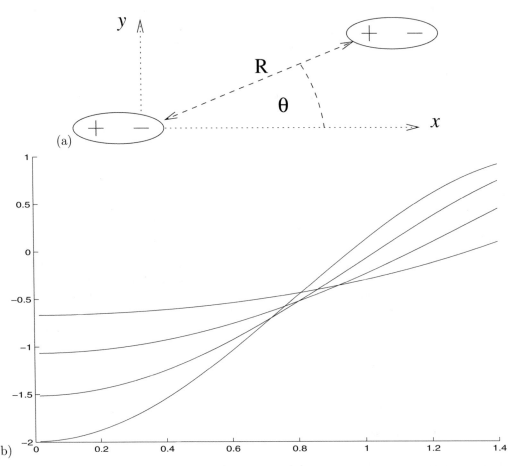

Fig. 10.4 Dipole–dipole interaction: rotated parallel configuration. (a) The negative (resp., positive) charge center of the dipole on the left is at $(0,0)$ (resp., at $(-1,0)$). The positive (resp., negative) charge center of the dipole on the right is at $R(\cos\theta, \sin\theta)$ (resp., at $(1 + R\cos\theta, \sin\theta)$). (b) Dipole–dipole (parallel) interaction energy (vertical axis), scaled by R^3, for $R = 2, 4, 10, 1000$. Horizontal θ-axis measured in radians.

A plot of the interaction energy (10.9) is given in Figure 10.4(b) as a function of θ for various values of R. Since we know (cf. (3.11)) that the interaction energy will decay like R^{-3}, we have scaled the energy in Figure 10.3(b) by R^3 to keep the plots on the same scale. The value of $R = 1000$ indicates the asymptotic behavior; see Exercise 10.2 for the analytical expression. Again, there is little difference between $R = 100$ (not shown) and $R = 1000$. The flatter curve is the smallest value of R ($=2$) and shows only limited angular dependence.

10.2.2 Two-Parameter Interaction Configuration

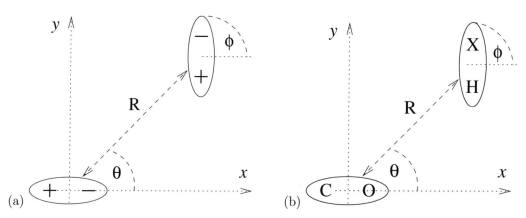

Fig. 10.5 Dipole–dipole (two-angle) interaction configuration. (a) The abstract case. The negative (resp., positive) charge center of the dipole on the left is at $(0,0)$ (resp., at $(-1,0)$). The positive (resp., negative) charge center of the dipole on the right is at $R(\cos\theta, \sin\theta)$ (resp., at $R(\cos\theta, \sin\theta) + (\cos\phi, \sin\phi)$). (b) Examples: in the serine-hydrogen placement problem, the atom X is oxygen (O), and for a carbonyl-amide backbone hydrogen bond, the atom X is nitrogen (N).

Now we consider the effect of a dual angular orientation on the strength of interaction of two dipoles. We do this as a first step to understanding the problem of hydrogen placement for serine, cf. Section 6.3. It also has direct bearing on the energy associated with different orientations of the charge groups (e.g., amide and carbonyl) in hydrogen bonds, cf. Section 6.4.

Suppose we have two dipoles as indicated in Figure 10.5. The exact positions of the charges are as follows. We take as origin the dipole center between the positive and negative charges on the left, and we assume the separation distance of the charges is two. Thus the $\pm$ charges on the left are at $(\mp 1, 0)$. The distance from the origin to the midpoint between the positive and negative charges on the right is R. Thus the charge centers of the dipole on the right are at $R(\cos\theta, \sin\theta) - (\cos\phi, \sin\phi)$ (positive charge) and $R(\cos\theta, \sin\theta) + (\cos\phi, \sin\phi)$ (negative charge). The distance between the negative charge on the right and the negative charge on the left is

$$|(1,0) - R(\cos\theta, \sin\theta) - (\cos\phi, \sin\phi)| = \sqrt{(1 - R\cos\theta - \cos\phi)^2 + (-R\sin\theta - \sin\phi)^2}$$
$$= \sqrt{2 + R^2 - 2R\cos\theta - 2\cos\phi + 2R(\cos\theta\cos\phi + \sin\theta\sin\phi)} = \sqrt{a - b - c + d},$$
(10.10)

where $a = 2 + R^2$, $b = 2R\cos\theta$, $c = 2\cos\phi$, $d = 2R(\cos\theta\cos\phi + \sin\theta\sin\phi)$. The distance between the positive charge on the right and the positive charge on the left is

$$|(-1,0) - R(\cos\theta, \sin\theta) + (\cos\phi, \sin\phi)| = \sqrt{(-1 - R\cos\theta + \cos\phi)^2 + (-R\sin\theta + \sin\phi)^2}$$
$$= \sqrt{2 + R^2 + 2R\cos\theta - 2\cos\phi - 2R(\cos\theta\cos\phi + \sin\theta\sin\phi)} = \sqrt{a + b - c - d}.$$
(10.11)

The distance between the positive charge on the left and the negative charge on the right is

$$|(-1,0) - R(\cos\theta, \sin\theta) - (\cos\phi, \sin\phi)| = \sqrt{(-1 - R\cos\theta - \cos\phi)^2 + (R\sin\theta + \sin\phi)^2}$$
$$= \sqrt{2 + R^2 + 2R\cos\theta + 2\cos\phi + 2R(\cos\theta\cos\phi + \sin\theta\sin\phi)} = \sqrt{a + b + c + d}.$$
(10.12)

The distance between the positive charge on the right and the negative charge on the left is

$$|(1,0) - R(\cos\theta, \sin\theta) + (\cos\phi, \sin\phi)| = \sqrt{(1 - R\cos\theta + \cos\phi)^2 + (R\sin\theta - \sin\phi)^2}$$
$$= \sqrt{2 + R^2 - 2R\cos\theta + 2\cos\phi - 2R(\cos\theta\cos\phi + \sin\theta\sin\phi)} = \sqrt{a - b + c - d}.$$
$$(10.13)$$

Thus the interaction energy for the dipole pair (assuming unit charges) is

$$V(R) = \frac{1}{\sqrt{a + b - c - d}} + \frac{1}{\sqrt{a - b - c + d}} - \frac{1}{\sqrt{a + b + c + d}} - \frac{1}{\sqrt{a - b + c - d}}, \quad (10.14)$$

where we recall that $a = 2 + R^2$, $b = 2R\cos\theta$, $c = 2\cos\phi$, and $d = 2R(\cos\theta\cos\phi + \sin\theta\sin\phi)$.

10.2.2.1 Minimum Energy Configuration

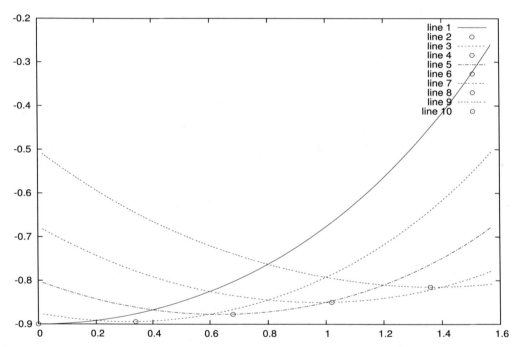

Fig. 10.6 Dipole–dipole (two-angle) interaction energy (vertical axis), scaled by R^3, for $R = 3$, as a function of ϕ for various fixed values of $\theta = 0, 0.2, 0.4, 0.6, 0.8$. Approximate minimum values of the energy are indicated by circles at the points $\phi = 1.7\,\theta$. The horizontal ϕ-axis is measured in radians.

Since there are now two angles to vary, it is not so clear how to display the energy in a useful way. But one question we may ask is: what is the minimum energy configuration if we allow ϕ to vary for a given θ? We might think that the dipole on the right would always point at the negative charge at the left. This would correspond to having the minimum energy configuration at $\phi = \theta$. This is clearly true at $\theta = 0$, but say at $\theta = \pi/2$, we might expect the minimum energy configuration to occur when the dipole on the right is flipped, that is at $\phi = \pi = 2\theta$. We plot the energy as a function of ϕ for various values of θ in Figure 10.6. As an aid to the eye, we plot a circle at a point close the minimum in energy, as a way to see how the optimum ϕ varies as a function of θ. In particular, we have plotted the point not at

$\phi = \theta$, nor at $\phi = 2\theta$, but rather $\phi = 1.7\,\theta$. This is convincing evidence that the relationship between the optimum value of ϕ for a fixed value of θ is complicated.

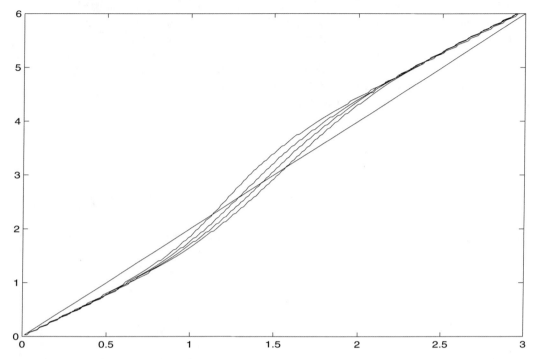

Fig. 10.7 Optimal ϕ angle (vertical axis) corresponding to the minimum interaction energy as a function of θ (horizontal axis) for the dipole–dipole (two-angle) interaction, for $R = 3, 5, 10, 1000$ (the left-most curve corresponds to $R = 3$, and they move to the right with increasing R). Both the horizontal θ-axis and the vertical ϕ-axis are measured in radians. The line $\phi = 2\theta$ has been added as a guide.

In the case that $\theta = \pi/2$, the expression (10.14) simplifies, with $b = 0$ and $d = 2R\sin\phi$, so that

$$
\begin{aligned}
V(R) &= \frac{1}{\sqrt{a-c-d}} + \frac{1}{\sqrt{a-c+d}} - \frac{1}{\sqrt{a+c+d}} - \frac{1}{\sqrt{a+c-d}}, \\
&= \frac{1}{\sqrt{2+R^2-2\cos\phi-2R\sin\phi}} + \frac{1}{\sqrt{2+R^2-2\cos\phi+2R\sin\phi}} \\
&\quad - \frac{1}{\sqrt{2+R^2+2\cos\phi+2R\sin\phi}} - \frac{1}{\sqrt{2+R^2+2\cos\phi-2R\sin\phi}}.
\end{aligned}
\tag{10.15}
$$

Then if $\phi = \pi$, this further simplifies to

$$
V(R) = \frac{1}{\sqrt{4+R^2}} + \frac{1}{\sqrt{4+R^2}} - \frac{1}{\sqrt{R^2}} - \frac{1}{\sqrt{R^2}} = \frac{2}{\sqrt{4+R^2}} - \frac{2}{R},
\tag{10.16}
$$

as we would expect. We leave as Exercise 10.4 to plot (10.15) as a function of ϕ for various values of R to see the behavior and to determine the minimum of the expression (10.15).

When R is large, we might expect that $\phi_{opt} \approx \theta$, since the dipole should point in the general direction of the other dipole. However, this is not the case; rather there is a limiting behavior that is different. In Figure 10.7, the optimal ϕ is plotted as a function of θ, and we note that it is very nearly equal to 2θ, but not exactly. For θ small, it behaves more nearly like $\phi \approx 1.7\,\theta$, but for larger values of θ the optimal ϕ increases to, and then exceeds, 2θ, before returning to the value of 2θ near $\theta = \pi$.

The minimum ϕ has been determined by computing the energies for discrete values of ϕ and then interpolating the data by a quadratic around the discrete minimum. Necessary adjustments at the ends of the computational domain are evident. Limited resolution in the computations contributes to the visible jaggedness of the curves in the plot. We leave as an exercise to produce smoother plots, as well as to explore the asymptotic behavior as $R \to \infty$.

The energy, again scaled by R^3, at the optimal value of ϕ is plotted as a function of θ in Figure 10.8. Since the curves in this figure are not horizontal, the dipole system has a torque that would tend to move them to the $\theta = 0$ position if θ were not fixed (as we assume it is, due to some external geometric constraint).

10.2.2.2 Limiting Expression

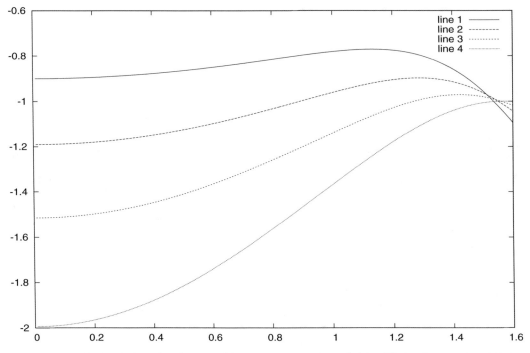

Fig. 10.8 Dipole–dipole (two-angle) interaction energy minimum, scaled by R^3, for $R = 3, 5, 10, 1000$ (top to bottom), as a function of θ. Plotted is the energy (vertical axis) at the optimal value of ϕ that minimizes the energy as a function of ϕ for fixed θ. Horizontal θ-axis measured in radians.

To examine the limiting behavior of the energy minimum in detail, we use a general formula valid in three dimensions. We return to the two-angle case in Section 10.2.4. We will show there that the limiting case for large R gives

$$\phi = \cot^{-1}\left(\frac{\frac{1}{3} + \cos 2\theta}{\sin 2\theta} \right). \tag{10.17}$$

10.2.3 Three Dimensional Interactions

We now consider the interaction between two dipoles in arbitrary orientation. For simplicity, we place the center of the two dipoles along the x-axis, one at the origin and the other at $\mathbf{r} = (r, 0, 0)$. Thus assume that there are charges ± 1 at $\pm \mathbf{u}$, and correspondingly charges ± 1 at points $\mathbf{r} \pm \mathbf{v}$ where $\mathbf{u}$ and $\mathbf{v}$ are unit vectors, that is, $u_1^2 + u_2^2 + u_3^2 = v_1^2 + v_2^2 + v_3^2 = 1$. The interaction potential is then

$$V(\mathbf{r}, \mathbf{u}, \mathbf{v}) = \frac{1}{|\mathbf{u} - \mathbf{r} - \mathbf{v}|} + \frac{1}{|-\mathbf{u} - \mathbf{r} + \mathbf{v}|} - \frac{1}{|\mathbf{u} - \mathbf{r} + \mathbf{v}|} - \frac{1}{|-\mathbf{u} - \mathbf{r} - \mathbf{v}|}. \tag{10.18}$$

We use the relation

$$|\mathbf{w} + \mathbf{x} + \mathbf{y}|^2 = |\mathbf{w}|^2 + |\mathbf{x}|^2 + |\mathbf{y}|^2 + 2(\mathbf{w} \cdot \mathbf{x} + \mathbf{w} \cdot \mathbf{y} + \mathbf{y} \cdot \mathbf{x}) \tag{10.19}$$

to get

$$V(\mathbf{r}, \mathbf{u}, \mathbf{v}) = \frac{1}{\sqrt{2 + r^2 - 2\mathbf{u} \cdot \mathbf{v} - 2r(u_1 - v_1)}} + \frac{1}{\sqrt{2 + r^2 - 2\mathbf{u} \cdot \mathbf{v} + 2r(u_1 - v_1)}}$$
$$- \frac{1}{\sqrt{2 + r^2 + 2\mathbf{u} \cdot \mathbf{v} - 2r(u_1 + v_1)}} - \frac{1}{\sqrt{2 + r^2 + 2\mathbf{u} \cdot \mathbf{v} + 2r(u_1 + v_1)}} \tag{10.20}$$
$$= \frac{1}{\sqrt{2 + r^2}} \left(\frac{1}{\sqrt{1 - a - b}} + \frac{1}{\sqrt{1 - a + b}} - \frac{1}{\sqrt{1 + a - c}} - \frac{1}{\sqrt{1 + a + c}} \right)$$

where

$$a = \frac{2\mathbf{u} \cdot \mathbf{v}}{2 + r^2}, \quad b = \frac{2r(u_1 - v_1)}{2 + r^2} = \frac{2\mathbf{r} \cdot (\mathbf{u} - \mathbf{v})}{2 + r^2}, \quad \text{and} \quad c = \frac{2r(u_1 + v_1)}{2 + r^2} = \frac{2\mathbf{r} \cdot (\mathbf{u} + \mathbf{v})}{2 + r^2}, \tag{10.21}$$

where we note that $a = \mathcal{O}\left(r^{-2}\right)$, $b = \mathcal{O}\left(r^{-1}\right)$, and $c = \mathcal{O}\left(r^{-1}\right)$. Expanding using the expression

$$\boxed{(1 - \epsilon)^{-1/2} = 1 + \tfrac{1}{2}\epsilon + \tfrac{3}{8}\epsilon^2 + \tfrac{5}{16}\epsilon^3 + \mathcal{O}\left(\epsilon^4\right),} \tag{10.22}$$

we find for large r that

$$\begin{aligned} V(\mathbf{r}, \mathbf{u}, \mathbf{v})\sqrt{2 + r^2} &= \frac{1}{\sqrt{1 - a - b}} + \frac{1}{\sqrt{1 - a + b}} - \frac{1}{\sqrt{1 + a - c}} - \frac{1}{\sqrt{1 + a + c}} \\ &\approx \tfrac{1}{2}\left(a + b + a - b + a - c + a + c\right) \\ &\quad + \tfrac{3}{8}\left((a + b)^2 + (a - b)^2 - (a - c)^2 - (a + c)^2\right) \\ &\quad + \tfrac{5}{16}\left((a + b)^3 + (a - b)^3 + (a - c)^3 + (a + c)^3\right) + \mathcal{O}\left(r^{-4}\right) \\ &= 2a + \tfrac{3}{4}\left(b^2 - c^2\right) + \tfrac{5}{16}\left(4a^3 + 6a(b^2 + c^2)\right) + \mathcal{O}\left(r^{-4}\right) \\ &= 2a + \tfrac{3}{4}\left(b^2 - c^2\right) + \mathcal{O}\left(r^{-4}\right) \end{aligned} \tag{10.23}$$

because $a = \mathcal{O}\left(r^{-2}\right)$, $b = \mathcal{O}\left(r^{-1}\right)$, and $c = \mathcal{O}\left(r^{-1}\right)$. But

$$b^2 - c^2 = \frac{4r^2}{(2 + r^2)^2}\left((u_1 - v_1)^2 - (u_1 + v_1)^2\right) = \frac{-16u_1 v_1 r^2}{(2 + r^2)^2} = \frac{-16(\mathbf{u} \cdot \mathbf{r})(\mathbf{v} \cdot \mathbf{r})}{(2 + r^2)^2} \tag{10.24}$$

157

Therefore

$$V(\mathbf{r}, \mathbf{u}, \mathbf{v})\sqrt{2 + r^2} = \frac{4\mathbf{u} \cdot \mathbf{v}}{2 + r^2} - \frac{12(\mathbf{r} \cdot \mathbf{u})(\mathbf{r} \cdot \mathbf{v})}{(2 + r^2)^2} + \mathcal{O}\left(r^{-4}\right). \tag{10.25}$$

Dividing, we get

$$\begin{aligned} V(\mathbf{r}, \mathbf{u}, \mathbf{v}) &= \frac{4\mathbf{u} \cdot \mathbf{v}}{(2 + r^2)^{3/2}} - \frac{12(\tilde{\mathbf{r}} \cdot \mathbf{u})(\tilde{\mathbf{r}} \cdot \mathbf{v})}{(2 + r^2)^{3/2}(2r^{-2} + 1)} + \mathcal{O}\left(r^{-5}\right) \\ &= \frac{4\mathbf{u} \cdot \mathbf{v}}{r^3(2r^{-2} + 1)^{3/2}} - \frac{12(\tilde{\mathbf{r}} \cdot \mathbf{u})(\tilde{\mathbf{r}} \cdot \mathbf{v})}{r^3(2r^{-2} + 1)^{5/2}} + \mathcal{O}\left(r^{-5}\right). \end{aligned} \tag{10.26}$$

where $r = |\mathbf{r}|$ and $\tilde{\mathbf{r}} = (1/r)\mathbf{r}$ is the unit vector in the direction of $\mathbf{r}$. Note that

$$\frac{1}{\sqrt{2r^{-2} + 1}} \approx 1 - r^{-2} + \mathcal{O}\left(r^{-4}\right) \tag{10.27}$$

by (10.22). Thus

$$\boxed{V(\mathbf{r}, \mathbf{u}, \mathbf{v}) \approx \frac{4\mathbf{u} \cdot \mathbf{v} - 12(\tilde{\mathbf{r}} \cdot \mathbf{u})(\tilde{\mathbf{r}} \cdot \mathbf{v})}{r^3} + \mathcal{O}\left(r^{-5}\right).} \tag{10.28}$$

10.2.4 Application to Two-Angle Problem

As an application, we consider the limiting behavior of the two-angle problem considered in Section 10.2.2. We write the coordinates in Figure 10.5 using the notation of Section 10.2.3. Thus we take $\mathbf{u} = (1, 0)$, $\mathbf{v} = (\cos\phi, \sin\phi)$, and $\tilde{\mathbf{r}} = (\cos\theta, \sin\theta)$. Therefore

$$\begin{aligned} r^3 V(\mathbf{r}, \phi, \theta) \approx U(\phi, \theta) : &= 4\cos\phi - 12\cos\theta(\cos\phi\cos\theta + \sin\phi\sin\theta) \\ &= 4\cos\phi(1 - 3\cos^2\theta) - 12\cos\theta\sin\phi\sin\theta \\ &= 4\cos\phi(1 - 3\cos^2\theta) - 6\sin\phi\sin 2\theta. \end{aligned} \tag{10.29}$$

We are interested in finding, for each θ, the value of ϕ that minimizes $U(\phi, \theta)$. Thus we compute

$$\frac{\partial U}{\partial\phi}(\phi, \theta) = -4\sin\phi(1 - 3\cos^2\theta) - 6\cos\phi\sin 2\theta. \tag{10.30}$$

Therefore, at a minimum, we have

$$\tan\phi = \frac{\sin 2\theta}{2\cos^2\theta - \frac{2}{3}} = \frac{\sin 2\theta}{\frac{1}{3} + \cos 2\theta}, \tag{10.31}$$

from which we get the expression $\phi(\theta)$ in (10.17). Expanding for small θ, we find $\phi = \frac{3}{2}\theta + \mathcal{O}\left(\theta^2\right)$. It also follows that when $\theta = \pi/2$, then $\phi = \pi$. We leave to Exercise 10.17 further investigation of this curious function.

10.3 Charged Interactions

Interactions involving charged groups may seem conceptually simpler than those involving only dipoles. The interaction between two charges is given in (3.1); it depends only on the distance between them. We will see that the interaction between a single charge and a dipole is not much more complex. But there are more complex interactions between charged groups that require careful analysis. These have direct bearing on the expected orientation of neighboring charged sidechains (e.g., Asp-Glu) in particular.

10.3.1 Charge-Dipole Interactions

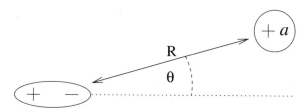

Fig. 10.9 Charge-dipole interaction configuration. The dipole on the left has its negative charge at the origin and its positive charge center at $(-1, 0, 0)$. The positive charge $(+a)$ center on the right is at $R(\cos\theta, \sin\theta, 0)$.

Charge-dipole interactions are simpler to analyze, and we have already anticipated their asymptotic strength in (3.8). On the other hand, this forms a very important class of interactions. Although mainchain–mainchain (hydrogen bond) interactions do not involve such pairs, all of the three other interactions among sidechains and mainchains can involve charge-dipole interactions. In addition, more complex interactions, such as cation-π interactions (Section 12.3) are of this form. Thus we develop the basics of charge-dipole interactions in some detail.

By choosing coordinates appropriately, we can assume that the positive and negative sites of the dipole align on the x-axis, and that the charge is located in the x, y plane, as depicted in Figure 10.9. Assume that the negative charge of the dipole is at the origin and that the isolated charge is positive, located at $r(\cos\theta, \sin\theta, 0)$. We choose scales such that the charges of the dipole are of unit size (± 1), located at ($\mp 1, 0, 0$), that is, the positive charge of the dipole is at $(-1, 0, 0)$. If a is the charge of the isolated charge, then the interaction energy of the system is

$$V(r, \theta) = -\frac{a}{\sqrt{(1 - r\cos\theta)^2 + r^2 \sin^2\theta}} + \frac{a}{\sqrt{(1 + r\cos\theta)^2 + r^2 \sin^2\theta}}. \qquad (10.32)$$

We leave as Exercise 10.8 to show that

$$V(r, \theta) \approx -\frac{2a\cos\theta}{r^2} \qquad (10.33)$$

for large r and fixed θ.

We will be interested in the force field that the dipole exerts on the charge as well. It is easier to compute the gradient of V in Cartesian coordinates (note that we can ignore the z direction in our computations):

$$V(x,y) = -\frac{a}{\sqrt{(1-x)^2+y^2}} + \frac{a}{\sqrt{(1+x)^2+y^2}}$$

$$= a\frac{\sqrt{(1-x)^2+y^2} - \sqrt{(1+x)^2+y^2}}{\sqrt{(1-x)^2+y^2}\,\sqrt{(1+x)^2+y^2}}$$

$$= a\frac{\sqrt{1-2x+r^2} - \sqrt{1+2x+r^2}}{\sqrt{1-2x+r^2}\,\sqrt{1+2x+r^2}} \tag{10.34}$$

$$= \frac{a}{r}\frac{\sqrt{(1-2x)r^{-2}+1} - \sqrt{(1+2x)r^{-2}+1}}{\sqrt{(1-2x)r^{-2}+1}\,\sqrt{(1+2x)r^{-2}+1}}$$

$$\approx -\frac{2a\cos\theta}{r^2}$$

for large $r = \sqrt{x^2+y^2}$.

To improve readability, we will use the notation $[x,y]$ to denote the vector with components x and y. Thus we find

$$\nabla V(x,y) = a((1-x)^2+y^2)^{-3/2}[x-1,y] - a((1+x)^2+y^2)^{-3/2}[x+1,y]$$

$$= a\frac{(1+2x+r^2)^{3/2} - (1-2x+r^2)^{3/2}}{(1-2x+r^2)^{3/2}(1+2x+r^2)^{3/2}}[x,y]$$

$$- a\frac{(1+2x+r^2)^{3/2} + (1-2x+r^2)^{3/2}}{(1-2x+r^2)^{3/2}(1+2x+r^2)^{3/2}}[1,0] \tag{10.35}$$

$$= \frac{a}{r^2}\frac{(r^{-2}+2r^{-1}\cos\theta+1)^{3/2} - (r^{-2}-2r^{-1}\cos\theta+1)^{3/2}}{(r^{-2}-2r^{-1}\cos\theta+1)^{3/2}(r^{-2}+2r^{-1}\cos\theta+1)^{3/2}}[\cos\theta,\sin\theta]$$

$$- \frac{a}{r^3}\frac{(r^{-2}+2r^{-1}\cos\theta+1)^{3/2} + (r^{-2}-2r^{-1}\cos\theta+1)^{3/2}}{(r^{-2}-2r^{-1}\cos\theta+1)^{3/2}(r^{-2}+2r^{-1}\cos\theta+1)^{3/2}}[1,0]$$

$$\approx \frac{6a\cos\theta}{r^3}[\cos\theta,\sin\theta] - \frac{2a}{r^3}[1,0]$$

for large r, where we used $(1+\epsilon)^{3/2} \approx 1 + \frac{3}{2}\epsilon$. Finally, we recall that these calculations are fully valid in three dimensions, so we have derived expressions valid for all z as well. In all cases, the z-component of ∇V is zero.

10.3.2 Charge-Charge Interactions

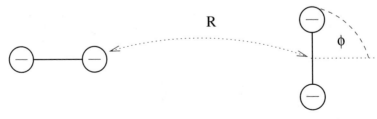

Fig. 10.10 Charge-charge interaction configuration similar to what is found in an interaction between Asp and Glu. The pair of negative charges on the right is at $\pm(\cos\phi,\sin\phi)$ and the pair of negative charges on the left is at $(-R\pm1,0)$.

We now consider the preferred angular orientation for two like charged groups as one finds in residues such as Asp and Glu. Suppose we have two charge groups as indicated in Figure 10.10. The exact positions of the charges are as follows. We assume the separation distance of the charges is two, and we assume that the origin is the center of the two negative charges on the right. Thus there are negative charges at $(\cos\phi, \sin\phi)$ and $(-\cos\phi, -\sin\phi)$. The separation between the charge groups is R; the negative charges on the left are fixed at $(-R\pm1, 0)$. Thus the interaction energy for the charged pairs (assuming unit charges) depend on the distances

$$
\begin{aligned}
r_{++} &= |+(\cos\phi, \sin\phi) - (R+1, 0)|, \\
r_{-+} &= |-(\cos\phi, \sin\phi) - (R+1, 0)|, \\
r_{+-} &= |+(\cos\phi, \sin\phi) - (R-1, 0)|, \quad \text{and} \\
r_{--} &= |-(\cos\phi, \sin\phi) - (R-1, 0)|.
\end{aligned}
\tag{10.36}
$$

With the denotations $C = \cos\phi, S = \sin\phi$, we find

$$
\begin{aligned}
r_{++}^2 &= (C - R - 1)^2 + S^2 = 2 + R^2 - 2\cos\phi(R+1) + 2R, \\
r_{-+}^2 &= (C + R + 1)^2 + S^2 = 2 + R^2 + 2\cos\phi(R+1) + 2R, \\
r_{+-}^2 &= (C - R + 1)^2 + S^2 = 2 + R^2 - 2\cos\phi(R-1) - 2R, \quad \text{and} \\
r_{--}^2 &= (C + R - 1)^2 + S^2 = 2 + R^2 + 2\cos\phi(R-1) - 2R.
\end{aligned}
\tag{10.37}
$$

We can write these expressions succinctly as

$$
r_{\pm_1 \pm_2} = \sqrt{2 + R^2 \pm_1 2(\cos\phi(R \pm_2 1)) \pm_2 2R}.
\tag{10.38}
$$

The energy (of repulsion) for the charge groups is

$$
\frac{1}{r_{++}} + \frac{1}{r_{-+}} + \frac{1}{r_{+-}} + \frac{1}{r_{--}},
\tag{10.39}
$$

and we seek to find the value of ϕ that minimizes it. We leave as Exercise 10.5 to plot the expression in (10.39) which is symmetric around $\phi = \pi/2$ and has a simple minimum there.

A more realistic model of the charge group for Asp and Glu is depicted in Figure 10.11. The configuration is now three-dimensional, with the carbon joining the oxygens below the plane. We assume a positive charge on the left at $(-R, 0, -z)$ and on the right at $(0, 0, -z)$. The negative charges are at $(\cos\phi, \sin\phi, 0)$, $(-\cos\phi, -\sin\phi, 0)$, and $(-R\pm1, 0, 0)$. We leave as Exercise 10.6 to investigate the minimum energy configuration.

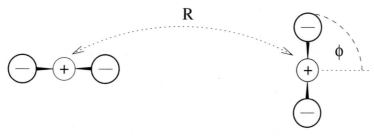

Fig. 10.11 Charge-charge interaction configuration similar to what is found in an interaction between Asp and Glu, but with a more refined model. The pair of negative charges on the right are at $(\cos\phi, \sin\phi, 0)$ and $(-\cos\phi, -\sin\phi, 0)$ and the associated positive charge is at $(0, 0, -z)$. The pair of negative charges on the left are at $(-R\pm1, 0, 0)$, and the associated positive charge is at $(-R, 0, -z)$.

10.4 General Form of a Charge Group

We now develop some technology to allow us to analyze charge groups in general contexts. Our main objective will be to understand the asymptotic decay rate of the corresponding potential (or resulting force field, the derivative of the potential). We have seen in Section 10.1 that the exponent of the decay rate is crucial in determining the global effect of the charge group. Thus our main objective is to develop only qualitative comparisons of different charge groups. Some quantitative comparisons are presented in Figure 10.12, but even these are intended only to clarify the qualitative relationship between different types of charge groups.

The general form of a potential for a charged system can be written as a sum of point charge potentials

$$V(\mathbf{r}) = \sum_{k=1}^{K} \frac{q_k}{|\mathbf{r} - \mathbf{r}_k|}, \tag{10.40}$$

where the charges q_k are at $\mathbf{r}_k$. When the net charge of the system is zero, we can interpret V as being defined by a difference operator applied to the fundamental charge potential

$$W(\mathbf{r}) = 1/|\mathbf{r}| \tag{10.41}$$

as follows. Define a translation operator $T_{\mathbf{x}}$ by

$$(T_{\mathbf{x}}f)(\mathbf{r}) = f(\mathbf{r} - \mathbf{x}) \tag{10.42}$$

for any function f. Then we can interpret the expression (10.40) as

$$V = \sum_{k=1}^{K} q_k T_{\mathbf{r}_k} W, \tag{10.43}$$

where W is defined in (10.41). In view of (10.43), we define the operator

$$\mathcal{D} = \sum_{k=1}^{K} q_k T_{\mathbf{r}_k}. \tag{10.44}$$

We will see that this corresponds to a difference operator when the net charge of the system is zero.

10.4.1 Asymptotics of General Potentials

The decay of $V(\mathbf{r})$ for simple dipoles can be determined by algebraic manipulations as in Section 3.3. However, for more complex arrangements, determining the rate is quite complicated. Multipole expansions become algebraically complex as the order increases. Here we offer an alternative calculus to determine asymptotic behavior of general potentials. We begin with some more precise notation.

Let as assume that there is a small parameter ϵ that defines the distance scale between the charge locations. That is, we define

$$V_\epsilon(\mathbf{r}) = \sum_{k=1}^{K} \frac{q_k}{|\mathbf{r} - \epsilon \mathbf{r}_k|}, \tag{10.45}$$

where we now assume that the $\mathbf{r}_k$'s are fixed and are of order one in size. There is a dual relationship between the asymptotics of $V_\epsilon(\mathbf{r})$ as $\mathbf{r} \to \infty$ and $\epsilon \to 0$, as follows:

$$\boxed{V_\epsilon(\mathbf{r}) = |\mathbf{r}|^{-1} V_{\epsilon/|\mathbf{r}|}\left(|\mathbf{r}|^{-1}\mathbf{r}\right).}$$

(10.46)

The proof just requires changing variables in (10.45):

$$V_\epsilon(\mathbf{r}) = \frac{1}{|\mathbf{r}|} \sum_{k=1}^{K} \frac{q_k}{|\mathbf{r}|^{-1}|\mathbf{r} - \epsilon \mathbf{r}_k|} = \frac{1}{|\mathbf{r}|} \sum_{k=1}^{K} \frac{q_k}{|(|\mathbf{r}|^{-1}\mathbf{r}) - (\epsilon/|\mathbf{r}|)\mathbf{r}_k|} = |\mathbf{r}|^{-1} V_{\epsilon/|\mathbf{r}|}\left(|\mathbf{r}|^{-1}\mathbf{r}\right).$$

(10.47)

Given a general potential V of the form (10.40), we can think of this as having $\epsilon = 1$, that is, $V = V_1$. Using (10.47), we can derive the asymptotic form

$$V(\mathbf{r}) = V_1(\mathbf{r}) = |\mathbf{r}|^{-1} V_{|\mathbf{r}|^{-1}}\left(|\mathbf{r}|^{-1}\mathbf{r}\right) = \epsilon V_\epsilon(\omega),$$

(10.48)

where we now define $\epsilon = |\mathbf{r}|^{-1}$ and $\omega = |\mathbf{r}|^{-1}\mathbf{r}$ satisfies $|\omega| = 1$. This says that we can determine asymptotics of V as $\mathbf{r} \to \infty$ by considering instead the behavior of V_ϵ on bounded sets (e.g., ω with $|\omega| = 1$) as $\epsilon \to 0$.

The reason that V_ϵ is useful is that we can write it in terms of a difference operator applied to W. Recalling (10.44), we define

$$\mathcal{D}_\epsilon = \sum_{k=1}^{K} q_k T_{\epsilon \mathbf{r}_k},$$

(10.49)

and observe from (10.40) and (10.41) that

$$V_\epsilon = \mathcal{D}_\epsilon W.$$

(10.50)

We will see in typical cases that, for some $\kappa \geq 0$,

$$\lim_{\epsilon \to 0} \epsilon^{-\kappa} \mathcal{D}_\epsilon = \mathcal{D}_0,$$

(10.51)

where $\mathcal{D}_0$ is a differential operator of order κ. The convergence in (10.51) is (at least) weak convergence, in the sense that for any smooth function f in a region $\Omega \subset \mathbb{R}^3$,

$$\lim_{\epsilon \to 0} \epsilon^{-\kappa} \mathcal{D}_\epsilon f(\mathbf{x}) = \mathcal{D}_0 f(\mathbf{x})$$

(10.52)

uniformly for $\mathbf{x} \in \Omega$. In particular, we will be mainly interested in sets Ω that exclude the origin, where the potentials are singular. Thus we conclude that

$$\lim_{\epsilon \to 0} \epsilon^{-\kappa} V_\epsilon(\omega) = \mathcal{V}(\omega), \qquad |\omega| = 1,$$

(10.53)

where the limiting potential is defined by

$$\mathcal{V}(\mathbf{r}) = \mathcal{D}_0 W(\mathbf{r}).$$

(10.54)

Applying (10.48), (10.53), and (10.54), we conclude that

$$V(\mathbf{r}) \approx \frac{1}{|\mathbf{r}|^{\kappa+1}} \mathcal{D}_0 W(|\mathbf{r}|^{-1}\mathbf{r}),$$

(10.55)

for large $\mathbf{r}$. More precisely, we will typically show that

$$\epsilon^{-\kappa} \mathcal{D}_\epsilon \phi(\mathbf{r}) = \mathcal{D}_0 \phi(\mathbf{r}) + \mathcal{O}(\epsilon) \tag{10.56}$$

in which case we can assert that

$$V(\mathbf{r}) = \frac{1}{|\mathbf{r}|^{\kappa+1}} \mathcal{D}_0 W(|\mathbf{r}|^{-1}\mathbf{r}) + \mathcal{O}(|\mathbf{r}|^{-\kappa-2}). \tag{10.57}$$

10.4.2 Application of (10.57)

Let us show how (10.57) can be used in practice by considering a known situation, that of a dipole. Thus take $\mathbf{r}_1 = (\frac{1}{2}, 0, 0)$ and $\mathbf{r}_2 = (-\frac{1}{2}, 0, 0)$. We can compute the action of $\mathcal{D}_\epsilon$ on smooth functions via

$$\mathcal{D}_\epsilon \phi(x, y, z) = \phi(x + \tfrac{1}{2}\epsilon, y, z) - \phi(x - \tfrac{1}{2}\epsilon, y, z). \tag{10.58}$$

By Taylor's theorem, we can expand a function ψ to show that

$$\psi(x + \xi) - \psi(x - \xi) = 2\xi \psi'(x) + \tfrac{1}{3}\xi^3 \psi^{(3)}(x) + \mathcal{O}(\xi^5). \tag{10.59}$$

Applying (10.59) to $\psi(x) = \phi(x, y, z)$, we have

$$\mathcal{D}_\epsilon \phi(x, y, z) = \epsilon \frac{\partial}{\partial x} \phi(x, y, z) + \mathcal{O}(\epsilon^3). \tag{10.60}$$

Taking limits, we see that

$$\epsilon^{-1} \mathcal{D}_\epsilon \rightarrow \frac{\partial}{\partial x} \tag{10.61}$$

as $\epsilon \rightarrow 0$. Thus we conclude that the potential for a dipole is $\mathcal{O}(|\mathbf{r}|^{-2})$ for large $\mathbf{r}$, in keeping with the derivation in Section 3.3. More precisely, applying (10.57) we have

$$V(\mathbf{r}) = |\mathbf{r}|^{-2} \left(\frac{\partial}{\partial x} W \right) (|\mathbf{r}|^{-1}\mathbf{r}) + \mathcal{O}(|\mathbf{r}|^{-3}). \tag{10.62}$$

10.5 Quadrapole Potential

The most important potential after the dipole is the quadrapole. As the name implies, it typically involves four charges. For this reason, the geometry can be quite complex. This provides an opportunity to apply the techniques developed in Section 10.4. We begin with a simple case.

10.5.1 Opposing Dipoles

Two opposing dipoles tend to cancel each other out, but the result is not zero, rather it is a quadrapole. For example, suppose there unit negative charges at $\pm(a, 0, 0)$, where a is some (positive) distance parameter, with unit positive charges at $\pm(a + 1, 0, 0)$, viz.:

+   -   -   +

These four charges can be arranged as two dipoles, one centered at $a + \frac{1}{2}$ and the other centered at $-a - \frac{1}{2}$. Thus the separation distance S between the two dipoles is

$$S = 2a + 1. \tag{10.63}$$

The partial charges for a benzene ring as modeled in Table 12.1 consist of three sets of such paired dipoles, arranged in a hexagonal fashion.

The potential for such a charge group can be estimated by algebraic means, as we did in Chapter 3, or we can utilize the technology of Section 10.4. We define

$$\mathcal{D}_\epsilon = T_{\epsilon(a+1,0,0)} - T_{\epsilon(a,0,0)} + T_{\epsilon(-a-1,0,0)} - T_{\epsilon(-a,0,0)}. \tag{10.64}$$

In evaluating $\mathcal{D}_\epsilon \phi$, we may as well assume that ϕ is a function only of x, cf. Section 10.4.2. Applying (10.59) to ϕ and ϕ' we find that

$$
\begin{aligned}
\mathcal{D}_\epsilon \phi(x) &= \phi(x - \epsilon(a+1)) - \phi(x - \epsilon a) + \phi(x + \epsilon(a+1)) - \phi(x+\epsilon a)\\
&= \epsilon \phi'(x - \epsilon(a + \tfrac{1}{2})) - \epsilon \phi'(x + \epsilon(a + \tfrac{1}{2})) + \mathcal{O}(\epsilon^3)\\
&= \epsilon^2 (2a + 1)\phi''(x) + \mathcal{O}(\epsilon^3)\\
&= \epsilon^2 S \phi''(x) + \mathcal{O}(\epsilon^3),
\end{aligned}
\tag{10.65}
$$

where S is the separation distance between the dipoles. Thus

$$\lim_{\epsilon \to 0} \epsilon^{-2} \mathcal{D}_\epsilon = (2a + 1)\frac{\partial^2}{\partial x^2} = S \frac{\partial^2}{\partial x^2}, \tag{10.66}$$

where $S = 2a + 1$ is the separation distance (10.63) between the dipoles. Applying (10.57), we find

$$V(\mathbf{r}) = |\mathbf{r}|^{-3} S \frac{\partial^2}{\partial x^2} W(|\mathbf{r}|^{-1}\mathbf{r}) + \mathcal{O}(|\mathbf{r}|^{-4}) \tag{10.67}$$

for large $\mathbf{r}$, where $W \ (= 1/r)$ is defined in (10.41).

The potential for opposing dipoles is depicted in Figure 10.12 for two separation distances, $S = 2$ (a) and $S = 4$ (b). For the larger value of the separation, there is little difference between the dipole and quadrapole potentials near the right-most charge. There is a much greater difference between the potentials for a single charge and that of a dipole. Thus the separation distance affects substantially the cancellation of the second dipole, at least locally. If the distance units are interpreted as Ångstroms, then the separation $S = 4$ (b) is roughly comparable to the partial charge model of a benzene ring (cf. Table 12.1) consisting of three sets of such paired dipoles.

10.5.2 Four-Corner Quadrapole

The four-corner arrangement provides a two-dimensional arrangement of opposing dipoles, as follows:

$$
\begin{matrix}
+ & & - \\
\\
- & & +
\end{matrix}
$$

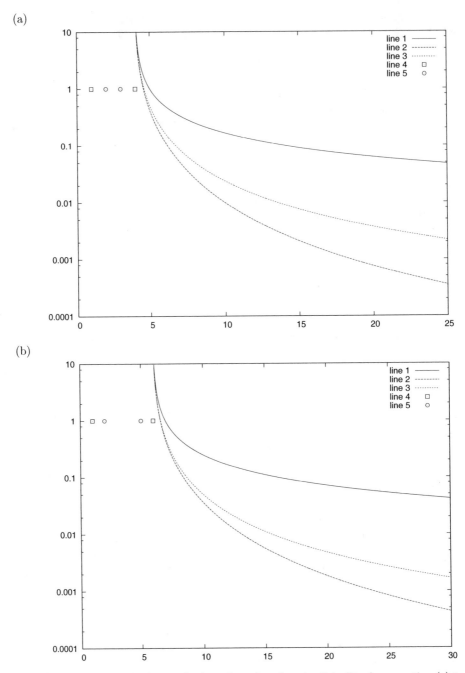

Fig. 10.12 Comparison of single charge, dipole and quadrapole potentials. Dipole separation (a) two units and (b) four units. The locations of the negative charges are indicated by circles and the locations of the positive charges are indicated by squares. The upper solid line is the potential for a single positive charge whose horizontal position is indicated by the right-most square. The middle, short-dashed line is the potential for the dipole corresponding to the right-most dipole. The lower, longer-dashed line is the potential for the dipole corresponding to the quadrapole formed by the pair of dipoles.

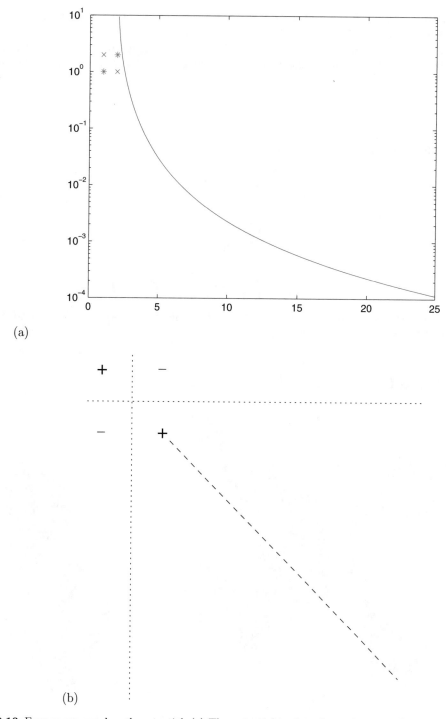

(a)

(b)

Fig. 10.13 Four corner quadrapole potential. (a) The potential is plotted as a function of distance s along the line $(x(x), y(s)) = ((2 + s/\sqrt{2}, 1 - s/\sqrt{2})$ which emanates from the lower-right corner of the quadrapole. The locations of the negative charges are indicated by circles and the locations of the positive charges are indicated by squares. (b) Schematic representation. The line used for the plot in (a) is indicated as a dashed line. The potential vanishes, by symmetry, on the dotted lines.

This quadrapole system has positive charges $q_1 = q_2 = 1$ at $\mathbf{r}_1 = (-1, 1, 0)$ and $\mathbf{r}_2 = (1, -1, 0)$ and negative charges $q_3 = q_4 = -1$ at $\mathbf{r}_3 = (1, 1, 0)$ and $\mathbf{r}_4 = (-1, -1, 0)$. A plot of the potential along a diagonal where it is maximal is given in Figure 10.13. Note that it dies off a bit more rapidly than the potential for the opposing dipoles (cf. Figure 10.12). Defining

$$\mathcal{D}_\epsilon = \sum_{k=1}^{K} q_k T_{\epsilon r_k} \tag{10.68}$$

and applying (10.59) twice, we see that

$$\mathcal{D}_\epsilon \phi(\mathbf{r}) = \sum_{k=1}^{K} q_k \phi(\mathbf{r} - \epsilon \mathbf{r}_k)$$
$$= 4 \frac{\partial}{\partial x} \frac{\partial}{\partial y} \phi(0) \epsilon^2 + \mathcal{O}(\epsilon^3). \tag{10.69}$$

Thus

$$V(\mathbf{r}) = |\mathbf{r}|^{-3} 4 \frac{\partial}{\partial x} \frac{\partial}{\partial y} W(|\mathbf{r}|^{-1} \mathbf{r}) + \mathcal{O}(|\mathbf{r}|^{-4}). \tag{10.70}$$

It is not hard to generalize these results to the case where the opposing charges are located at the four corners of any parallelogram.

10.5.3 Quadrapole Example

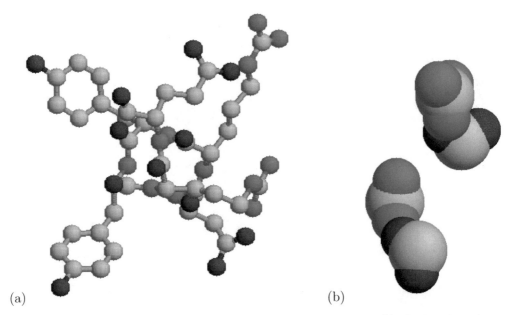

(a)   (b)

Fig. 10.14 A near quadrapole found in the PDB file 1I4M of the human prion. (a) The four charged groups are nearly aligned on the right side of the figure. Shown is the residue sequence DRYYRE. (b) Detail of the charged groups indicating the alignment of the opposing dipoles.

An example of a (near) quadrapole is found in the human prion (PDB file 1I4M) in the motif DRYYRE [180]. This is shown in Figure 10.14. The charges closely approximate the 'four corner' arrangement for a suitable parallelogram. The DRYYRE residue group forms a helical structure. Note that the four charged sidechains are nearly planar, with the tyrosines transverse to this plane. The detail Figure 10.14(b) shows the skewness of the two opposing dipoles.

10.5.4 Water: Dipole or Quadrapole?

Water can be written as a combination of two dipoles, following the general pattern of Section 10.4. So is water a quadrapole or just a dipole? The answer is crucial to determine the locality or globality of water–water interaction.

We can approximate the electronic structure of water as a system with positive charges

$$q_1 = q_2 = a \quad \text{at} \quad \mathbf{r}_{1,2} = (\pm c, -1, 0) \tag{10.71}$$

and negative charges

$$q_3 = q_4 = -b \quad \text{at} \quad \mathbf{r}_{3,4} = (0, y^0, \pm d), \tag{10.72}$$

where $y^0 > 0$ denotes the position above the x-axis of the lone-pair oxygen charges. Note that we have chosen the spatial unit so that the hydrogens are exactly one unit below the x-axis (and the charge center is the origin), but otherwise all positions are arbitrary. This is exactly the model of water that is known as Tip5P [330], with $a = b$. We would like to show that this system is a dipole; by that, we mean two things, one of which is that it is *not* a quadrapole.

To discover the exact multipole nature of the water model encoded in (10.71) and (10.72), we modify it to form a quadrapole. We extend the system (10.71–10.72) to involve two more charges:

$$q_5 = -2a \quad \text{at} \quad \mathbf{r}_5 = (0, -1, 0) \quad \text{and}$$
$$q_6 = 2b \quad \text{at} \quad \mathbf{r}_6 = (0, y_0, 0). \tag{10.73}$$

The configuration of charges is depicted in Figure 10.15.

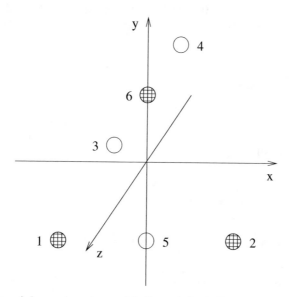

Fig. 10.15 Configuration of charges in water model. Open circles indicate negative charge locations; shaded circles indicate locations of positive charge.

The extended system (10.71),(10.72),(10.73) is a quadrapole due to the cancellations leading to an expression such as (10.69). More precisely, note that the charges at locations 1, 2 and 5 correspond to a second difference stencil centered at point 5 for approximating

$$\frac{\partial^2 \phi}{\partial x^2}(0, -1, 0) \tag{10.74}$$

(with suitable scaling). Similarly, the charges at locations 3, 4 and 6 correspond to a second difference stencil centered at point 6 for approximating

$$\frac{\partial^2 \phi}{\partial z^2}(0, y^0, 0) \tag{10.75}$$

(with suitable scaling). Therefore

$$\begin{aligned}
\mathcal{D}_\epsilon \phi(0) &= \sum_{k=1}^{6} q_k \phi(\epsilon \mathbf{r}_k) \\
&= ac^2 \epsilon^2 \frac{\partial^2 \phi}{\partial x^2}(0, -1, 0) - \\
&\quad bd^2 \epsilon^2 \frac{\partial^2 \phi}{\partial z^2}(0, y^0, 0)\epsilon^2 + \mathcal{O}(\epsilon^4),
\end{aligned} \tag{10.76}$$

and a similar result would hold when expanding about any point $\mathbf{r}$.

Let V^D denote the potential of the system with charges as indicated in (10.73). We leave as Exercise 10.10 to show that this is a dipole provided $a = b$. Let V_Q denote the quadrapole potential associated with (10.76), and let V^W be the water potential using the model (10.71–10.72). Thus we have written the water potential as

$$V^W = V^D + V^Q \tag{10.77}$$

for an explicit dipole potential V^D, with charges at $\mathbf{r}_5$ and $\mathbf{r}_6$, and a quadrapole. Thus the water model (10.71–10.72) is asymptotically a dipole, and not a quadrapole. Moreover, we see that the axis of the dipole is the y-axis, the bisector of the angle $\angle HOH$.

10.6 Multipole Summary

Let us summarize the asymptotic behavior of the various potentials that can arise. We have seen in (10.57) that the order of decay of the potential can be determined by the arrangement of the charges (10.49). When the net charge is non-zero, we have $k = 0$, but when the net charge is zero then $k \geq 1$. The dipole is the case $k = 1$, and $k = 2$ is called a quadrapole. Similarly, $k = 3$ is an octapole, and so on. We summarize in Table 10.2 the different powers for the potential of these different charge groups.

charge group	nonzero net charge	dipole	quadrapole	octapole
decay rate	r^{-1}	r^{-2}	r^{-3}	r^{-4}
κ	0	1	2	3

Table 10.2 Asymptotic decay rates for potentials of various charge groups.

The interaction energy between different charge groups has been worked out in specific cases. We summarize the general case in Table 10.3. We leave as Exercise 10.11 to verify the additional cases not already covered.

charge group	nonzero	dipole	quadrapole	octapole
nonzero	r^{-1}	r^{-2}	r^{-3}	r^{-4}
dipole	r^{-2}	r^{-3}	r^{-4}	r^{-5}
quadrapole	r^{-3}	r^{-4}	r^{-5}	r^{-6}
octapole	r^{-4}	r^{-5}	r^{-6}	r^{-7}

Table 10.3 Asymptotic decay rates for interaction energies of various charge groups.

10.7 Further Results

We collect here some further results about electrostatic interactions.

10.7.1 Dipole Induction by Dipoles

Water has both a fixed dipole and an inducible dipole. That is, water is both polar and polarizable. The dipole moment of water in the gas phase $\mu \approx 0.5$eÅ (cf. Section 18.7.2), and the polarizability $\alpha \approx 1.2$Å^3. Thus an electric field strength of only one-tenth of an electron per square Ångstrom (0.1eÅ^{-2}) could make a substantial modification to the polarity of water, since the change in polarity is approximated by the product of the polarizability and the electric field strength (see (3.27)).

10.7.2 Modified Dipole Interaction

Since the dipole–dipole interaction does not reproduce the sort of angular dependence we expect for certain bonds, e.g., hydrogen bonds, it is reasonable to try to modify the model. We ask the question: if the hydrogen charge density is represented in a more complex way, will a stronger angular dependence appear? To address this question, we introduce a negative charge to represent the electron density beyond the hydrogen. The exact positions of the charges are as follows. The position of the negative charge on the right we take as the origin, and we assume the separation distance between the charges is one. The separation of the positive charge on the left and the negative charge on the right is R. Thus the charge centers of the multipole on the left are at $(-R-1, 0)$ (negative charge $-\alpha$), $(-R, 0)$ (positive charge $+\beta$) and $(-R+\delta, 0)$ (negative charge $-\gamma$). The positive charge on the right is at $(\cos\phi, \sin\phi)$. The original model depicted in Figure 10.3 corresponds to the choices $\alpha = 1$, $\beta = 1$, $\gamma = 0$, in which case the value of δ does not matter.

The distances between the various charges are easy to compute. The distance between the positive charge on the left and the negative charge on the right is R, and the distance between the main (α) negative charge on the left and the negative charge on the right is $R + 1$. The distance between the minor (γ) negative charge on the left and the negative charge on the right is $R - \delta$.

The distance between the positive charge on the right and the minor (γ) negative charge on the left is

$$
\begin{aligned}
|(\cos\phi, \sin\phi) - (-R+\delta, 0)| &= \sqrt{(R-\delta+\cos\phi)^2 + \sin^2\phi} \\
&= \sqrt{1 + (R-\delta)^2 + 2(R-\delta)\cos\phi}.
\end{aligned}
\tag{10.78}
$$

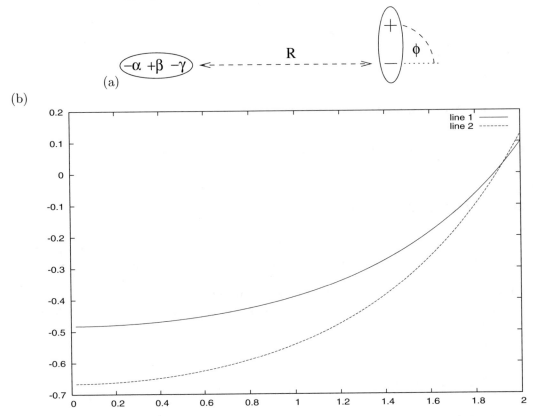

Fig. 10.16 (a) A modified model for dipole–dipole interaction. The negative (resp., positive) charge center of the dipole on the right is at $(0,0)$ (resp., at $(\cos\phi, \sin\phi)$). The charge centers of the multipole on the left are at $(-R-1, 0)$ (negative charge $-\alpha$), $(-R, 0)$ (positive charge $+\beta$) and $(-R+\delta, 0)$ (negative charge $-\gamma$). (b) Dipole–dipole (in-line) interaction energy, scaled by R^3, for $R = 2$ for the modified model versus the conventional model, depicted in Figure 10.3. Horizontal ϕ-axis measured in radians. The solid curve corresponds to $\alpha = 0.8, \beta = 1.0, \gamma = 0.2, \delta = 0.2$, whereas the flattest (dashed) curve corresponds to $\alpha = 1.0, \beta = 1.0, \gamma = 0.0$, which is the same as the model depicted in Figure 10.3.

The distance between the two positive charges is

$$
\begin{aligned}
|(\cos\phi, \sin\phi) - (-R, 0)| &= \sqrt{(R+\cos\phi)^2 + \sin^2\phi} \\
&= \sqrt{1 + R^2 + 2R\cos\phi},
\end{aligned}
\tag{10.79}
$$

and the distance between the main (α) negative charge on the left and the positive charge on the right is

$$|(\cos\phi, \sin\phi) - (-R-1, 0)| = \sqrt{(1 + R + \cos\phi)^2 + \sin^2\phi}$$
$$= \sqrt{1 + (R+1)^2 + 2(R+1)\cos\phi}. \tag{10.80}$$

Thus the interaction energy for the dipole pair (assuming unit charges) is

$$\frac{\alpha}{R+1} - \frac{\beta}{R} + \frac{\gamma}{R-\delta} - \frac{\gamma}{\sqrt{1 + (R-\delta)^2 + 2(R-\delta)\cos\phi}}$$
$$+ \frac{\beta}{\sqrt{1 + R^2 + 2R\cos\phi}} - \frac{\alpha}{\sqrt{1 + (R+1)^2 + 2(R+1)\cos\phi}}. \tag{10.81}$$

A plot of the interaction energy (10.81) is given in Figure 10.16 as a function of ϕ for $R = 2$, scaled by $R^{-3} = 8$. The flatter curve corresponds to the new model with a more complex dipole. Thus we see that this does not produce an improved model of the angular dependence of a hydrogen bond.

10.7.3 Hydrogen Placement for Ser and Thr

Let us consider the problem of determining the angular orientation of the hydrogen in serine and threonine, depicted in Figure 6.5. We choose coordinates so that the x, y plane contains the terminal carbon and oxygen from the sidechain of Ser/Thr and the negative site of the partial charge of the moiety that is forming the hydrogen bond, as depicted in Figure 10.17. In the special case that the positive charge in the dipole forming the hydrogen acceptor is also in this plane, then we can argue by symmetry that the hydrogen must lie in this plane as well, at one of the solid dots indicated at the intersection of the circle with the plane of the page. But in general, we must assume that the location of the positive partial charge is outside of this plane.

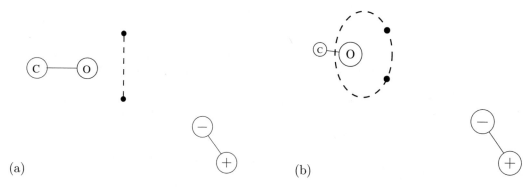

(a)   (b)

Fig. 10.17 Configuration for the placement of hydrogen at the end of the sidechain of serine or threonine in response to a nearby dipole. The dashed line indicates the circle of possible hydrogen placements. (a) The plane of the circle is orthogonal to the plane of the page. (b) The plane of the circle is skew to the plane of the page.

In Figure 10.17(b), we indicate the view from the plane defined by the positions of the oxygen and the negative and positive partial charges of the dipole. The circle of possible locations for the hydrogen (see Figure 6.5) is now clearly visible, and the intersection points with the plane of the page are again indicated by black dots. Now we see it is not obvious what the optimal position for the hydrogen would be.

To determine the optimal hydrogen position, let us assume that the coordinates are as in Figure 10.17, with the origin chosen to be at the center of the circle. Thus, the plane of the page is the x, y plane, and the coordinates of the circle are $(0, \cos \theta, \sin \theta)$. The position of the negative partial charge is then $(x_0, y_0, 0)$ and the positive partial charge is (x_1, y_1, z_1). The interaction potential between the dipole and the hydrogen is thus

$$\frac{-1}{\sqrt{x_0^2 + (y_0 - \cos \theta)^2 + \sin^2 \theta}} + \frac{1}{\sqrt{x_1^2 + (y_1 - \cos \theta)^2 + (z_1 - \sin \theta)^2}} \tag{10.82}$$

For given x_0, y_0, x_1, y_1, z_1, this expression can be minimized to find the optimal θ.

We can also use the expression (10.35) to find the optimum θ. In coordinates determined so that the hydrogen and the dipole lie in a plane, the interaction field (10.35) has a zero component orthogonal to the plane. For the hydrogen position on the circle to be correct, the tangent to the circle must be orthogonal to the gradient of the interaction potential at that point. Suppose that we write the circle as $(x(\phi), y(\phi), z(\phi))$ in these coordinates. Then a necessary condition is that

$$\nabla V(x(\phi), y(\phi), z(\phi)) \cdot (x'(\phi), y'(\phi), z'(\phi)) = 0. \tag{10.83}$$

10.8 Exercises

Exercise 10.1. Show that the interaction energy (10.6) tends to the asymptotic form

$$\frac{-2 \cos \phi}{R^3}. \tag{10.84}$$

Exercise 10.2. Show that the interaction energy (10.9) tends to the asymptotic form

$$\frac{-\frac{1}{2} - \frac{3}{2} \cos \phi}{R^3}. \tag{10.85}$$

Exercise 10.3. Verify that the second term in the energy expression in (10.14) is indeed the same as (10.7). Also verify that the fourth term in the energy expression in (10.14) is correct.

Exercise 10.4. Plot the expression in (10.15) and determine if it is symmetric around $\phi = \pi$ for finite R. Where does the minimum occur as a function of ϕ? Determine the limiting expression as $R \to \infty$ after scaling by R^3.

Exercise 10.5. Plot the expression in (10.39) and verify that it is symmetric around $\phi = \pi/2$ and has a simple minimum there.

Exercise 10.6. Carry out the calculations leading to the expression in (10.39) in the case that the charge group has a positive charge as well as the negative charges, as shown in Figure 10.11. Take the charges to be appropriate for Asp or Glu. Investigate the minimum energy configuration. Also consider three-dimensional configurations in which the positive charge is located below the negative charges.

Exercise 10.7. Investigate the optimal (minimum energy) configuration for charge-dipole pairs in which the charge is fixed at a distance r from the center of the dipole, which is free to rotate by an angle ϕ. Determine the value of ϕ at the minimum.

Exercise 10.8. Prove that the asymptotic expression (10.33) is valid for fixed θ and large r. (Hint: show that

$$V(r,\theta) = \frac{a}{r}\left(\frac{\sqrt{1-2r^{-1}\cos\theta+r^{-2}}-\sqrt{1+2r^{-1}\cos\theta+r^{-2}}}{\sqrt{1-2r^{-1}\cos\theta+r^{-2}}\sqrt{1+2r^{-1}\cos\theta+r^{-2}}}\right) \qquad (10.86)$$

and expand the expression in the numerator. Is this asymptotic approximation uniformly valid for all θ?)

Exercise 10.9. Determine the percentage error in the approximation (10.33) when $\theta = \pi/4$ and $r = 3$.

Exercise 10.10. Show that a charge system with only the charges as indicated in (10.73) forms a dipole provided $a = b$ and examine its asymptotic behavior.

Exercise 10.11. Verify the interaction energies listed in Table 10.3 for the cases not already covered. (Hint: develop technology similar to Section 10.4.1. The interaction introduces an additional difference operator that multiplies the one associated with the potential. The order of the product of the two limiting differential operators is equal to the sum of the orders of the individual operators.)

Exercise 10.12. Plot the forces corresponding to the potentials in Figure 10.12. That is, compute and compare the forces on a single charge from an isolated charge, a dipole and a quadrapole.

Exercise 10.13. Suppose that the units for the horizontal axis in Figure 10.12 are chosen to be Ångstroms, and that the unit of charge corresponds to one electron. Determine the units of the vertical axis (see Chapter 18).

Exercise 10.14. Compare the asymptotic expression (10.28) for the general dipole–dipole interaction with the special cases considered earlier in the text, e.g., (3.11), (3.15), (10.84) and (10.85).

Exercise 10.15. Determine whether the force between two neutral groups as depicted in Figure 10.18 is attractive or repulsive.

Exercise 10.16. Show that the interaction energy for two quadrapoles as depicted in Figure 10.18 is

$$\frac{1}{r-2} - \frac{4}{r-1} + \frac{6}{r} - \frac{4}{r+1} + \frac{1}{r+2} \approx \left(\frac{\partial^4}{\partial r^4}\right)\left(\frac{1}{r}\right) = \frac{24}{r^5} \qquad (10.87)$$

as $r \to \infty$ by considering the error in the difference operator represented by the left-hand side, cf. Exercise 3.7.

Fig. 10.18 Interaction between two quadrapoles.

Exercise 10.17. Plot $\phi(\theta)$ given by (10.17). Be careful about what happens when the denominator vanishes.

Exercise 10.18. Determine the operator $\mathcal{D}_0$ defined in (10.51) in the case of the Na-Cl crystal, which has a charge $(-1)^{i+j+k}$ on integer lattice points (i,j,k).

Exercise 10.19. Investigate the potential energy of the system indicated in Figure 10.19. Determine the value of ϕ that minimizes the energy.

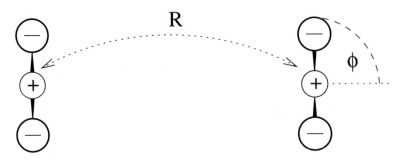

Fig. 10.19 Charge-charge interaction configuration similar to what is found in an interaction between Asp and Glu, but with a rotated sidechain. The pair of negative charges on the right are at $(\cos\phi, \sin\phi, 0)$ and $(-\cos\phi, -\sin\phi, 0)$ and the associated positive charge is at $(0, 0, -z)$. The pair of negative charges on the left are at $(-R, \pm 1, 0)$, and the associated positive charge is at $(R, 0, -z)$.

Exercise 10.20. The strength of a dipole is the product of the charge difference times the separation distance. Prove that the separation distance does not affect the conclusions of Section 10.2.4. More precisely, let the separation vector $\mathbf{u} = (\epsilon, 0)$ and the charge strength be $1/\epsilon$, so that the dipole strength is constant. Show that (10.29) is still obtained for all $\epsilon > 0$.

Chapter 11
Dehydrons in Protein Interactivity

Howard Walter Florey (1898–1968) won the Nobel Prize in Physiology or
Medicine in 1945 (with two others) for the discovery of penicillin [299].

We have established that dehydrons are sticky binding sites for protein ligands. Here we
examine several cases where these play an important biologic role. We begin by considering
diverse examples in the literature where specific dehydrons can be seen to be critical in
specific interactive sites. The preformed dehydrons in one protein are shown to be wrapped
intermolecularly upon binding. The protein associations of interest include different types
of interactions, involving signaling, structural roles, and enzymatic activity. Subsequently, we
study protein interactions from a more high-level view, showing that the number of dehydrons
in proteins correlates positively with protein interactivity.

11.1 Basic Cases

Several examples of the role of dehydrons in protein associations were given in [174]. We
review them here briefly. They illustrate the way that dehydrons play a role in signaling and
in protein structure formation.

11.1.1 Dehydrons in Antibody–Antigen Associations

One striking example of a dehydron involved in protein–protein association is the binding of
the light chain of antibody FAB25.3 with the HIV-1 capsid protein P24 [174], as depicted in
the PDB file 1AFV. Antibodies are proteins that bind to antigens; they can in some cases
directly neutralize the antigen, but in most cases they tag things for attack by the immune
system. In this role, we can think of antibodies as playing the role of signaling.

Antibodies consist of two ("light" and "heavy") chains. The smaller light chain has an
important region that resembles a finger reaching out to touch the antigen at a particular
location. It was found that this finger points directly at a dehydron in the antigen [174]. Thus,
it appears that the dehydron provides an important part of the recognition process for this
antibody–antigen complex.

© Springer International Publishing AG 2017
L.R. Scott and A. Fernández, *A Mathematical Approach to Protein Biophysics*,
Biological and Medical Physics, Biomedical Engineering,
DOI 10.1007/978-3-319-66032-5_11

In Figure 11.1(a), a cartoon of [174, Figure 2] depicts a portion of the capsid protein P24 (on the left) with the dehydron depicted as a gray strip. The light chain of the antibody FAB25.3 is depicted on the right, pointing directly at the dehydron on the protein P24 [174]. The binding interface for this complex is quite small, and yet the single dehydron in this helix of the HIV-1 capsid protein lies precisely at the interface with the extended finger of the light chain. The intermolecular wrapping of the dehydron is apparent and becomes a critical factor for the association.

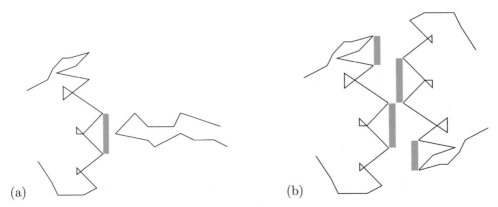

(a)   (b)

Fig. 11.1 Cartoon of two fragments of proteins binding (a) at a dehydron on one of the proteins and (b) along interfaces with multiple dehydrons on both of the proteins. The dehydrons are depicted as a gray strip.

The tip of the finger in the light chain in this case is the residue Tyr 32 in chain L. Five of the nonpolar carbons in this residue are in a desolvation domain (radius 6.5 Å) of the hydrogen bond between Thr 108 (donor) and Ser 102 (acceptor) in the antigen (chain A), which is one of a sequence of hydrogen bonds in a helix in the HIV-1 capsid. The desolvation domain (consisting of spheres centered at the CA atoms of the donor and acceptor) contains only nine nonpolar groups natively available from chain A. The wrapping of this dehydron by Tyr 32 is thus quite significant.

It should be noted that the desolvation domain of this dehydron also contains four nonpolar groups from the B chain in 1AFV. The PDB file 1AFV consists of two copies of the antibody–antigen complex. Chains A, H, L represent one copy, and chains B, K, M represent the other. Thus, atoms from chain B nearby atoms in chain A represent a crystallization artifact. It is not surprising (Section 11.1.3) that a crystal contact would appear at a dehydron, but it could be of concern that the region around this underwrapped dehydron only becomes structured due to the artificial dimer. This dehydron is part of a short helix involving only residues 101 to 104 in the A chain in the HIV-1 capsid.

11.1.2 Dehydrons in Quaternary Structural Assemblages

Dehydrons also appear more broadly in structural roles [174]. In some cases, they guide formation of quaternary structures, such as dimers, through intermolecular wrapping. For example, dehydrons (G49, G52), (G78, T80), and (T91, G94) in the HIV-1 protease guide the formation of the dimer structure [169]. In other cases, the structures can be quite complex. A cartoon of this kind of behavior is presented in Figure 11.1(b).

Virus capsids can be viewed as a model for protein–protein interaction [549–551]. The formation of the capsid in picornaviruses [198, 286] was shown to be essentially determined by the distribution of dehydrons in the individual virus peptide (VP) subunits [174]. This distribution concentrates at the symmetry centers of the capsid and edge-to-edge subunit positioning.

A84-LEU-(H)–A81-TYR-(O)	A98-GLU-(H)–A96-GLY-(O)
A96-GLY-(H)–A102-GLU-(O)	B162-LYS-(H)–A96-GLY-(O)
A101-SER-(H)–A98-GLU-(O)	A205-HIS-(H)–B208-GLN-(O)
A210-ASN-(H)–A207-ARG-(O)	A212-GLY-(H)–A210-ASN-(O)
B252-VAL-(H)–B101-ARG-(O)	B161-ARG-(H)–B159-THR-(O)
B228-ALA-(H)–B225-SER-(O)	C72-VAL-(H)–C70-THR-(O)
C141-GLN-(H)–C138-SER-(O)	C180-ILE-(H)–C178-ALA-(O)
C204-SER-(H)–C202-PRO-(O)	

Table 11.1 List of dehydrons for the protein unit of Mengo encephalomyocarditis virus (PDB code 2MEV) [286] engaged in known crystal contacts (their wrapping hydrophobic groups belong to the sidechains of residues known to form crystal contacts [268]). The proton-donor residue is marked as (H) and the electron-donor residue supplying the carbonyl group is marked (O).

There are three individual VP (virus peptide) subunits that assemble into a virus unit for the foot-and-mouth disease virus (FMDV) [198]. The atomic coordinates correspond to PDB entry 1BBT. This type of assembly of a virus capsid from several copies of a small number of VP subunits is typical [549, 551, 552]. Regions of high concentration of dehydrons are found at the symmetry centers of the capsid, and the exposure of these dehydrons to solvent is eliminated upon the associations of the units ([174, Figure 5a]). Approximately two-thirds of the dehydrons involved in domain-swapping and unit assembly become "desensitized" upon formation of the unit. Moreover, the capsid has four remaining regions with a high concentration of dehydrons.

A comparison between [198, Figures 5a and 5b] reveals that the four regions in the unit with a high dehydron density can be readily associated either with the centers of symmetry of the capsid or to the interunit edge-to-edge assembly of VP2 and VP3 [198]. Thus, the FMDV unit presents two loop regions highly sensitive to water removal on VP1 and VP3 which are directly engaged in nucleating other units at the pentamer and hexamer (with three-fold symmetry) centers of the capsid. Furthermore, a helical region on VP2 is severely under-wrapped and occupies the central dimeric center (shown in [198, Figure 5b]).

The FMDV capsid protein VP2 has a "handle" region on the β-hairpin near the N-terminus (12-27 region) that is known to be structurally defective, and it is also underwrapped. This region is also present in the VP2 subunits of all the picornaviruses. Because of its high density of dehydrons, this region is a strong organizing center in the capsid. It has been suggested as a potential drug target [174].

Dehydrons in the human rhinovirus (PDB code 1R1A [268]) can be seen to have similar properties [174]. While most dehydrons can be attributed to the unit assembly, there are three particular sites with high dehydron density which do not become well wrapped either after the formation of the unit or after the assembly of the whole capsid. Two of them correspond to antibody binding sites, but the other site lies under the so-called canyon of the VP1 structure and has been known to be the target region for the drug WIN 51711 used to treat the common cold [268].

The pattern of dehydrons in the virus is shown in [174, Figure 7]. The viral unit has only two very strong dehydron centers on its rim: the pentamer center located in VP1, and the part of VP2 involved in the VP2–VP3 edge-to-edge contact. The remaining 15 dehydrons are listed in Table 11.1 are not involved in the organization of the capsid. Of the 60 residues known to be engaged in crystal contacts for this virus [286], 54 of them have sidechain carbonaceous groups in the desolvation domains of the 15 dehydrons marked in Table 11.1. Thus, the dehydrons not associated with the structural organization of the capsid can be seen to correlate with the crystal packing.

11.1.3 Crystal Contacts

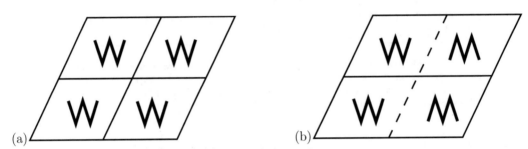

Fig. 11.2 Crystal contacts. In (a), a single copy of a protein W is repeated in a regular pattern. In this case, only one copy (chain) will appear in the PDB file. In (b), a pair with one copy transformed in some way appear together, again in a regular pattern. In this case, both copies W and M, where M is W rotated by some amount, will appear as separate chains in the PDB file.

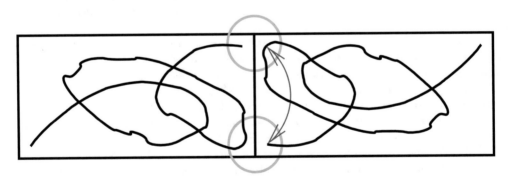

Fig. 11.3 False contacts can occur between two copies that appear as a pair of chains in a PDB file.

When proteins are crystallized for the purpose of imaging, the individual protein units contact each other at points called **crystal contacts** [135]. These frequently occur at dehydrons, since these promote complexation. Many PDB files involve multiple copies of proteins due the manner in which they are crystallized, as shown in Figure 11.2(b). Thus, the crystal contacts may appear in the PDB file as interactions between different chains, as shown in Figure 11.3. These are points at which each individual chain makes contact(s) with another chain in the formation of the crystal used in the process of X-ray imaging. Thus, they may be thought of as a type of artifact of the protein–protein interactions related to the imaging process. One example of a crystal contact was observed in Section 11.1.1.

Another view of this type of artifact can be seen in the PDB file 1P2C (see Section 13.1). This consists of two copies of an antibody–antigen complex. These complexes consist of two antibody chains and the antigen chain. In 1P2C, one of the antibody chains (chain A) makes several hydrogen bonds with the antigen (chain F) in the other copy of the complex. These are not biologically relevant.

11.1.4 Enzymatic Activity

Enzymes are proteins that **catalyze** (i.e., enhance) chemical reactions. We can think of them as machines, since they start with an input resource, a molecule called the **substrate**, and convert the substrate into a different molecule, the product. The process is called **catalysis**. The chemical reaction that is facilitated by an enzyme would occur naturally without a catalyst, but at a much slower rate.

Enzymes can be quite large proteins [520, 521], but the "active site" in which catalysis takes place is localized. There are thousands of enzymes that play a role in metabolic pathways [441]. Many drugs are designed to be inhibitors of enzymes [169], disrupting the role of an enzyme in the metabolic system.

The active site of an enzyme will contain water until the substrate enters. Moreover, water removal is critical for the success of enzymatic process [169, 520]. The process of water removal can be enhanced energetically by the presence of dehydrons near the active site [169]. The target of the catalysis is frequently a small molecule, or a small protruding part of a larger molecule, and in this case,

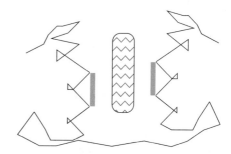

Fig. 11.4 Cartoon of an enzyme with a small molecule (herringbone shading) in the active site. The dehydrons are depicted as gray strips.

the dehydrons are found on the enzyme, as shown in the cartoon in Figure 11.4. Although most drugs were designed via high-throughput screening and not specifically intended as wrappers, we can see retrospectively that particular drugs play a significant role in wrapping dehydrons [169, 176]. Moreover, drugs can be redesigned to enhance the interaction between hydrophobic groups on the drug and dehydrons on the target protein [114, 171, 176].

11.2 Debunking Hydrophobic Contacts

In a recent paper [545], certain conserved residues (see Figure 11.5) were highlighted that make "hydrophobic contacts" upon binding of two proteins. The paper introduced two new PDB files, 2XS1 and 2XS8, depicting the binding of Alix protein to Gag proteins of simian immunodeficiency viruses SIVmac239 and SIVagmTan-1. The authors observed conserved hydrophobic residues critical to binding, also found in Gag proteins of other viruses [545].

Since hydrophobic residues involved in binding often play the role of wrappers stabilizing preformed hydrogen bonds, we decided to analyze this motif in detail. There are several software systems that analyze wrapping. We chose to use Wrappa [193, 194] to examine these systems. We confirmed that the conserved hydrophobic residues were indeed wrapping candidate dehydrons [174] in the Gag proteins in 2XS1 and 2XS8. Moreover, the use of Wrappa in this analysis required exercising several of Wrappa's features. We have written this in part as a tutorial to describe how Wrappa can be used in such a study.

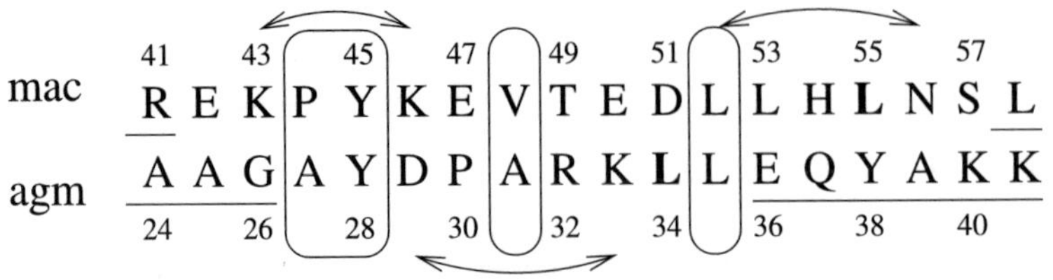

Fig. 11.5 Alignment of Gag proteins (chains B) of simian immunodeficiency viruses SIVmac239 (PDB file 2XS1, denoted by `mac` in the figure) and SIVagmTan-1 (PDB file 2XS8, denoted by `agm` in the figure). Conserved residues are enclosed in rounded rectangles. All of these are wrappers of hydrogen bonds in the Alix (A) chains in PDB files 2XS1 and 2XS8, respectively. Two additional Leucines which provide wrappers are shown in bold face. The arrows indicate underwrapped mainchain hydrogen bonds in the Gag proteins listed in Table 11.4. The initial residue R and the last two residues S and L are not represented in the PDB file 2XS1; the initial residues AAG and the final residues EQYAKK are not represented in the PDB file 2XS8 (missing residues underlined).

11.2.1 Hydrogen Bond Definitions

Wrappa computes standard distances and angles used to define hydrogen bonds Section 6.4. The relevant angles are shown in Figure 6.7. The angles ∠NHO, ∠NOC, and ∠HOC are all assumed to be at least 110 degrees by default in Wrappa.

The distances of interest are d_{NO}, the distance between O and N in the mainchain, and d_{HO}, the distance between O and H in the mainchain. The default assignments in Wrappa include the requirements that

$$d_{NO} \leq 3.5\,\text{Å} \quad \text{and} \quad d_{HO} \leq 2.5\,\text{Å}, \tag{11.1}$$

for the distances.

All of these default values are easily changed by the user of Wrappa. We describe one example of how such flexibility is crucial for analyzing PDB files. All of the data presented here were obtained using the default parameter choices in Wrappa except as noted in Section 11.2.3 to detect imperfect hydrogen bonds.

11.2.2 Interchain Wrapping and Protein Association

When the hydrogen bonds in a protein are not well wrapped, their strength and stability can be enhanced by the approach of a ligand that enhances water removal [174]. Wrappa was designed to find hydrogen bonds automatically and unambiguously by using rigorous criteria (albeit user-controllable) for wrapping and hydrogen bond definition. However, using Wrappa with the default settings may mask interchain wrapping for the following reason.

Suppose that chain A donates n wrappers to a hydrogen bond desolvation sphere [174, 194] as shown in Figure 11.6(a), and that chain B donates m wrappers to the same hydrogen bond desolvation sphere as shown in Figure 11.6(b). Then Wrappa will (correctly) assess that this hydrogen bond is surrounded by $m + n$ wrappers. If $m + n$ exceeds the wrapping threshold (the default is 19), then this hydrogen bond will not be reported as underwrapped. But if $n < 19$, the hydrogen bond may have been underwrapped before association, and the further contribution of $m > 0$ wrappers could have contributed significantly to the enthalpy of binding.

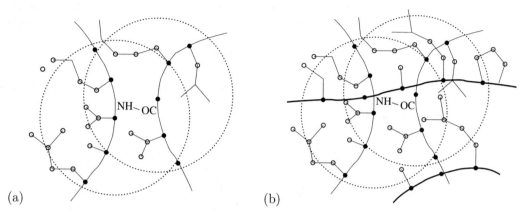

(a)   (b)

Fig. 11.6 Caricatures of an underwrapped mainchain hydrogen bond (a) that becomes further wrapped upon association with another protein (b). The backbone of the other protein is shown as a darker line for emphasis. The dark dots indicate the C-alpha carbons, and the open circles represent nonpolar carbonaceous groups C_n, $n = 0, 1, 2, 3$, on sidechains. The latter determine the wrapping number. The dashed-line circles represent the boundaries of the spherical components of the desolvation domain [174, 194].

There are two ways of addressing this with Wrappa. One way is to declare on the "upload" page that only chain A should be analyzed. This will then provide a list of hydrogen bonds whose intramolecular wrapping is deficient. The other way is to increase the wrapping threshold so that all hydrogen bonds are analyzed and reported. By examining the corresponding "wrappers" file, it is easy to find hydrogen bonds that are within chain A and have wrappers coming from chain B. A combination of these two approaches is often useful to make sure that all the appropriate candidate dehydrons have been identified. In this way, the data in Tables 11.2 and 11.3 were identified.

donor	acceptor	d_{NO}	d_{HO}	∠NHO	∠NOC	∠HOC	AW	BW	B chain residues
ASN 494	ALA 490	3.30	2.39	150.83	143.75	137.16	16	1	Leu 55
THR 497	THR 493	3.05	2.17	145.08	142.72	134.17	15	0	
LYS 501	THR 497	3.29	2.36	153.97	149.37	143.97	18	3	Val 48
ALA 505	LYS 501	2.88	1.91	162.85	160.34	157.13	17	11	Pro 44, Tyr 45, Val 48
GLN 508	GLN 504	3.13	2.25	146.21	152.29	144.22	17	2	Pro 44
CYS 512	GLN 508	3.03	2.09	155.11	149.99	142.71	17	2	Pro 44
GLU 672	ALA 668	3.00	2.05	156.70	154.41	147.93	16	4	Tyr 45
GLU 679	LYS 675	2.77	1.79	166.50	167.01	165.28	17	3	Leu 52

Table 11.2 Candidate dehydrons in the PDB file 2XS1. The first two columns give the donors and acceptors of the hydrogen bonds from the A chain. The distance between O and N in the mainchain is denoted by d_{NO}, and d_{HO} is the distance between O and H in the mainchain. The angles are indicated in Figure 6.7. The numbers listed under AW are the numbers of wrappers coming from the A chain, and the numbers listed under BW are the numbers of wrappers coming from the B chain (both are based on desolvation radii of 6.5 Å). The final column lists the residues providing wrapping coming from the B chain. The first group of dehydrons comes from the α-helix 18, which extends from PRO A 480 to SER A 515. The second group of dehydrons comes from the α-helix 25, which extends from ASN A 643 to ARG A 698.

Table 11.2 identifies seven candidate dehydrons in the A chain (corresponding to Alix) of PDB file 2XS1 that are wrapped by sidechains in chain B (a fragment of the Gag protein of SIVmac239), five in one helix in Alix and two in another. Such a large number and dense concentration of dehydrons could contribute significantly to the enthalpy of binding.

Similarly, Table 11.3 identifies analogous candidate dehydrons in the A chain of PDB file 2XS8 that are wrapped by sidechains in chain B (a fragment of the Gag protein of SIVagmTan-1). As is typical in homologous complexes, the wrapping patterns are different even though the hydrogen bond patterns are quite similar [176]. In addition, fine details of the hydrogen bonds differ between the two PDB files, with hydrogen bond distances and angles varying by more than 10% (compare Tables 11.2 and 11.3). Although data in PDB files have significant error bars due to the challenge of interpreting image data, the systematic differences of the data in Tables 11.2 and 11.3 are consistent with an interpretation that there are substantial environmental differences in the two Gag–Alix systems. Thus, the observed conservation of wrappers is quite significant.

donor	acceptor	d_{NO}	d_{HO}	$\angle$NHO	$\angle$NOC	$\angle$HOC	AW	BW
ASN A 494	ALA A 490	3.17	2.19	166.26	150.87	148.64	16	0
THR A 497	THR A 493	3.09	2.19	149.63	148.27	141.47	17	1
LYS A 501	THR A 497	3.36	2.47	148.43	138.21	133.18	17	2
ALA A 505	LYS A 501	3.05	2.20	142.00	155.03	145.26	15	8
GLN A 508	GLN A 504	3.36	2.53	140.25	142.11	132.97	17	0
CYS A 512	GLN A 508	3.12	2.20	152.24	148.66	141.21	18	0
GLU A 672	ALA A 668	2.94	1.95	174.08	153.64	152.12	16	4
GLU A 679	LYS A 675	2.93	1.93	170.33	159.76	158.22	18	0

Table 11.3 Candidate dehydrons in the PDB file 2XS8. See the caption for Table 11.2 for definitions of terms. The wrappers for these hydrogen bonds were ALA B 27, TYR B 28, ALA B 31, LEU B 34, LEU B 35.

11.2.3 Finding Imperfect Hydrogen Bonds

There is one significant difference between the hydrogen bonds in Tables 11.2 and 11.3. The hydrogen bond GLN A 508 — GLN A 504 in 2XS8 is less well formed than in 2XS1. In fact, the default settings of Wrappa require $d_{HO} \leq 2.5$, and this is violated for 2XS8, as shown in Table 11.3. Thus, an application of Wrappa to 2XS8 with the default settings will not identify this as a hydrogen bond. It is worth noting that there is no wrapping of this hydrogen bond by Gag residues in 2XS8, and this could potentially explain why this hydrogen bond is less well formed.

It is easy to extract information on a wide range of potential interactions using Wrappa. In the case of the hydrogen bond GLN A 508 — GLN A 504 in 2XS8, the needed information can be obtained by relaxing the requirements for the definition of a hydrogen bond. In this case, relaxing the H–O distance to $d_{HO} \leq 2.6$ is sufficient to get Wrappa to provide the requisite information. Of course, initially we would not know why this hydrogen bond did not appear in an assessment by Wrappa, so it is often necessary to relax all of the distance and angle constraints substantially in order to discover the potential interaction. The only consequence of doing so is larger output data files and slower execution time for Wrappa. Relaxing angles and distances in the Wrappa configuration page can be used generally to study the effects of mutations on conserved hydrogen bonds.

11.2.4 Reciprocal Wrapping

When two proteins form an association, each can wrap hydrogen bonds in the other. We see examples of this in the PDB files presented in [545]. The B chains in both 2XS1 and 2XS8 have significant structure as represented by intramolecular hydrogen bonds. Some of these are underwrapped based on the nonpolar groups found in the B chain alone, but they appear further wrapped from the A chain when viewing the association of the chains in the PDB file. These hydrogen bonds are detailed in Table 11.4.

PDB	donor	acceptor	d_{NO}	d_{HO}	∠NHO	∠NOC	∠HOC	BW	AW
2XS1	LYS B 46	LYS B 43	2.93	1.95	169.67	131.20	131.03	18	2
2XS1	ASN B 56	LEU B 52	2.82	1.83	170.48	166.38	166.67	15	9
2XS8	LYS B 33	ASP B 29	3.07	2.21	143.56	160.98	152.19	17	1

Table 11.4 Intramolecular hydrogen bonds in the B chains in the PDB files 2XS1 and 2XS8 that are wrapped by nonpolar groups in the A chain. See the caption for Table 11.2 for definitions of terms. The second dehydron in the first pair of dehydrons comes from the α-helix 26, which extends from TYR B 45 to ASN B 56 in the B chain in PDB file 2XS1. The remaining dehydron comes from the short α-helix 27 in the B chain in PDB file 2XS8, which extends from TYR B 28 to LYS B 33.

Note that the underwrapped hydrogen bonds in the B chains, which are depicted by the arrows in Figure 11.5, are not conserved. They connect distinctly different pairs of residues. Thus, it is only the structure of the wrappers in the B chains that appear to be conserved.

11.2.5 Caveats

It is not possible to infer binding events by looking at a single PDB file that shows the bound proteins. It is tempting to assume that the structure of the two ligands remains the same when they are separated, but structural changes upon binding often occur. Thus, our analysis based on looking only at the bound Gag–Alix systems can only suggest possible explanations consistent with possible binding scenarios. In particular, the regions of the Gag proteins depicted in Figure 11.5 could be unstructured in the unbound state, only becoming structured upon binding to Alix, and conversely. If this is the case, an even stronger enthalpic contribution may be associated with the water removal by the conserved hydrophobic residues in Figure 11.5. Thus, understanding the full impact suggested by the role of these residues as wrappers requires a more extensive analysis together with confirming experiments. We have emphasized this by referring to the hydrogen bonds identified by Wrappa as candidate dehydrons, since only experimental confirmation can assure that these hydrogen bonds actually benefit from dehydration upon protein–protein association.

11.2.6 Polar Interactions

It is not our intention to say that the candidate dehydrons identified here are the whole story regarding enthalpic interactions between the Gag and Alix proteins. As observed in [545], the conserved tyrosines in the Gag proteins make polar interactions with atom OD1 in ASP A 506 in the Alix chain. The oxygen–oxygen distances are 2.45 Å in 2XS1 and 2.55 Å in 2XS8.

Moreover, with hydrogens added by the MolProbity Web site, the corresponding hydrogen–oxygen distances are 1.68 Å in 2XS1 and 1.57 Å in 2XS8. Here, we mean the HH hydrogen attached to the OH oxygen in the tyrosine sidechains (TYR B 45 in 2XS1 and TYR B 28 in 2XS8). Thus, these tyrosines are both wrapping candidate dehydrons and making polar contacts, something not uncommon for amphiphilic residues.

There are few other polar interactions. In 2XS1, the closest heavy atoms that could form a salt bridge are at a distance 4.42 Å, beyond the distance 4 Å considered stabilizing for salt bridge [289, 290]. These atoms are NZ LYS A 501 and OE2 GLU B 47. No other complementary heavy polar atoms, coming from the A and B chains, are closer than 4.5 Å in 2XS1. On the other hand, in 2XS8, atom NH1 ARG B 32 is 2.61 Å from OE1 and 3.65 Å from OE2 in GLU A 679. The distances for atom NH2 ARG B 32 are 4.43 Å from OE1 and 4.44 Å from OE2 in GLU A 679. The distances for atom NE ARG B 32 are 4.68 Å from OE1 and 4.71 Å from OE2 in GLU A 679. Thus, the distances between ARG B 32 and GLU A 679 suggest a monodentate, backside interaction [122].

11.2.7 Conclusions about ALIX

There is a substantial number of mainchain–mainchain hydrogen bonds in the PDB files 2XS1 and 2XS8 that appear underwrapped based on a count of nonpolar groups within their respective chains. Many of these become further wrapped by nonpolar groups in the complementary chain upon association of the two chains. Thus, there is a potential enthalpic gain caused by the enhancement of these hydrogen bonds due to the removal of water by the nonpolar groups. This could explain the observed "hydrophobic contacts" described in [545].

The wrappers of the candidate dehydrons in the A chains in the PDB files 2XS1 and 2XS8 coming from the B chains include all of the conserved residues identified in [545] and shown in Figure 11.5. Note that the underwrapped hydrogen bonds in the B chain, indicated by the arrows in Figure 11.5, are not conserved. Thus, it appears that wrapping is the primary conservation variable.

The Wrappa code can be an effective tool to reveal subtle relationships between hydrogen bonds and hydrophobic residues in protein–ligand association. These observations can suggest further analysis and experiments.

11.3 Neurophysin/Vasopressin Binding

In [61], a polar/aromatic interaction is suggested for the residue B-Tyr99 of the hormone vasopressin, a ligand of the bovine protein neurophysin (BNP)-II, a transporter of hormones along axons. Aromatic rings can in principle interact with any polar moiety (cf. Section 3.1.4 and Section 12.4). We analyze the interaction between neurophysin and vasopressin represented in PDB file 2BN2 discussed in [61] and exhibit some additional candidate interactions to explain the particular details of the interaction zone.

Part of the reason for the suggestion [61] of a polar/aromatic interaction is the large number of disulfide bonds in neurophysin [437]. There are seven such bonds in a protein with only 79 residues; nearly a fifth of all of the residues are involved in one. Moreover, there is a disulfide bond in the ligand vasopressin. But potential polar interacting partners of B-Tyr99 are not made clear in [61]. For clarity, we designate the protein neurophysin as the A chain, and the ligand fragment of vasopressin the B chain, in keeping with the notation of the PDB file

2BN2. This fragment consists of only two residues, B-Phe98 and B-Tyr99. Thus, B-Tyr99 denotes the tyrosine residue of vasopressin that is found in the active site of neurophysin.

One corollary of the large number of disulfide bonds in neurophysin is the expectation of a large number of underwrapped hydrogen bonds [151, 163]. There are 19 intramolecular mainchain hydrogen bonds in neurophysin with less than 20 wrappers using a desolvation domain radius of 6.6 Å. We will see that there is correspondingly a large number of hydrogen bonds in neurophysin that are wrapped by the ligand vasopressin.

11.3.1 The Role of Tyrosine-99

Donors	Acceptors	AW	Phe	Tyr
A-Cys 21 NH	A-Leu 11 OC	24	0	3
A-Phe 22 NH	A-Ile 26 OC	25	0	4
A-Gly 23 NH	A-Ile 26 OC	20	0	6
A-Ile 26 NH	A-Gly 23 OC	20	0	6
A-Cys 44 NH	A-Ala 41 OC	24	0	2
A-Gln 45 NH	A-Leu 42 OC	19	0	2
A-Asn 48 NH	A-Gln 45 OC	16	3	3
A-Gln 55 NH	A-Arg 8 OC	12	0	1
A-Cys 54 NH	B-Phe 98 OC	17	7	6

Table 11.5 Mainchain hydrogen bonds wrapped by B-Phe 98 and B-Tyr 99 in PDB file 2BN2. The column "AW" lists the number of wrappers coming from the A chain. The columns "Phe" and "Tyr" indicate the number of wrappers contributed by B-Phe 98 and B-Tyr 99, respectively, to the desolvation domain for the hydrogen bond. The desolvation domain radius chosen was 6.6 Å.

Donors	Acceptors	AW	Phe	Tyr	OBW
A-Cys 21 NH	A-Leu 11 OC	25	0	3	0
A-Phe 22 NH	A-Ile 26 OC	22	0	5	0
A-Ile 26 NH	A-Gly 23 OC	18	0	6	0
A-Cys 44 NH	A-Ala 41 OC	23	0	2	0
A-Gln 45 NH	A-Leu 42 OC	20	0	1	0
A-Glu 47 NH	A-Cys 44 OC	26	0	5	1
A-Asn 48 NH	A-Gln 45 OC	19	0	3	2
A-Leu 50 NH	A-Glu 47 OC	19	0	5	1
A-Gln 55 NH	A-Arg 8 OC	12	3	1	0
A-Cys 54 NH	B-Cys 1 OC	14	3	6	3
B-Phe 3 NH	A-Cys 54 OC	8	7	3	5
B-Asn 5 NH	B-Tyr 2 OC	6	2	6	5
B-Cys 6 NH	B-Tyr 2 OC	6	2	6	4

Table 11.6 Mainchain hydrogen bonds wrapped by B-Phe 3 and B-Tyr 2 in PDB file 1KJ4. The column "AW" lists the number of wrappers coming from the A chain. The columns "Phe" and "Tyr" indicate the number of wrappers contributed by B-Phe 3 and B-Tyr 2, respectively, to the desolvation domain for the hydrogen bond. The column "OBW" lists the number of other wrappers coming from the B chain. The desolvation domain radius chosen was 6.6 Å.

Examination of the PDB file 2BN2 shows that B-Phe98 and B-Tyr99 wrap several under-wrapped mainchain–mainchain hydrogen bonds listed in Table 11.5. The active site, into which B-Tyr99 inserts, has a large water-exposed area without the ligand. They provide wrapping in a sector that is otherwise exposed to water attack.

The paper [61] reports binding constants for various mutations of the native B-Phe98, B-Tyr99 pair. A comparable binding constant, only 15% smaller, is obtained for B-Phe98, B-Phe99. The terminal O-H group in B-Tyr99 also forms a sidechain–mainchain hydrogen bond with A-Cys44(O). This presumably accounts for the additional binding strength that this ligand has compared to the B-Phe98, B-Phe99 variant [61]. B-Phe98, B-Phe99 is the corresponding residue pair at the binding site for the ligand phenypressin, a hormone in Australian macropods.

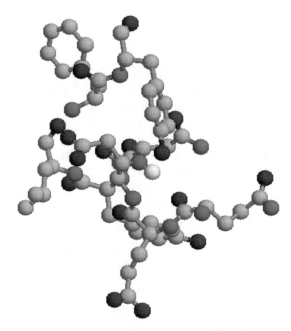

Fig. 11.7 Wrapping of hydrogen bond Asn 48 N—Gln 45 O in PDB file 2BN2. The hydrogen (in white) attached to Asn48N has been added to indicate the bond, which is in the middle of the figure. The ligand pair B-Tyr99 and B-Phe98 is in the upper left of the figure. Residues 45 to 50 in the A chain are shown to indicate the bulk of the wrapping.

It is also interesting to note that the mutation B-Tyr→B-Leu would both modify wrapping patterns and eliminate the sidechain–mainchain hydrogen bond. This is reported [61] to have a significantly lower binding constant.

The CD1 and CE1 hydrophobic groups of B-Tyr99 are also in close proximity to the sidechain–mainchain bond A-Ser56(OG)—A-Cys21, which otherwise has a small number of nearby hydrophobic groups.

In summary, B-Tyr99 is involved in an intermolecular hydrogen bond and in wrapping several dehydrons. The mutation B-Tyr99→B-Leu99 would both eliminate the intermolecular hydrogen bond and reduce significantly the wrapping of some of these bonds.

11.3.2 The Role of Phenylalanine-98

The role of B-Phe98 is less critical. It makes the mainchain–mainchain bond A-Cys54—B-Phe98 and wraps only the mainchain–mainchain bond A Asn 48–A Gln 45, shown in

Figure 11.7. The mutation Leu98 binds with about half the affinity of Phe98 [61]. Presumably, the mainchain bond to A-Cys54 would be preserved, and the amount of wrapping only slightly decreased.

On the other hand, the combination B-Phe98, B-Leu99 has significantly reduced affinity compared with the native B-Phe98, B-Tyr99. This could be due to two different reasons. One could be a polar–aromatic interaction of some sort [308], but it could also be due to the factors discussed here. Compared with Tyr, Leu lacks the sidechain–mainchain bond with Cys44. Moreover, it is positions CD and CE on residue 99 that provide the wrapping. The CE positions are absent on Leu, and the CD positions are the end of the Leu sidechain, so its ability to wrap, compared with Phe, would be reduced. Similarly, two other second-position sidechains (Met and His) show minimal affinity, consistent with their lack of wrappers at the end of the sidechain and lack of ability to form a hydrogen bond at the end with Cys44.

The role of residue 48 in chain A was explored in [61] by a natural "mutation" arising due to the fact that bovine (Asn 48) and ovine (Ile 48) differ at this location. The ovine affinity is slightly higher, but of the same order. The "mutation" Asn 48→Ile 48 is isosteric, so it is plausible that the mainchain–mainchain hydrogen bond to Gln 45 O is maintained.

11.3.3 2BN2 Versus 1JK4

The PDB structure 1JK4 (Figure 11.8) also represents neurophysin bound to a larger fragment of vasopressin [528], but with the orientation of the fragment is reversed, that is, the sequence of the vasopressin fragment is B-Cys1, B-Tyr2, B-Phe3, B-Gln4, and so forth. The residues B-Tyr2, B-Phe3 in 1JK4 correspond to B-Phe98, B-Tyr99 in 2BN2. The individual bonds, listed in Table 11.6, are quite similar, and the general picture is the same. One new ingredient is the fact that dehydrons in the vasopressin fragment can now be seen that are wrapped in part by residues from neurophysin. Both B-Asn 5 — B-Tyr 2 and B-Cys 6 — B-Tyr 2 have these complex-wrapping patterns.

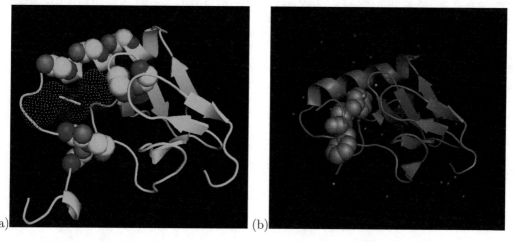

(a)   (b)

Fig. 11.8 (a) The active site of the bovine protein neurophysin bound to a fragment of the hormone vasopressin, in the PDB file 2BN2. The fragment B-Phe 98, B-Tyr 99 is shown as a cloud, and several of the hydrogen bonds in Table 11.5 are indicated by showing their backbone atoms as spheres. (b) The active site of the bovine protein neurophysin bound to a fragment of the hormone vasopressin, in the PDB file 1JK4. The two residues B-Tyr 2, B-Phe 3 of the vasopressin fragment are shown as spheres.

The main difference that becomes apparent in 1JK4 is the decreased role of B-Phe3. Indeed, upon examination, it is no longer in the active site but rather sticking out of it. B-Tyr 2 is found in 1JK4 in the same place as B-Tyr 99 in 2NB2, and in 1JK4 its role as a wrapper is more pronounced, as indicated in Table 11.6.

11.4 Sheets of Dehydrons

It is possible for protein systems to form using only indirect dehydration forces. One example is given by associations of β-sheets [149, 360].

11.5 Exercises

Exercise 11.1. Examine the PDB file 1AFV in a protein viewer. Identify Tyr 42 in the L chain and the nearby hydrogen bond in the A chain. In addition to wrapping the hydrogen bond, what other interactions could Tyr 42 be having with the A chain? Which way do you think the hydrogen attached to the terminal oxygen in Tyr 42 would be pointing?

Exercise 11.2. Scan the PDB for instances of crystal contacts. Plot a distribution of wrapping for hydrogen bonds that have atoms in their desolvation domains from artificial contacts in other chains.

Exercise 11.3. Determine the residues most likely to be involved in catalytic activity in the active site of an enzyme.

Exercise 11.4. The residue Asp is often involved in catalytic activity in the active site of an enzyme. It is often found to make local sidechain–mainchain bonds [446] in an underwrapped environment. Explore the possible correlation of these two observations.

Chapter 12
Aromatic Interactions

Partha Niyogi (1967–2010) was the Louis Block Professor in Computer
Science and Statistics at the University of Chicago. He is known for his
work in artificial intelligence, especially in the field of manifold learning
and evolutionary linguistics.

Fig. 12.1 Side view of an aromatic ring showing three slabs containing alternating negative and net positive
charges, providing a directional quadrupole perpendicular to the face of the ring. The + charges correspond
to eight times as much charge as each − charge.

Aromatic sidechains are special for many reasons. First of all, they are just big relative
to other sidechains. But they also have a subtle dual role as hydrophobic groups and as
acceptors in polar interactions (e.g., as hydrogen bond acceptors [2]). Their delocalized π-
electronic cloud is highly polarizable and enables interactions with cations, so-called cation-π
interactions. On the one hand, the benzene-like rings in them are largely hydrophobic, and
they have very small dipole moments. However, the charge distribution of these rings creates
a significant quadrupole moment (Section 10.5), as we will explain.

12.1 Classical Representation of Electron Distribution

The center of a surface above each side of the face of the aromatics is negatively charged, and
the C–H groups in the ring form positive charge centers [123, 327]. This can be easily visualized
since all of the protons lie in the plane of the ring, whereas the electrons are distributed all
around them. The fact that the positive charges are in the plane means that there is a net
positive charge at points in this plane, with a net negative charge in planes above and below
this. More precisely, we should define three thin slabs, as shown in Figure 12.1, one around
the plane of the aromatic ring, and two above and below this slab. The upper and lower
slabs contain electrons but no protons, thus have a net negative charge. Since the sum of the

© Springer International Publishing AG 2017
L.R. Scott and A. Fernández, *A Mathematical Approach to Protein Biophysics*,
Biological and Medical Physics, Biomedical Engineering,
DOI 10.1007/978-3-319-66032-5_12

charges in the three slabs is zero, the middle slab is positively charged. Moreover, the positive charge is twice the charge of each of the other two slabs, so we have a typical arrangement of a quadrapole.

12.2 Partial Charge Model

Partial charges (Figure 12.2) are frequently used to model aromatics [75, 308] as shown in Table 12.1. However, this model is planar and thus does not directly represent the three-dimensional charge distribution depicted in Figure 12.1. In fact, it replaces this polarity that is orthogonal to the face with one that is within the plane of the aromatic ring.

In Figure 12.5(a), we plot the electrostatic potential corresponding to the partial charges in Table 12.1 in a plane parallel to the plane of the ring at a distance of 1.0 Å from the plane of the ring. We do see that the face of the ring has a negative charge, as required.

However, the planar partial charges cause a large polar behavior in the plane of the ring near the locations of the hydrogens. In Figure 12.5(b), we plot the electrostatic potential corresponding to the partial charges in Table 12.1 in a plane near the plane of the ring (at a distance from the plane of the ring of only 0.1 Å). In this plane, there is a strong polarity, one that might lead to hydrophilic behavior.

A more sophisticated approximation of the aromatic ring might be to put negative charges at positions near the carbons but in the direction normal to the ring. This improvement would be similar to current models of water, such as Tip5P [330].

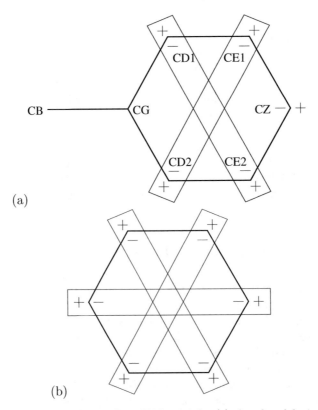

(a)

(b)

Fig. 12.2 Arrangement of partial charges from Table 12.1 for (a) the phenylalanine sidechain (only carbons are shown). Corresponding partial charges are depicted for (b) a benzene ring. The boxes indicate the counterbalancing dipoles which combine to form quadrupoles.

12.3 Cation-π Interactions

Residue	atom type	PDB code	charge
PHE	C	CDi, CEi, $i = 1, 2$, CZ	−0.1
	HC	HDi, HEi, $i = 1, 2$, HZ	0.1
TYR	C	CDi, CEi, $i = 1, 2$	−0.1
	HC	HDi, HEi, $i = 1, 2$	0.1
	C	CZ	0.15
	OA	OH	−0.548
	H	HH	0.398
TRP	C	CG	−0.14
	C	CD1, CE3, CZi, $i = 2, 3$, CH2	−0.1
	HC	HD1, HE3, HZi $i = 2, 3$, HH2	0.1
	NR	NE1	−0.05
	H	HE1	0.19

Table 12.1 Partial charges from the Gromos force field for aromatic amino acids; cf. also [308].

Among pair interactions at interfaces (Section 7.5), the Arg–Trp interaction has the fourth highest log-odds ratio. This pair is an example of what is known as a cation-π pair [53, 103, 200, 343, 390, 400, 423, 519, 538, 547]. It has a strength comparable to that of a hydrogen bond. It is based on an interaction between the negative charge on the face of aromatic residues and positively charged (cation) residues (Lys, Arg, His). The cation-π motifs play a special role in protein interfaces [103, 538]. The cation-π interaction also has a significant role in α-helix stabilization [454].

Cation-π interactions can take place with other residues, such as a phosphorylated tyrosine, cf. Figure 12.3. The aromatic ring still provides a distributed negative charge, unaffected by the addition of the phosphate group. And the two-sided nature of the polarity of an aromatic ring means that it can interact with two cations at one time, one on each side. This is depicted in Figure 12.3(a) which shows a phosphorylated tyrosine flanked by an arginine and a lysine in the SH2 domain in the PDB file 1JYR. SH2 domains [316] specifically recognize phosphorylated tyrosines and bind proteins containing them, and the cation-π interaction is presumably important in the binding process.

Correspondingly, a cation could be sandwiched between two aromatics, as shown in Figure 12.3(b) which depicts part of the PDB file 1BMB. The arginine (A67) interacts with both a phosphorylated tyrosine (PTR-I3) and a phenylalanine (Phe-I4). In addition, Arg-A67 is hydrogen bonded with the backbone oxygen on Phe-I4 and one of the terminal oxygens on PTR-I3.

Finally, we show a cation-π grouping involving a complex of two cations and two aromatics, as shown in Figure 12.4. The arginine is interacting with both of the aromatics, whereas the lysine is interacting only with the phosphorylated tyrosine.

12.4 Aromatic-Polar Interactions

Aromatic rings can in principle interact with any polar moiety [69, 70, 308]. We will consider different classes of polar interactions. The most familiar will be a type of hydrogen bond, in which the face of the aromatic ring forms the acceptor of a hydrogen bond. Cation-π

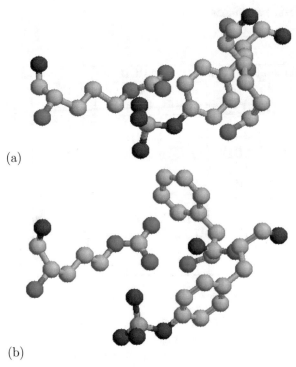

(a)

(b)

Fig. 12.3 Cation-π groups found in the SH2 domains in the PDB files (a) 1JYR and (b) 1BMB. (a) Shown are a phosphorylated tyrosine (PTR1003) flanked by Arg67 (upper left) and Lys109 (lower right). (b) Shown are a phosphorylated tyrosine (PTR-I3, lower right) and a phenylalanine (Phe-I4, upper right) flanking an Arg-A67 (middle left).

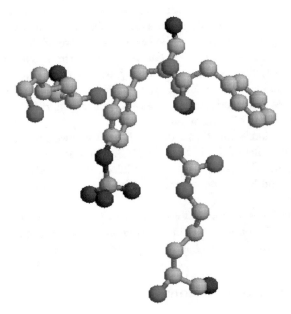

Fig. 12.4 Cation-π group found in the SH2 domain in the PDB file 1TZE. Shown are a phosphorylated tyrosine (PTR-I4, middle left) and a phenylalanine (Phe-I3, upper right) together with Arg-E67 (lower right) and Lys-E109 (upper left).

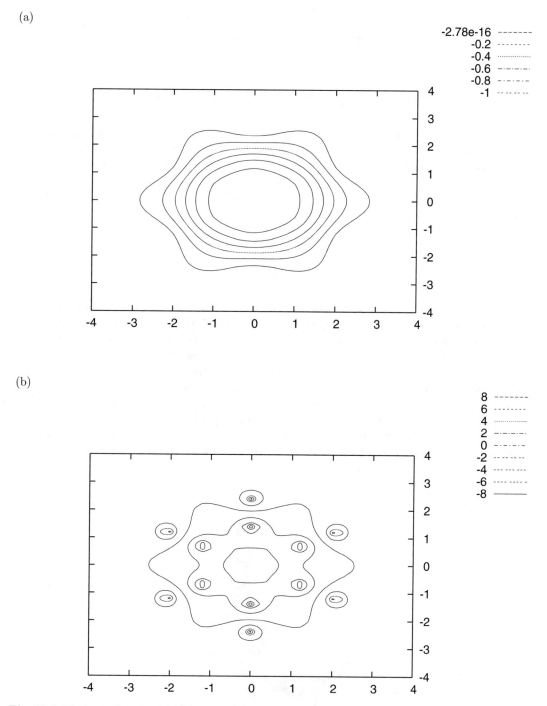

Fig. 12.5 Electrostatic potential of the partial charge model of a benzene ring [308] in parallel planes above the ring, using the partial charges in Table 12.1: (a) 1.0 Ångstrom above and (b) 0.1 Ångstrom above. The dimensions of the axis scale are Ångstroms.

interactions can take place with the NH groups on Asn and Gln [53]. However, there are also other types of interactions that appear to be possible.

12.4.1 Aromatics as Acceptors of Hydrogen Bonds

In [250, Figure 6], a hydrogen bond with the polar face of an aromatic ring is described. The aromatic face can interact with typical hydrogen bond donors [73, 308, 391, 468, 469, 475]. One criterion for such a hydrogen bond involves the distance from the center of the aromatic face, with the angle formed between the donor and the aromatic face also taken into account (cf. Figure 1 in [308]). See Exercises 12.3 and 12.4 for a further exploration of interactions of this type.

12.5 Aromatic–Aromatic Interactions

It is possible for two aromatics to interact via more complex interactions, including edge-to-face interactions [78, 303, 540]. In Figure 12.6, the relative orientations of two phenylalanines are shown [7] as found in the PDB file 1TLA. A vector along the line formed by the C_γ—C_ζ carbons in Phe117 (upper) is pointing toward the face of Phe153.

The interaction depicted in Figure 12.6 is consistent with a model of Phe in which there is a small positive partial charge, or positive polarity, at the end of the sidechain, that is near the C_ζ carbon. The partial charge model in Table 12.1 is precisely of that type. There is a dipole formed between C_ζ and H_ζ, with no opposing dipole related to C_γ. The other dipoles in the ring are always counterbalanced, forming quadrapoles. Thus, there is a net dipolar behavior to the carbonaceous ring of Phe in the partial charge model in Table 12.1, consistent with the Phe–Phe interaction in Figure 12.6.

See Exercises 12.5 and 12.6 for a further exploration of interactions of this type.

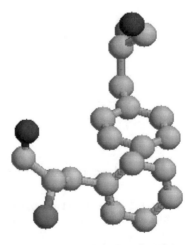

Fig. 12.6 Pairwise interactions of two phenylalanines (117 and 153) in a mutant S117F (Ser 117 → Phe) of T4 lysozyme detailed in the PDB file 1TLA [7].

12.6 Exercises

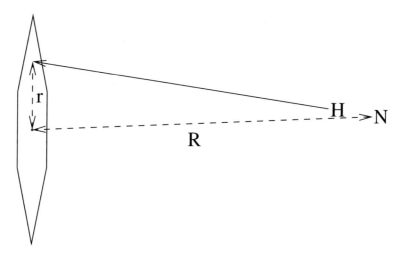

Fig. 12.7 Coordinate definitions for Exercise 12.3.

Exercise 12.1. Are the aromatic rings planar? Three points determine a plane, but there are six points in an aromatic ring. Determine the plane that best fits six points in three dimensions by least squares. Test this on Phe, Tyr, and Trp in several PDB files. How big is the deviation of the six points from the plane (compute the squared deviation)?

Exercise 12.2. Where is the center of an aromatic ring? The three lines connecting opposite carbons shown in Figure 12.2 should intersect at the center of the plane formed by the aromatic ring. The midpoint of each line segment should be at the center of the ring. But if the carbon positions do not lie in a plane, or if the hexagon is not regular, these lines may not intersect. Determine an algorithm for finding a point that best approximates the near intersection of these three lines. For example, you could just average the three midpoints of each of the above line segments. Does this point lie on the plane determined in Exercise 12.1? Test this on Phe, Tyr, and Trp in several PDB files. For each aromatic ring, give the sum of squares of distances from each line-segment midpoint to the center point you determine.

Exercise 12.3. Look for instances separately for Phe, Tyr and Trp forming acceptors for hydrogen bonds where the hydrogen bond donors are either mainchain (NH) donors or NH groups on Asn and Gln. For each aromatic, look for the closest candidate hydrogen bond donor. Plot the distance from the nitrogen donor to the center of the aromatic face (Exercise 12.2) and the distance from this center point to the point of intersection of the vector generated by NH to the plane (Exercise 12.1) of the aromatic. See Figure 12.7 for a depiction of these coordinates. Use (6.1) to determine hydrogen placement for the mainchain donors, and use (6.2) and (6.3) for the sidechains. Use a large subset of the PDB, such as the nonredundant PDB, for your experiments. Give the distributions for the mainchain and sidechain donors separately, as well as for each aromatic sidechain. (This means six different distributions.)

Exercise 12.4. Catalog all hydrogen bonds having aromatics as an acceptor in the PDB. Determine the distribution of distances from the donor to the face of the other aromatic, as well as the angles formed between the donor, hydrogen, and the plane of the aromatic. Compare this with the two distances used in Exercise 12.3.

Exercise 12.5. Investigate the interactions of two phenylalanine residues as depicted in Figure 12.6. Determine the distribution of distances from the terminal carbon of one of the residues to the face of the other. Also record the angle between the two faces.

Exercise 12.6. Investigate interactions of a phenylalanine residue and another aromatic (Tyr or Trp) similar to the Phe–Phe interaction depicted in Figure 12.6. Determine the distribution of distances from the terminal carbon of one the Phe residues to the face of the other aromatic. Also record the angle between the two faces.

Exercise 12.7. Investigate the Trp-Pro 'sandwich' [408] by examining the geometry of Trp–Pro pairs that appear in PDB files in close proximity.

Exercise 12.8. Investigate the interaction geometry for all planar groups in sidechains [80].

Chapter 13
Antibody–Antigen Complexes

Janet Davison Rowley (1925–2013), a professor at the University of Chicago, was the first scientist to identify a chromosomal translocation as the cause of leukemia and other cancers. In 1998, she received the National Medal of Science and was one of three scientists awarded the Lasker Award. In 2009, Dr. Rowley was awarded the Presidential Medal of Freedom, and in 2012, she won the Japan Prize with two other scientists for her role in the creation of the cancer drug Imatinib, whose commercial name in the U. S. is Gleevec.

Antibody–antigen complexes have a natural protein-protein interface whose study illuminates how one molecule can be designed to bind to another. Moreover, antibodies evolve naturally to improve specificity and affinity, so we can observe the effect of different choices in protein sequence. We will see that we must use all of the ideas developed so far to assess such interfaces completely.

In order to make comparisons between related protein interactions, we need some quantitative measures. For hydrogen bonds, we use the metrics introduced in Section 6.4 based on donor-acceptor distances and angles. Although these do not give a single "quality" value, they are useful to visualize differences between good and bad hydrogen bonds. For other interactions, we introduce a simple quality measure based on the electrostatic potential given by the partial charges of the interactants:

$$-\sum_{i \neq j} \frac{q_i q_j}{|\mathbf{r}_i - \mathbf{r}_j|},$$

(13.1)

where the partial charges q_i are given, e.g., in Tables 8.4 and 8.5, and x_i denotes the position of this partial charge. We make no claim for the efficacy of this model of multipolar interaction, but it may give some way of comparing like interactions in different systems. The negative sign in (13.1) allows us to interpret bigger positive numbers as better (more favorable energetically). A negative quality estimate would correspond to a repulsive force.

13.1 Comparing F10.6.6 and D44.1 Antibodies

In the paper [79], the structures of an antibody in two different maturation states were compared. Here we review their analysis. We find some differences in interpretation of the

© Springer International Publishing AG 2017
L.R. Scott and A. Fernández, *A Mathematical Approach to Protein Biophysics*,
Biological and Medical Physics, Biomedical Engineering,
DOI 10.1007/978-3-319-66032-5_13

changes in bonding structure. Our main conclusion that differs substantially from [79] is that the change in strength of the individual bonds can account for the difference in binding affinity of the two complexes.

The structural comparisons are based on the PDB structures 1P2C and 1MLC. They pertain to antibodies denoted by F10.6.6 and D44.1, respectively. Each PDB file contains two, asymmetric copies of the antigen–antibody complexes. These provide some redundancy in the analysis, but as we will see they tend to be somewhat contradictory in certain details. Thus one is not sure which one of the asymmetric copies is native, and perhaps neither represents the state achieved in a context different from the process of crystallization, cf. Section 11.1.3.

The chains in these files are denoted by different descriptors. The light chains in D44.1 are denoted by A,C in 1MLC, whereas in F10.6.6 they are denoted by A,D in 1P2C. The heavy chains in D44.1 are B,D and in F10.6.6 are B,E. The HEL chains (C,F) are numbered 6xy in the 1P2C PDB file, and this corresponds to xy in the HEL chain (E,F) in the 1MLC PDB file. This is summarized in Table 13.1 for reference.

PDB file/antibody	Light	Heavy	HEL
1P2C/F10.6.6	A,D	B,E	C,F
1MLC/D44.1	A,C	B,D	E,F

Table 13.1 Naming conventions for different PDB files for closely related antigen–antibody structures. The numbering of residues within each chain is also different. In 1MLC, the residues are numbered starting with 1 in each case. But in 1P2C, the j-th chain begins with $1 + 300(j - 1)$. Thus Leu C 684 corresponds to Leu F 1584. In 1MLC, this residue is Leu E 84.

PDB	Type	Donor	Acceptor	Distance
1P2C	M-M	F-Arg 1521 N	A-Asp 151 O	2.85
1P2C	S-M	A-Ser 153 OG	F-Asn 1519 O	2.55
1P2C	M-S	F-GLY1522 N	A-Ser 153 OG	2.60
1P2C	S-S	A-Arg 188 NH2	F-Asp 1518 OD1	2.78
1MLC	S-S	F-Arg 128 NE	A-Asn 76 OD1	2.66
1MLC	S-S	F-Arg 128 NH2	A-Asn 76 OD1	2.74

Table 13.2 Nonphysical crystal artifacts in PDB files 1P2C and 1MLC. Hydrogen bonds between the A and F chains, which are in different copies of the antibody–antigen systems.

Thus the interfaces of interest in 1MLC are AE, BE, CF, DF; the interfaces of interest in 1P2C are AC, BC, DF, EF. However, one also finds significant other interactions that cross between the two copies of the antigen–antibody structures (Table 13.2). However, these are likely artifacts of the crystallization process, cf. Section 11.1.1.

The definition of hydrogen bonds is subjective to a certain extent. Thus it is natural that one definition would provide a different set from another. For example, in [79], the mainchain-sidechain hydrogen bond in 1MLC, E-Thr 47-O — A-Asn 92-ND2, is claimed between HEL and the light chain. However, the angle θ between the N–H vector and the C–O vector (see 6.4) is fairly small, about 116 degrees. On the other hand, there is a water (HOH1601) in 1MLC which provides a plausible link between E-Thr 47 and A-Asn 92, although more likely to A-Asn 92-OD1, as shown in Figure 13.1(a).

The number of hydrogen bonds that pass a rigorous test is actually quite small. These are enumerated in Table 13.3. For the most part, the number of intermolecular hydrogen bonds for F10.6.6 and D44.1 are the same, except that some hydrogen bonds in one structure appear as water-mediated bonds in the other. One can argue that F10.6.6 has more, or better, hydrogen bonds to HEL that D44.1 does, although the advantage is not overwhelming. But the two S-M bonds in F10.6.6 do not have comparable analogues in D44.1, except for one case in the second chain.

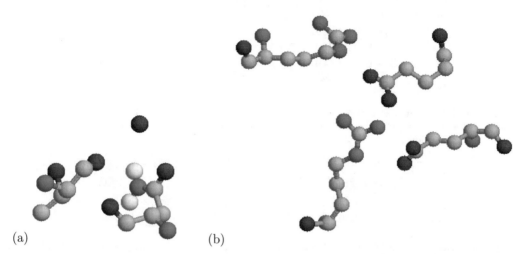

(a)   (b)

Fig. 13.1 (a) Configuration of E-Thr 47 (left), A-Asn 92 (right), and HOH1601 (single red dot at top) in 1MLC. The hydrogen positions on the terminal N in Asn have been determined synthetically. (b) Configuration of the double salt bridge in the PDB file 1P2C. Note that the positive and negative charges are nearly coplanar, and closely approximate opposing corners of a rhombus, a configuration consistent with a quadrapole.

type	donor : acceptor	F10.6.6(1)	F10.6.6(2)	D44.1(1)	D44.1(2)
M-M	hel-Gly649-N : L-Gly/Asn92-O	2.88	2.90	2.83	not found
S-M	hel-Ser681-OG : H-Thr330-O	3.02	2.77	not found	2.97
S-S	H-Trp333-NE1 : hel-Tyr653-OH	2.84	2.77	2.87	2.91
S-S	hel-Thr643-OG1 : H-Ser357-OG	2 WB	2 Ws	3.48	not found
S-M	H-Ser357-OG : hel-Gln641-O	1 WB	3.31	not found	not found

Table 13.3 Intermolecular hydrogen bonds between antibody and antigen. S = sidechain, M = mainchain. W indicates a water molecule hydrogen bonded instead of the S or M hydrogen bond; B indicates that an apparent water bridge exists. The numbers given are the distance in Ångstroms between the donor and acceptor (heavy) atoms in the hydrogen bond. The notation L-Gly/Asn92 in the mainchain–mainchain bond with hel-Gly649 refers to the alignment D44.1/Asn92 → F10.6.6/Gly92.

The mutation of D44.1/Asn92 to F10.6.6/Gly92 has only a minor effect, since the mainchain–mainchain bond hel:Gly49-light:Asn92 in D44 becomes the mainchain–mainchain bond hel:Gly49-light:Gly92 in F10. The angular arrangement in F10.6.6 is slightly better than D44.1, whereas the bond distance is comparable.

In the heavy chains, there is a mutation of D44.1/Ser330 to F10.6.6/Thr330. Correspondingly, the strong sidechain-mainchain bond between hel-Ser81 and H-Thr330 in the F10.6.6 complex replaces a rather weak bond in the D44.1 complex involving H-Ser330, and it appears in only one copy. Surprisingly, B-Ser30 (in the first copy of the complex in 1MLC) does not form any sort of bond with other atoms.

There is a sidechain-sidechain hydrogen bond that appears in the D44.1 complex that does not have a direct analogue in the F10.6.6 complex. It is between hel-Thr643 and H-Ser357, a residue that is preserved in both complexes. However, there is a neighboring mutation D44.1/H-Gly356 to F10.6.6/H-Asp356 which presumably affects the local environment. The (hel-Thr643, H-Ser357) bond in the D44.1 complex, which is not noted in [79], is replaced by a water-mediated hydrogen bond (hel-Thr643, HOH2545, H-Ser357).

cation	aromatic	F10.6.6(1)	F10.6.6(2)	D44.1(1)	D44.1(2)
hel-Arg 668	L-Trp 94	12.3	11.9	12.2	9.93
hel-Arg 668	H-Trp 333	18.7	18.4	17.2	19.0
hel-Arg 645	H-Trp 333	8.84	8.66	8.41	8.92
hel-Arg 645	H-Tyr 359	20.1	18.7	18.8	18.0
L-Arg 96	L-Trp 94	none	15.3	16.6	16.7

Table 13.4 Cation-π interactions in D44.1 and F10.6.6. The letter 'H' refers to the 'heavy' chain, and the letter 'L' refers to the 'light' chain. The numbers given are the quality estimate (13.1) of the bonds modeled as multipole interactions.

cation	anion	F10.6.6(1)	F10.6.6(2)	D44.1(1)	D44.1(2)
hel-Arg 668	H-Glu 335	71.5	75.5	73.4	70.0
hel-Arg 645	H-Glu 350	69.6	71.5	68.7	80.1
hel-Arg 668	H-Glu 350	113.	110.	102.	94.2

Table 13.5 Salt bridges in D44.1 and F10.6.6. The letter 'H' refers to the 'heavy' chain, and the letter 'L' refers to the 'light' chain. The numbers given are the quality estimate (13.1) of the bonds modeled as multipole interactions.

In the analysis in [79], cation-π interactions were omitted. However, there are significant cation-π interactions as detailed in Table 13.4 between the light and heavy chains and HEL. Intermolecular salt bridges are also detailed in Table 13.5. Like the cation-π interactions, the salt bridges are similar for both F10.6.6 and D44.1. It is significant that the salt bridges form a complex, and the positive and negative charges lie very nearly in a plane (in fact, they are almost co-linear). Moreover, their arrangement, at the vertices of a rhombus, is consistent with a quadrapole structure.

What is even more striking is the fact that the cations are involved in both salt bridges and interactions with aromatics. The aromatics form a cage around the salt bridges, but it is not clear that there are strong cation-pi interactions in all cases. However, the unique combination of residues in both F10.6.6 and D44.1 that form bonds with the two arginines in HEL is presumably a critical part of the recognition and binding strategy for this antigen–antibody pair.

	bond details	PDB/antibody	d_{NO}	d_{HO}	DHA	DAB	HAB	raW	erW
donor	hel-Leu (6)84 N	1P2C/F10(1)	2.96	2.12	166	119	118	16	4
acceptor	hel-Ser (6)81 O	1P2C/F10(2)	2.95	2.13	158	118	115	16	4
interface	H/hel	1MLC/D44(1)	3.02	2.08	156	103	98	16	3
SS type	helix	1MLC/D44(2)	3.57	2.68	148	101	104	17	4
donor	H-Thr (3)31 N	1P2C/F10(1)	3.13	2.27	171	128	126	19	2
acceptor	H-Thr (3)28 O	1P2C/F10(2)	2.99	2.14	172	129	127	19	2
interface	H/hel	1MLC/D44(1)	3.05	2.09	160	126	121	19	2
SS type	helix	1MLC/D44(2)	3.04	2.04	175	140	138	18	2

Table 13.6 Underwrapped mainchain (M-M) intramolecular hydrogen bonds in two pairs of antibody–antigen complexes. d_{NO} is the distance is between the donor N and acceptor O (heavy) atoms in the hydrogen bond, in Ångstroms, and d_{HO} is the hydrogen-oxygen distance. Angles DHA, DAB, and HAB are in degrees as described in Section 6.4. Numbers of nonpolar carbons in a 6.5Å desolvation shell: erW, for intermolecular wrappers, and raW, for intramolecular wrappers. In 1MLC/D44(2), the pair Leu F 84 N–Ser F 81 O may not qualify as a hydrogen bond.

Further differentiation between 1P2C/F10.6.6 and 1MLC/D44.1 emerges when the under-wrapped intramolecular hydrogen bonds are considered. The dehydrons are listed in Table 13.6.

13.2 Four Antibody Complexes

cation	aromatic	1NDM	1DQJ	1C08	1NDG
hel-ARG 621	L-TRP 94	11.4		11.4	11.2
hel-ARG 621	L-TYR 96	38.4	39.5	41.5	39.4
hel-ARG 621	H-TYR 358	27.3	27.2	27.2	1.82*
hel-ARG 621	H-TYR 350	55.9	52.1	56.1	51.9
hel-LYS 696	L-TYR 50	25.3	24.6	26.0	25.1
hel-LYS 697	H-TRP 398		9.50	9.27	9.71
hel-LYS 697	H-TYR 333	21.7	19.7	20.4	20.5
cation	anion	1NDM	1DQJ	1C08	1NDG
hel-LYS 697	H-Asp 332	118	109	112	110
hel-LYS 697	H-GLU/Asp 399	97.4		122	108

Table 13.7 Quality estimates (13.1) for cation-π interactions and salt bridges in four antibody-lysozyme complexes. In the cation-π interactions, note the change in position 358 in 1NDG from Tyr to Phe and the corresponding change in quality estimate (13.1). In the salt bridges, note that H-GLU 399 in 1NDM aligns with H-Asp 399 in the other three structures. Note that position 399 in 1DQJ is a glycine which explains the absence of this salt bridge in 1DQJ. In all of these, the light chain is chain A, the heavy chain is chain B, and the antigen (HEL) is chain C.

In another paper [312], the structures of four HEL antibodies were studied. These antibodies are very close in affinity, but they can be ranked in order of affinity, and we have followed this ranking in our presentation of the data.

The antibody complexes are found in four PDB files: 1NDM, 1DQJ, 1C08, and 1NDG. In all of these, the light chain is chain A, the heavy chain is chain B, and the antigen (HEL) is chain C. Unlike the antibody complexes in the previous section, these appear as single complexes in each PDB file, with no crystal doubles. If there is any artifact due to crystallization in these complexes, we do not have the same direct evidence as in the previous section.

Again we see a wide range of interactions represented between the antigen and the antibody light and heavy chains. Both cation-π interactions and salt bridges appear, as indicated in Table 13.7. One of the cations (Lys 697) in the antigen interacts with two aromatics and two negatively charged residues in the antibody. This suggests a complex recognition pattern for this residue. Another cation (Arg 621) interacts with four aromatics in the antibody, again suggesting a complex signal.

Intermolecular hydrogen bonds are listed in Table 13.8. We see a diverse set of mainchain and sidechain intermolecular hydrogen bonds, connecting both the light and heavy chains to the antigen. The pattern of bonds among the four antibodies provides a signature for them. Overall, there are fourteen different bonds represented. But in each antibody–antigen system, a range of 7 to 10 of these appears. Only four of the bonds are preserved in all antibody–antigen complexes, including the only mainchain–mainchain hydrogen bond. Curiously, the antibody with the highest affinity has the lowest number of intermolecular hydrogen bonds. Underwrapped intramolecular mainchain–mainchain hydrogen bonds are listed in Table 13.9.

type	donor	acceptor	1NDM	1DQJ	1C08	1NDG
M-M	hel-ARG 621 N	L-Asn 92 O	2.68 4.13	2.81 3.32	2.89 3.31	2.80 3.47
M-S	H-Ser 354 N	hel-Asp 701 OD1	3.06 7.24	3.15 6.43	3.20 6.16	(3.59 4.59)
M-S	hel-GLY 702 N	H-Ser 356 OG	2.88 6.46	2.82 6.33	3.00 6.21	
S-M	H-Ser 331 OG	hel-ARG 673 O			3.42 1.28	(4.04 0.483)
S-M	H-TYR 350 OH	hel-Ser 700 O	2.55 5.35	2.60 5.02	2.64 4.63	2.66 4.46
S-M	H-TYR 358 OH	hel-Asp 701 O	3.17 0.32	3.38 0.52		
S-M	hel-ARG 673 NH2	H-Thr 330 O		2.47 0.81		
S-S	hel-ARG 673 NH1	H-Thr 330 OG1	3.16 3.91			
S-S	H-Ser 352 OG	hel-Asp 701 OD1	2.71 8.07	2.70 7.98	2.50 10.2	2.88 2.60
S-S	H-Ser 354 OG	hel-Asp 701 OD1	2.85 5.56	2.49 11.5	2.77 5.93	2.80 6.64
S-S	L-GLN 53 NE2	hel-Asn 693 OD1	2.85 0.56	2.84 0.85		2.81 0.82
S-S	hel-Asn 693 ND2	L-GLN 53 OE1		3.30 0.228	2.83 0.375	3.30 0.296
S-S	hel-LYS 696 NZ	L-Asn 31 OD1			2.95 4.76	2.83 4.64
S-S	hel-LYS 696 NZ	L-Asn 32 OD1			2.82 7.13	
		total bonds	9	10	10	7

Table 13.8 Intermolecular hydrogen bonds in four antibody-lysozyme complexes. S = sidechain, M = mainchain. The numbers given are (1) the distance between the donor and acceptor (heavy) atoms in the hydrogen bond and (2) the quality estimate of the hydrogen bond modeled as a dipole-dipole interaction as defined in (13.1). Note that the two bonds involving hel-LYS 696 NZ in 1C08 are in conflict, in the sense that one would not normally think of the N-H group represented by NZ as capable of forming two hydrogen bonds. However, this ambiguity reflects the geometry involving this group and the two 'acceptor' atoms (OD1 of L-Asn 31 and L-Asn 32). In 1NDG(H8), the donor for the S-M bond with hel-Arg673-O changes from H-Ser331-OG to H-Arg331-NE. Data in parentheses are for reference only. By relaxing the definition of hydrogen bond, we can determine the data not called a hydrogen bond. In all of these, the light chain is chain A, the heavy chain is chain B, and the antigen (HEL) is chain C.

IC	donor	acceptor	1NDM	1DQJ	1C08	1NDG
AC	A-GLY 68 N	A-Ser 30 O	2.89 12+0	2.80 12+0	2.82 12+0	2.85 14+0
BC	B-Ser (3)31 N	B-Ser (3)28 O	3.20 11+8	2.99 10+6	3.15 11+6	2.95 18+6
CB	C-GLY 104 N	C-Asp 101 O	3.13 15+10	3.21 14+11	2.86 13+11	3.36 15+14

Table 13.9 Underwrapped intramolecular mainchain–mainchain hydrogen bonds in four antibody-lysozyme complexes. Key: IC = interface chains. Distances in Ångstroms between acceptor and donor (heavy) atoms for each dehydron are listed, followed by the intramolecular+intermolecular wrapping for a desolvation radius of 6.5 Å. In 1NDG, B-Ser 31 N is replaced by B-Arg 331 N. In general in 1NDG, the residue numbering in the B chain is incremented by 300 and in the C chain is incremented by 600. Although there are no intermolecular wrappers for the A-GLY 68 N — A-Ser 30 O dehydron, constituents of C-Gly 16 are within 6.5Å of A-Ser 30 CA.

13.3 Exercises

Exercise 13.1. Examine the hapten-antibody complex in the PDB file 2RCS [87, 513]. Determine which bonds are involved and how they change with antibody maturation.

Exercise 13.2. Examine the antibody–antigen complexes in the PDB files 3hfm, 1c08, and 1bwh [461]. Determine if there are underwrapped hydrogen bonds are involved and, if so, how they change with antibody maturation.

Chapter 14
Peptide Bond Rotation

Gertrude Belle Elion (1918–1999) was a 1988 recipient of the Nobel Prize in Physiology or Medicine. Working alone as well as with George H. Hitchings, Elion developed many new drugs, using innovative research methods that would later lead to the development of the AIDS drug AZT.

We now consider an application [145] of data mining that links the quantum scale with the continuum level electrostatic field. In other cases, we have considered the modulation of dielectric properties and the resulting effect on electrostatic fields. Here we study the effect of the local electrostatic field on the electronic structure of the peptide covalent bond. Wrapping by hydrophobic groups plays a role in our analysis, but we are interested primarily in an indirect effect of the interaction.

Typically, changes in the dielectric environment will have no direct effect on a covalent bond. However, such changes can affect the local electric field, and in many cases this can change the covalent bond structure, in particular when the electronic structure is representative of a zwitteronic state. One example is the effect of the protonation state of a His sidechain, in which two different resonant states are possible as indicated in Figure 4.9. Another has to do with the peptide bond itself. We consider modulations in the local electric field that cause a significant change in the electronic structure of the peptide bond and lead to a structural change in the type of the covalent bond.

We first describe the results and then consider their implications for protein folding.

14.1 Peptide Resonance States

The peptide bond is characterized in part by the planarity [382] of the six atoms shown in Figure 14.1. The angle ω (see Section 5.3.1) quantifies the orientation of this bond, with planarity corresponding to $\omega = \pi$ radians in the trans case as depicted in Figure 4.2(a), or $\omega = 0$ in the cis form as depicted in Figure 4.2(b). The torsional rigidity of the ω angle is due to the tautomerization of the keto form (A) into the enol zwitteronic form (B), with the latter having a double bond character.

It has been known for some time that the variation in ω is much greater than the variation in other parameters describing the peptide structure [82, 133, 134, 243, 247, 328]. It has been known [382] even longer that the planar state is not the preferred vacuum state. What determines the variation in ω is the local electronic environment [145], as we will review here.

© Springer International Publishing AG 2017
L.R. Scott and A. Fernández, *A Mathematical Approach to Protein Biophysics*,
Biological and Medical Physics, Biomedical Engineering,
DOI 10.1007/978-3-319-66032-5_14

The peptide bond is what is known as a **resonance** [382] between two states, shown in Figure 14.1. The "keto" state (A) on the left side of Figure 14.1 is actually the preferred state in the absence of external influences [382]. However, an external polarizing field can shift the preference to the "enol" state (B) on the right side of Figure 14.1, as we illustrate in Figure 14.2 [382].

We have already seen several examples of atomic groups whose electronic structure is a resonance between two distinct states. The C–C bonding structure in benzene, which is similar to the aromatic rings in Phe, Tyr and Trp, is a combination of single and double covalent bonds. Similarly, the allocation of charge in the oxygens in the symmetric ends of Asp and Glu represents a resonance, which includes the bond configuration with their terminal carbons. The guanidine-like group at the end of the Arg sidechain represents an even more complex type of resonance, in that the positive charge on the NH_2 groups is shared as well by the NH group in the ϵ position.

Resonance theory assumes that a given (resonant) state ψ has the form

$$\psi = C_A \psi_A + C_B \psi_B, \tag{14.1}$$

where ψ_X is the electronic wavefunction (eigenfunction of the Schrödinger equation) for state X ($=A$ or B). Since these are eigenstates normalized to have L^2 norm equal to one, we must have $C_A^2 + C_B^2 = 1$, assuming that ψ_A and ψ_B are orthogonal, as is the assumption for the ideal case of resonance theory. We will take this as the basis for the following discussions.

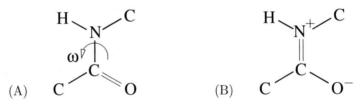

Fig. 14.1 Two resonance states of the peptide bond. The double bond between the central carbon and nitrogen keeps the peptide bond planar in the right state (B). In the left state (A), the single bond can rotate, cf. Figure 5.9(b).

The resonant state could be influenced by an external field in different ways, but the primary cause is hydrogen bonding. If there is a hydrogen bond to either the carbonyl or amide group in a peptide bond, this induces a significant dipole which forces the peptide bond into the (B) state shown in Figure 14.1. Such hydrogen bonds could be either with water, with sidechains, or with other backbone donor or acceptor groups, and such a configuration is indicated in Figure 14.2 in the representation (1) on the left side of the figure. An acceptor for the amide (NH) group is indicated by $\cdots$O, and a donor for the carbonyl (CO) group is indicated by H$\cdots$. Other acceptors could also be involved instead.

On the right side of Figure 14.2, in the representation (2), the polar environment is indicated by an arrow from the negative charge of the oxygen partner of the amide group in the peptide bond to the positive charge of the hydrogen partner of the carbonyl group. The strength of this polar environment will be less if only one of the charges (that is, only one of $\cdots$O or H$\cdots$, or equivalent) is available, but either one (or both) can cause the polar field.

However, if neither type of hydrogen bond is available, then the resonant state moves toward the preferred (A) state in Figure 14.1. The latter state involves only a single bond and allows ω rotation. Thus the electrostatic environment of peptides determines whether they are torsionally rigid or flexible. Since any hydrogen bond can enable the (B) state, it will prevail whenever a water molecule or a mainchain or sidechain donor or acceptor is appropriately located. Otherwise said, the (A) state persists only when water is removed and there are no other binding partners.

A strong electrostatic field could come from nearby charged residues or any of a number of other atom groups. For simplicity, we assume that a lack of hydrogen bond partners corresponds to a small local electrostatic field.

14.2 Measuring Variations in ω

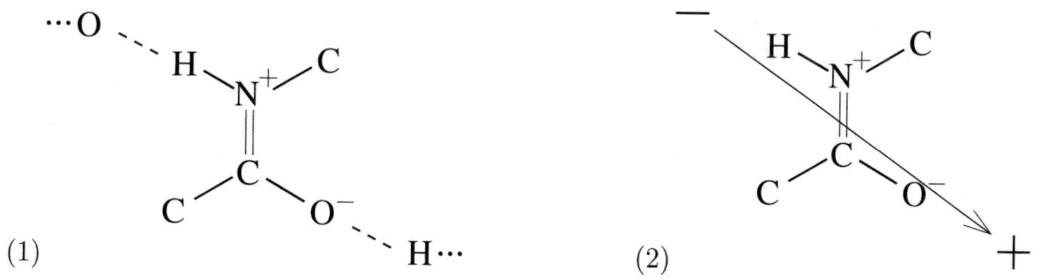

Fig. 14.2 The resonance state (B) of the peptide bond shown, on the left (1), with hydrogen bonds (dashed lines) to an acceptor (indicated by the oxygen preceded by dots) and a donor (indicated by the hydrogen followed by dots). These hydrogen bonds induce a polar field that reduces the preference for state (A). On the right (2), the (B) state is depicted with an abstraction of the dipole electrical gradient induced by hydrogen bonding indicated by an arrow.

There is no simple way to measure the flexibility of the peptide bond from typical structural data. If flexibility were linked with mobility of the ω bond, then it could potentially be inferred from the 'fuzziness' of the electron densities, e.g., as measured by the B-factors reported in the PDB. But such observations might be attributed to other factors, and there is no reason to believe that flexibility in ω would necessarily mean that the angle adopted would not be fixed and well resolved.

For any given peptide bond, the value of ω could correspond to the rigid (B) state even if it is fully in the (A) state. The particular value of ω depends not only on the flexibility but also on the local forces that are being applied. These could in principle be determined, but it would be complicated to do so. However, by looking at a set of peptide bonds, we would expect to see a range of values of ω corresponding to a range of local forces. It is reasonable to assume that these local forces would be randomly distributed in some way if the set is large enough. Thus, the *dispersion* $\Delta\omega$ for a set of peptide bond states would be proportional to the flexibility of the set of peptide bonds. Roughly speaking, $\Delta\omega$ is the width of the distribution of ω's.

The assessment of the flexibility of the peptide bond requires a model. We assume that the amount of rotation ω around the C–N bond is proportional to the applied force f, that is, $\omega - \omega_0 = \kappa f$, where we think of κ^{-1} as representing the strength of the bond to resist rotations, and $\omega_0 = \pi$ corresponds to the planar configuration. If $\kappa = 0$, then the bond is infinitely stiff. Thus if we have a set of rotations, due to a set of forces with dispersion Δf, then $\Delta\omega = \kappa\Delta f$.

Suppose that we assume that the bond adopts a configuration that can be approximated as a fraction C_B of the rigid (B) state and corresponding the fraction C_A of the (A) state.

Let us assume that the flexibility of the peptide bond is proportional to C_A, that is, the 'spring constant' κ of the bond depends linearly on the value of C_A: $\kappa = \kappa_0 C_A$. Thus we are assuming that the (B) state is infinitely rigid. For a set of bond configurations, we thus obtain $\Delta\omega = \kappa_0 C_A \Delta f$ relating the dispersion in ω's to the dispersion in applied forces f.

Let us summarize the assumptions of the model. We suppose that the peptide bond is subjected to a set of forces. We are presuming that Δf is the same for all proteins, and that we can measure $\Delta\omega$. From this we get $C_A = c\Delta\omega$, that is, C_A is proportional to $\Delta\omega$ Some of these forces will leave ω in the same value as for the rigid state. However, others will modify ω, and the extent of the modification will be proportional to the 'spring constant' κ of the bond and thus proportional to the value of C_A. Thus we can assert that the dispersion $\Delta\omega$ in ω is proportional to C_A:

$$\Delta\omega = \gamma C_A, \qquad (14.2)$$

where γ is a constant of proportionality.

The value of γ can be estimated [145] based on the observation that the value $C_B = 0.4$ [382] corresponds to the vacuum state of the peptide bond. In the peptide data, this state is approached with fully desolvated peptide groups that form no hydrogen bonds. Thus the constant will be determined as a by-product of the analysis. We are assuming that the resonant state ψ has the form (14.1), with $C_A^2 + C_B^2 = 1$. Therefore

$$\Delta\omega = \gamma C_A = \gamma\sqrt{1 - C_B^2}. \qquad (14.3)$$

We can now see how to determine γ. Suppose that $\Delta\omega^\infty$ is the observed value of dispersion in ω for fully desolvated peptide groups that form no hydrogen bonds. Then we must get $C_B = 0.4$ (and thus $C_A \approx 0.917$), which means that

$$\gamma \approx 1.09\Delta\omega^\infty. \qquad (14.4)$$

We can thus invert the relationship (14.3) to provide C_B as a function of $\Delta\omega$:

$$C_B = \sqrt{1 - (\Delta\omega/\gamma)^2}. \qquad (14.5)$$

The assumption (14.2) can be viewed as follows. The flexibility of the peptide bond depends on the degree to which the central covalent bond between carbon and nitrogen is a single bond. The (A) state is a pure single bond state and the (B) state is a pure double bond state. Since the resonance state is a linear combination, $\psi = C_A\psi_A + C_B\psi_B$, it follows that there is a linear relationship between C_A and flexibility provided that the single bond state can be quantified as a linear functional. To prove this relationship, we seek a functional L_A such that $L_A\psi_A = 1$ and $L_A\psi_B = 0$. If ψ_A and ψ_B are orthogonal, this is easy to do. We simply let L_A be defined by taking inner-products with ψ_A, provided (as we assume) that ψ_A and ψ_B are orthogonal.

14.3 Predicting the Electronic Structure

Since the preference for state (A) or (B) is determined by the local electrostatic environment, the easiest way to study the flexibility would be to correlate it with the gradient of the external electric field at the center of the peptide bond. However, this field is difficult to compute precisely due to the need to represent the dielectric effect of the solvent (Chapter 15) and to account for the polarizability of all the molecular groups. A dynamic simulation with an explicit water model and full representation of polarizability might be able to correctly estimate this accurately, but this is beyond current technological capability. However, it is possible to estimate the likelihood of a significant dipole moment based on the local environment [145].

The major contributor to a local dipole would be the hydrogen bonding indicated in Figure 14.2. These bonds can arise in two ways, either by backbone hydrogen bonding (or perhaps backbone-sidechain bonding, which is more rare) or by contact with water. The presence of backbone hydrogen bonds is indicated by the PDB structure, but the presence of water is not consistently represented. However, there is a proxy for the probability of contact with water: the wrapping of the local environment. So we can approximate the expected local electrostatic field by analyzing the backbone hydrogen bonding and the wrapping of these amide and carbonyl groups.

In [145], sets of peptide bonds were classified in two ways. First of all, they were separated into two groups, as follows. Group I consisted of peptides forming no backbone hydrogen bonds; in particular, this group excludes ones involved in either α-helices or β-sheets. Group II consisted of peptides forming at least one backbone hydrogen bond, thus including ones involved in either α-helices or β-sheets, but also including other residues which can be identified as forming hydrogen bonds in the PDB structure. Groups I and II are further subdivided according to wrapping. We will describe this in two steps, starting with a simple approach and then giving the full approach developed in [145].

A simplistic way of viewing the data is as follows based on dividing groups I and II into the under-wrapped subgroups (a) and the well-wrapped subgroups (b). We thus consider four groups:

- Group Ia consists of peptides forming no backbone hydrogen bonds and the amide and carbonyl groups are not well-wrapped,
- Group Ib consists of well-wrapped peptides forming no backbone hydrogen bonds,
- Group IIa consists of peptides capable of forming a backbone hydrogen bond but not well-wrapped, and
- Group IIb consists of well-wrapped peptides forming a backbone hydrogen bond.

Then we can estimate the electrostatic environment, and its impact on the resonant state, as depicted in Table 14.1. We make the simple assumption that well-wrapped peptides will not have water around, but under-wrapped ones will be solvated. With this assumption, we can classify the different groups according to whether or not water will be available to form hydrogen bonds with the peptide donor or acceptor. Thus in Table 14.1 we divide into four possibilities as above and assign the resonant state appropriate for the corresponding electrostatic environment. We see that only Group Ib can be in the A state.

Since wrapping can be defined using a count of nonpolar carbonaceous groups, it is possible [145] to give a more refined analysis than indicated in Table 14.1. In each major group (I and II), subsets can be defined based on the level ρ of wrapping in the vicinity of backbone [145]. For each subgroup, the dispersion of ω angles can be measured in the corresponding PDB files, and this data is plotted Figure 14.3 as a function of ρ. Figure 14.3 presents an even more refined analysis. It involves groups Iρ and IIρ for different values of ρ.

The left side of Figure 14.3 corresponds to Group Ia and Group IIa, and we see that the behavior is the same for the two groups. That is, the under-wrapped peptides have a similar dispersion $\Delta\omega$, corresponding to a dominant (B) state, whether or not they appear to be capable of hydrogen bonding, although the dispersion is decreasing with increasing wrapping. On the other hand, there is a difference between the Ib and IIb groups, as indicated on the right side of Figure 14.3. Group Ib prefers the vacuum state (A), whereas IIb tends to the (B) state.

Group number number of HB's	under-wrapped waters (a)	well-wrapped no water (b)	symbol used in Figure 14.3
Group I 0 HB's	$\Delta\omega \approx 0$ B state	$\Delta\omega >> 0$ A state	squares
Group II ≥ 1 HB's	$\Delta\omega \approx 0$ B state	$\Delta\omega \approx 0$ B state	x's

Table 14.1 Simple model of local electrostatic environment around a peptide bond and the resulting preferred resonant state.

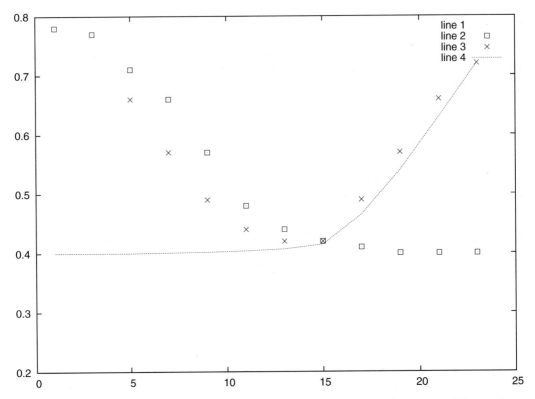

Fig. 14.3 Portion of the double-bond (planar) state in the resonance for residues in two different classes, adapted from [145]. The horizontal axis is the number of nonpolar carbonaceous groups inside two spheres of radius 6 Ångstroms centered at the midpoint of the peptide bond, cf. Section 8.4. The vertical axis is the quantity C_B indicating the prevalence of the (B) state. Small squares indicate the case when neither amide nor carbonyl group is engaged in a backbone hydrogen bond (group I). The x's indicate the case when at least one of the amide or carbonyl groups is engaged in backbone hydrogen bond (group II). The dotted line represents a hypothetical response due only to the dehydration of the mainchain bonds.

Figure 14.3 depicts the resulting observations for group I peptide bonds (small squares), using the model (14.5) to convert observed dispersion $\Delta\omega(\rho)$ to values of C_B, with a constant γ as given in (14.4). This value of γ was determined using the estimated value $C_B = 0.4$ [382] for the vacuum state of the peptide bond which is approached as ρ increases, in the absence of mainchain hydrogen bonds. Well-wrapped peptide bonds that do not form hydrogen bonds should closely resemble the vacuum state. However, poorly wrapped peptide amide and carbonyl groups would be strongly solvated, and thus strongly polarized, leading to a larger component of (B) as we expect and as Figure 14.3 shows.

Using the value (14.4) of γ allows an interesting assessment of the group II peptide bonds, as shown in Figure 14.3 (data represented by x's). These are bonds that, according to the PDB structures, are capable of participating in backbone hydrogen bonds. We see that these bonds also have a variable resonance structure depending on the amount of wrapping. Poorly wrapped backbone hydrogen bonds will likely be solvated, and thus the group I and group II peptides can be expected to behave similarly for small ρ. As with group I peptide bonds, we expect state (B) to be dominant for small ρ. Indeed, the two curves in Figure 14.3 (the squares and the x's) are quite similar for small ρ. As dehydration by wrapping improves, the polarity of the environment due to water decreases, and the proportion of state (B) decreases. But a limit occurs in this case, unlike with group I, due to the fact that wrapping now enhances the strength of the backbone hydrogen bonds, and thus increases the polarity of the environment.

The interplay between the decreasing strength of polarization due to one kind of hydrogen bonding (with water) and the increasing strength of backbone hydrogen bonding is quite striking. As water is removed, hydrogen bonds strengthen and increase polarization of the peptide bond. Figure 14.3 shows that there is a middle ground in which a little wrapping is not such a good thing. That is, small amounts of wrapping appear to remove enough water to decrease the polar environment. Moreover, with minimal wrapping, the backbone hydrogen bonds are screened, and therefore the resulting external polar environment of the peptide is weaker. But the effect of the hydrogen bonds increases as wrapping is increased. The solid line in Figure 14.3 represents a guess of the effect of wrapping of the backbone bonds alone.

It is striking that the group II data has a distinctive minimum. One might guess that there would be constant polarity in the transition from fully solvated peptides to fully desolvated peptides. But apparently there are two distinct behaviors. The fully solvated states appear to provide a strong dipole through the contact with water in a nearly bulk-like state. As wrapping is added, this polarity is disrupted as water becomes both excluded and disordered by the hydrophobic groups. On the other hand, the backbone hydrogen bonds can also be disrupted by just a few water molecules. Thus it takes a large amount of wrapping to establish a stable polar environment. Thus it is not just a simple exchange of one type of polarity for another. There are two different mechanisms. One has to do with the structure of water, and the resulting ability of water to establish a consistent polar environment around a peptide base. The other has to do with the requirements of backbone donors and acceptors to form stable attachments with other parts of the protein structure, unaffected by the presence of water. It is not surprising that these two disparate mechanisms would operate on their own scales and thus not cancel each other as the amount of wrapping is varied.

14.4 Implications for Protein Folding

After the "hydrophobic collapse" [548] a protein is compact enough to exclude most water. At this stage, few hydrogen bonds have fully formed. But most amide and carbonyl groups are protected from water. The data in Figure 14.3(a) therefore implies that many peptide bonds are flexible in the final stage of protein folding. This effect is not included in current models of protein folding. This effect buffers the entropic cost of hydrophobic collapse in the process of protein folding.

New models need to allow flexible bonds whose strengths depend on the local electrostatic environment [415]. Typical molecular dynamics (MD) models would either have peptide bonds fixed in the planar configuration or have a large spring constant for rotation in the ω angle. Here we need the spring constant to depend on the electrostatic field in the vicinity of the peptide bond.

The electrostatic field at a point $\mathbf{r}$ is given by

$$\sum_k q_k \frac{\mathbf{r} - \mathbf{r}_k}{|\mathbf{r} - \mathbf{r}_k|^3}, \tag{14.6}$$

where $\mathbf{r}_k$ denote the positions of the charges q_k, cf. (15.86). In particular, the quantity of interest is the strength of the dot product of the electrostatic field gradient and the vector $\overline{O - H}$ pointing from O to H in the peptide group. If $\mathbf{r}_0$ is the centroid of the peptide group, then one would seek a bending strength depending on the quantity

$$\sum_k q_k \frac{(\mathbf{r}_0 - \mathbf{r}_k) \cdot \overline{O - H}}{|\mathbf{r}_0 - \mathbf{r}_k|^3}. \tag{14.7}$$

Note that we are invoking a sum over all (charged) atoms in the system, and this type of global term will make the simulation much more costly. Using a cut-off radius to limit the number of charged atoms involved may be practical, but it introduces an approximation into the model whose effect would have to be assessed.

14.5 Exercises

Exercise 14.1. The dipole vector is twice as strong if both amide and carbonyl groups are involved in hydrogen bonds. Split group II into two groups, group II1 and II2 depending on the number of hydrogen bonds. How does the preference for the (B) state differ between groups II1 and II2?

Exercise 14.2. Scan some PDB files and form the groups Ia, Ib, IIa, IIb indicated in Section 14.3. Plot the distributions of ω angles for each group. Do the distributions for Ia and IIa look similar? How are the distributions for Ib and IIb different?

Exercise 14.3. Using a model of the dielectric effect (cf. Chapter 15), estimate the dipole vector at each peptide in a set of PDB files. Use this estimate to predict whether the peptide bond is in the (A) state or the (B) state. Compare this with the measured value of ω.

Exercise 14.4. Using a molecular dynamics model with explicit water, estimate the dipole vector at each peptide in a set of PDB files. Use this estimate to predict whether the peptide bond is in the (A) state or the (B) state. Compare this with the measured value of ω.

Exercise 14.5. Using a quantum chemistry model, calculate the flexibility of the ω bond as a function of an imposed dipole as indicated in Figure 14.2.

Exercise 14.6. Determine if wrapping correlates (or not) with the formation of mainchain-mainchain hydrogen bonds. Consider the average wrapping of carbonyls and amides that do not form hydrogen bonds and compare this with what is found in the figures in [194]. Note that there is a significant number of amide and carbonyl groups that are well wrapped and yet form no hydrogen bonds; these are the peptides in group Ib in Section 14.3.

Chapter 15
Continuum Equations for Electrostatics

Siméon Denis Poisson (1781–1840) conceived the central equation of electrostatics that relates the electric potential and the charge.

We began (Figure 1.1) by saying that the "dielectric modulation by the hydrophobic effect"
was a major factor in protein–ligand association. In this chapter, we derive continuum equations for dielectric materials to show how the hydrophobic effect can change the dielectric
environment, and how in a continuum model this leads to a change in energy. However, we
will also see that there is ambiguity in current dielectric models due to difficult modeling
issues. Thus, the continuum model, while useful, has limitations as a quantitative metric.

The dielectric properties of materials are important in many contexts [54, 337, 462]. A
dielectric medium [63] is often characterized by having charges organized in local groups
with net charge zero that are free to rotate to align with the ambient field. The properties of
a dielectric can be derived from first principles that take into account the polar properties of
the dielectric constituents.

15.1 Systems of Charged Particles

A dielectric medium is a collection of polar molecules which react to the electric field in
such a way as to moderate the overall electric field. Our case of interest is primarily water,
and we know water is a polar molecule; in fact, we know it can be represented as a dipole
(Section 10.5.4). We can model the dielectric effect by representing these polar molecules
as a collection of charges (cf. Section 10.5.4). Although this is just an approximation, we
can imagine making better and better approximations by larger collections of charges at
appropriate points to represent the polar nature of water. All such collections would of course
have net charge zero for each water molecule.

In Section 3.2, we describe the basic relationship between charge and electric potential and
force. Using (3.2) and assuming linearity, the basic equations of electrostatics for a collection
of charges of strength q_i at positions $\mathbf{r}_i$ in a vacuum can be derived from the simple expression

$$\nabla \cdot (\epsilon_0 \mathbf{e}) = \sum_i q_i \delta(\mathbf{r} - \mathbf{r}_i), \tag{15.1}$$

where ϵ_0 is the permittivity of the vacuum (see Section 18.2.3). Here, $\mathbf{e}$ is the resulting electric
field.

© Springer International Publishing AG 2017
L.R. Scott and A. Fernández, *A Mathematical Approach to Protein Biophysics*,
Biological and Medical Physics, Biomedical Engineering,
DOI 10.1007/978-3-319-66032-5_15

Using the ideas in Section 3.2, the equation (15.1) can be solved: $\mathbf{e} = -\nabla\phi$ where

$$\phi(\mathbf{r}) = \sum_i \frac{q_i}{|\mathbf{r} - \mathbf{r}_i|}, \tag{15.2}$$

where we have chosen physical units so that $\epsilon_0 = 1/4\pi$. But the ability to solve this so precisely is misleading. In interesting systems, the positions and orientations of many charge groups are unknown, and they must be deduced as part of the solution process.

15.2 Dielectric Materials

We think of our charged system as consisting of two parts: fixed charges (ρ) and dielectric charges (γ). We will split the total charge distribution $\sum_i q_i \delta(\mathbf{r} - \mathbf{r}_i)$ into two parts, $\gamma + \rho$, where γ is the part of the charge density corresponding to the dielectric, e.g., charge groups with net charge zero, and ρ denotes the remainder of the charge density (fixed charges). More precisely, we assume that the set of all charges is decomposed into two sets, I_ρ and I_γ, and

$$\rho = \sum_{i \in I_\rho} q_i \delta(\mathbf{r} - \mathbf{r}_i), \qquad \gamma = \sum_{i \in I_\gamma} q_i \delta(\mathbf{r} - \mathbf{r}_i). \tag{15.3}$$

Correspondingly, we define

$$\phi_\rho(\mathbf{r}) = \sum_{i \in I_\rho} \frac{q_i}{|\mathbf{r} - \mathbf{r}_i|}, \qquad \phi_\gamma(\mathbf{r}) = \sum_{i \in I_\gamma} \frac{q_i}{|\mathbf{r} - \mathbf{r}_i|}, \tag{15.4}$$

and $\phi = \phi_\rho + \phi_\gamma$. In particular,

$$\nabla \cdot (\epsilon_0 \nabla \phi_\rho)(\mathbf{r}) = \rho(\mathbf{r}), \qquad \nabla \cdot (\epsilon_0 \nabla \phi_\gamma)(\mathbf{r}) = \gamma(\mathbf{r}). \tag{15.5}$$

The key assumption is that the dielectric charges are free to self-organize in response to the overall electric field. For example, we imagine that water molecules can rotate so that their dipoles align with the electric field.

Define the **polarization** electric field $\mathbf{p} = -\nabla\phi_\gamma$, so that

$$\nabla \cdot (\epsilon_0 \mathbf{p}) = -\gamma. \tag{15.6}$$

Set $\mathbf{d} = \mathbf{e} + \mathbf{p}$. Then,

$$\nabla \cdot (\epsilon_0 \mathbf{d}) = \nabla \cdot (\epsilon_0 \mathbf{e}) + \nabla \cdot (\epsilon_0 \mathbf{p}) = \rho + \gamma + \nabla \cdot (\epsilon_0 \mathbf{p}) = \rho. \tag{15.7}$$

The electric field $\mathbf{e}$ thus satisfies $\mathbf{e} = \mathbf{d} - \mathbf{p}$, where $\mathbf{d}$ can be interpreted as the field generated by the fixed charges ρ in the absence of the charges associated with the dielectric effect.

Debye [110] postulated that the mobile charge groups would orient to oppose the resulting electric field. In a thermalized system, there will be fluctuations in the orientations of these charge groups, and we can talk only about mean orientations. Debye's postulate can thus be written

$$\widetilde{\mathbf{p}} = s\,\widetilde{\mathbf{e}}, \tag{15.8}$$

where s can be thought of as a constant of proportionality in the simplest case, but more generally we will think of s as an operator that maps $\widetilde{\mathbf{e}}$ to $\widetilde{\mathbf{p}}$. Here, $\widetilde{\mathbf{p}}$ (resp., $\widetilde{\mathbf{e}}$) can be interpreted as a temporal average over a timescale that is long with respect to the basic thermal motions. (More precisely, it is a thermodynamic quantity that represents an average over many realizations of the system.) The thermal motions occur on the order of fractions of picoseconds, so we could imagine a time average of the order of picoseconds.

It is useful to define $\epsilon = \epsilon_0 s + \epsilon_0$. If we are thinking of s as an operator, then the second term is just the identity operator scaled by the permittivity coefficient. Debye's postulate can now be written as

$$\epsilon_0 \widetilde{\mathbf{p}} = (\epsilon - \epsilon_0)\widetilde{\mathbf{e}}, \tag{15.9}$$

where we will see that ϵ can be thought of as an effective permittivity (which may be a constant in the simplest case, or more generally by an operator).

If one is uncomfortable with the Debye ansatz, we can define ϵ as follows. We decompose $\mathbf{p}$ into one part in the direction $\mathbf{e}$ and the other perpendicular to $\mathbf{e}$. That is, we write

$$\epsilon_0 \mathbf{p} = (\epsilon - \epsilon_0)\mathbf{e} + \zeta \mathbf{e}^{\perp}, \tag{15.10}$$

where ϵ is defined by

$$\epsilon = \epsilon_0 \left(1 + \frac{\mathbf{p} \cdot \mathbf{e}}{\mathbf{e} \cdot \mathbf{e}} \right), \tag{15.11}$$

with the appropriate optimism that $\mathbf{p} = 0$ when $\mathbf{e} = 0$. Here, we may assume that $\mathbf{e}^{\perp}$ is a unit vector. That is, $\epsilon - \epsilon_0$ reflects the correlation between $\mathbf{p}$ and $\mathbf{e}$. As defined, ϵ is a function of $\mathbf{r}$ and t and potentially singular. However, Debye postulated that a suitable average

$$\widetilde{\epsilon} = \epsilon_0 \left(1 + \left\langle \frac{\mathbf{p} \cdot \mathbf{e}}{\mathbf{e} \cdot \mathbf{e}} \right\rangle \right), \tag{15.12}$$

should be well behaved. Implicitly, it is assumed that averages of ζ defined in (15.10) may be ignored.

However we interpret the Debye ansatz, and we arrive at (15.8) as our model equation for the relationship between the electric field and the polarization field. For simplicity, we will drop the tildes and think from now on that everything represents temporal, or spatial, averages. Combining (15.9) (without the tildes) with the definition of $\mathbf{d}$, we conclude that

$$\epsilon_0 \mathbf{d} = \epsilon_0 \mathbf{e} + \epsilon_0 \mathbf{p} = \epsilon \mathbf{e}. \tag{15.13}$$

Thus (15.7) implies

$$\boxed{\nabla \cdot (\epsilon \nabla \phi) = \nabla \cdot (\epsilon \mathbf{e}) = \rho.} \tag{15.14}$$

Of course, we need to add boundary conditions to make this well posed; for example, we can demand that $\phi \to 0$ at infinity. This gives us an equation for ϕ (or $\mathbf{e}$) in terms of ρ, provided we have some way of estimating ϵ. Looking carefully at the derivation, we see that ϵ could be any operator, even a nonlinear one. We will see that it is appropriate to think of ϵ as just a simple constant in many cases, but in others it will be more complicated.

The expression (15.12) provides an operational definition for a computationally determined dielectric constant. That is, in a molecular dynamics computation, one can define a local dielectric constant by averaging the correlation coefficient

$$\widetilde{\epsilon} - \epsilon_0 = \epsilon_0 \frac{\mathbf{p} \cdot \mathbf{e}}{\mathbf{e} \cdot \mathbf{e}}, \tag{15.15}$$

over space, over certain molecules, and/or time. This correlation coefficient need not be positive, so it is conceivable that $\widetilde{\epsilon} < \epsilon_0$, and we could even have $\widetilde{\epsilon} < 0$.

Monitoring the other term $\zeta \mathbf{e}^{\perp}$ in (15.10) also allows us to quantify the extent to which the dielectric model holds. Using the orthogonality, we have

$$
\begin{aligned}
\zeta^2 &= \zeta \mathbf{e}^{\perp} \cdot \zeta \mathbf{e}^{\perp} = (\epsilon_0 \mathbf{p} - (\epsilon - \epsilon_0)\mathbf{e}) \cdot \zeta \mathbf{e}^{\perp} = \epsilon_0 \mathbf{p} \cdot \zeta \mathbf{e}^{\perp} \\
&= \epsilon_0 \mathbf{p} \cdot (\epsilon_0 \mathbf{p} - (\epsilon - \epsilon_0)\mathbf{e}) = \epsilon_0 \mathbf{p} \cdot \left(\epsilon_0 \mathbf{p} - \epsilon_0 \frac{\mathbf{p} \cdot \mathbf{e}}{\mathbf{e} \cdot \mathbf{e}} \mathbf{e} \right) \\
&= \epsilon_0^2 \mathbf{p} \cdot (\mathbf{p} - (\mathbf{p} \cdot \hat{\mathbf{e}})\hat{\mathbf{e}}) = \epsilon_0^2 \mathbf{p} \cdot (\mathbf{p} - \mathbf{p}_e) = \epsilon_0^2 \mathbf{p} \cdot \mathbf{p}_e^{\perp} \\
&= \epsilon_0^2 \mathbf{p}_e^{\perp} \cdot \mathbf{p}_e^{\perp}, = \epsilon_0^2 |\mathbf{p}_e^{\perp}|^2,
\end{aligned}
\tag{15.16}
$$

where we denote by $\hat{\mathbf{e}}$ the unit vector in the direction of $\mathbf{e}$, $\hat{\mathbf{e}} = |\mathbf{e}|^{-1}\mathbf{e}$, where

$$
\mathbf{p}_e = (\mathbf{p} \cdot \hat{\mathbf{e}})\hat{\mathbf{e}}
\tag{15.17}
$$

is the projection of $\mathbf{p}$ in the direction of $\mathbf{e}$, and where

$$
\mathbf{p}_e^{\perp} := \mathbf{p} - \mathbf{p}_e.
\tag{15.18}
$$

The quantity of interest would be the ratio of $|\zeta|$ and $|\epsilon - \epsilon_0| \, |\mathbf{e}|$, since these are the two terms in (15.10). The first one can be computed via (15.16). The second is also easily calculated. Taking the dot product of (15.10) with $\mathbf{e}$, we find

$$
|(\epsilon - \epsilon_0)| \, |\mathbf{e}|^2 = |(\epsilon - \epsilon_0)\mathbf{e} \cdot \mathbf{e}| = \epsilon_0 |\mathbf{p} \cdot \mathbf{e}| = \epsilon_0 |\mathbf{p} \cdot \hat{\mathbf{e}}| |\mathbf{e}| = \epsilon_0 |\mathbf{p}_e| \, |\mathbf{e}|.
\tag{15.19}
$$

Therefore,

$$
\frac{|\zeta|}{|\epsilon - \epsilon_0| \, |\mathbf{e}|} = \frac{|\mathbf{p}_e^{\perp}|}{|\mathbf{p}_e|}.
\tag{15.20}
$$

Thus, the ratio of the sizes of the terms in the Debye ansatz is easily monitored by the expression (15.20). Recall that $\mathbf{e}$ is just the electric vector resulting from all of the charge groups, both fixed and free to rotate, and $\mathbf{p}$ is the polarization vector resulting from the free charges. The terms on the right-hand side of (15.20) are defined in (15.17) and (15.18).

15.2.1 Bulk Water Dielectric

A very good approximation is obtained in bulk water by the assumption that ϵ is just a constant factor. The model

$$
\epsilon \approx 87.74 - 40.00\,\tau + 9.398\,\tau^2 - 1.410\,\tau^3,
\tag{15.21}
$$

where $\tau = T/100$ and T is the temperature in centigrade (for $T > 0$), is supported by extensive experiments for liquid water [220]. Note that the formula (15.21) is intended only to apply for water, hence $\tau \in [0,1]$.

This result is remarkable for many reasons. The fact that ϵ is much greater than one means that the strength of the opposing polarization field $\mathbf{p} = -\nabla \phi_\gamma$ is much greater than the field that induces it. The fact that ϵ increases with decreasing temperature implies that order is more important than mobility in the dielectric effect. When water freezes, it increases further: for ice at zero degrees centigrade, $\epsilon \approx 92$.

This simple model of the dielectric in bulk cannot hold when the spatial frequencies of the electric field $\nabla \phi$ are commensurate with the size of a water molecule, since the water molecules cannot orient appropriately to align with the field. Thus, frequency-dependent versions of ϵ have been proposed, and these are often called "nonlocal" models since the operator ϵ must be represented either as a Fourier integral (in frequency space), or as an integral in physical space with a nonlocal kernel [57, 447].

15.2.2 Wrapping Affects the Dielectric

Since the dipolar nature of water yields the dielectric effect, leading to the large coefficient ϵ in the bulk, it is natural to conclude that removal of water reduces the strength of the dielectric effect. Let us think for the moment of ϵ as a function for the sake of argument. Thus, removal of water by a nonpolar group reduces ϵ locally. Looking at the equation (15.14), this would mean that $\mathbf{e} = \nabla \phi$ would need to *increase* locally to keep a balance. Of course, partial differential equations are more complicated than this simple argument suggests, but one could look at the one-dimensional case for intuition. With suitable boundary conditions [60], we can solve an equation like (15.14) via

$$\phi(x) = \int^x \frac{R(t)}{\epsilon(t)} dt, \tag{15.22}$$

where $R' = \rho$. Thus, we see that a decrease in ϵ leads to an increase in ϕ, although the effect is spread out due the integral. In multiple dimensions, inverting an elliptic operator is even more complicated, but a similar principle holds, with the averaging involving the fundamental solution. We do not pursue the complicated details of such an argument since in the end we want to think of ϵ not just as a function but as an operator. In such a case, proving a relation analogous to (15.22) would be beyond the scope of this book. On the other hand, such a question can be addressed computationally with particular models for ϵ. However, we should note that a standard continuum model may not accurately capture the effect of the removal of water by a single nonpolar group. In particular, the enhancement of an underwrapped hydrogen bond due to the addition of a nonpolar group would require modeling accurately the potential energy of the hydrogen bond. That is, ρ would also need to change, as we have noted that a simple dipole–dipole interaction (that is, with fixed partial charges) does not accurately model a hydrogen bond.

15.2.3 Temporal Frequency Dependence

The relationship proposed by Debye between $\mathbf{p}$ and $\mathbf{e}$ depends on temporal frequency. In particular, many measurements have shown [261]

$$\epsilon(\nu) = \epsilon_0 + \frac{\epsilon_1 - \epsilon_0}{1 + \nu^2 \tau^2} \tag{15.23}$$

where τ is a characteristic time associated with the dielectric material.

This relationship has been verified extensively by experimental data, including recent work [261]. For plane waves, there is a simple relationship between spatial wave number ξ and temporal wave number ν given by $c\xi = \nu$ where c is the speed of light. Thus, plane waves with temporal frequency ν have spatial frequency $\xi = \nu/c$. This suggests using a form similar to (15.23) for the behavior of $\epsilon(\xi)$, although this form does not follow rigorously from current temporal experiments.

15.2.4 Spatial Frequency Dependence

We are interested in electric fields which are *not* time varying (i.e., $\nu = 0$) but rather spatially varying. It is easy to see that the driving electrical fields ρ generated by proteins have high frequencies. Salt bridges involve charge alternations on the order of a few Ångstroms, and polar sidechains such as Asn and Gln correspond to even higher frequencies, although at a smaller amplitude. Thus, the fixed charge density ρ will have significant frequency content on the order of the size of a water molecule. These high frequencies in ρ may be primarily in directions along the surface of the protein, but there also must be high frequencies normal to the surface as well resulting from the jump in dielectric, as we now explain.

The dielectric coefficient varies by a factor of nearly one hundred from inside the protein to the bulk dielectric value away from the surface. This forces a kink in the electric field in the vicinity of the boundary in the direction normal to the surface of the protein. Kinks (discontinuities in the derivative) imply high-frequency components in the electric field. Let us explain how such kinks can arise.

Suppose that the fixed charges are all within a domain Ω containing the protein and that water is excluded from this domain. Then, we expect $\epsilon = \epsilon_0$ inside Ω, and $\epsilon = \epsilon_1$ outside Ω. But the equation $\nabla\cdot(\epsilon\mathbf{e}) = \rho$ forces the quantity $\epsilon\mathbf{e}$ to have a continuous normal component across $\partial\Omega$, assuming there are no charges on $\partial\Omega$. Therefore, $\mathbf{n} \cdot \mathbf{e}$ has to be discontinuous across $\partial\Omega$ to compensate for the discontinuity of ϵ (see Exercise 15.7).

Thus, there are high-frequency components in the electric field both in the direction normal to the surface of the protein as well as in directions along the surface.

On the other hand, it is clear that the dielectric response has to go to zero for high frequencies. If the electric field varies at a spatial frequency whose wavelength is smaller than the size of a water molecule, the water molecule feels a diminished effect of that field component. Therefore, the dielectric coefficient must be a function of spatial wave number and go to zero for high frequencies.

Thus, we adopt the ansatz that the dielectric properties depend on spatial wave number ξ proportional to a factor κ given by

$$\kappa(\xi) = \epsilon_0 + \frac{\epsilon_1 - \epsilon_0}{1 + |\lambda\xi|^2}, \tag{15.24}$$

where λ is a length scale determined by the size of the dielectric molecules. In general, λ could be a matrix, allowing for anisotropy. But for the time being, we will think of it as a scalar.

We have anticipated that the dielectric coefficient may depend on space (and time), so we will be interested in cases where λ depends on the spatial variable $\mathbf{r}$, and in some cases, the length scale will tend to infinity. For this reason, we introduce $\mu = 1/\lambda$, and write

$$\kappa(\mathbf{r}, \xi) = \epsilon_0 + \mu(\mathbf{r})^2 \frac{\epsilon_1 - \epsilon_0}{\mu(\mathbf{r})^2 + |\xi|^2} \tag{15.25}$$

where $\mu(\mathbf{r})$ is a spatial frequency scale. However, for simplicity, we assume that the model (15.24) is sufficient for the moment.

15.2.5 Poisson-Debye Equation: Bulk Case

We can expand **e** and **p** in a Fourier series and use the Debye-like relationship (15.24) to relate the resulting coefficients in the series. That is, we have (using the inverse Fourier transform)

$$\frac{1}{(2\pi)^3} \int_{\mathbb{R}^3} e^{-i\mathbf{r}\cdot\xi} \kappa(\xi)\widehat{\mathbf{e}}(\xi)\,d\xi = \mathbf{p}(\mathbf{r})\,, \tag{15.26}$$

where here and subsequently we use the notation $\widehat{u}$ to denote the Fourier transform of a function u:

$$\widehat{u}(\xi) := \int_{\mathbb{R}^3} e^{i\xi\cdot\mathbf{r}} u(\mathbf{r})\,d\mathbf{r}\,. \tag{15.27}$$

Therefore, the basic equation is

$$\nabla\cdot\left(\frac{1}{(2\pi)^3} \int_{\mathbb{R}^3} e^{-i\mathbf{r}\cdot\xi} \kappa(\xi)\widehat{\mathbf{e}}(\xi)\,d\xi\right) = 4\pi\rho(\mathbf{r})\,. \tag{15.28}$$

We can write $\mathbf{e} = \nabla\phi$ using Maxwell's equations. Therefore, $\widehat{\mathbf{e}}(\xi) = i\xi\widehat{\phi}(\xi)$. Inserting this into (15.28) yields

$$\nabla\cdot\left(\frac{1}{(2\pi)^3} \int_{\mathbb{R}^3} e^{-i\mathbf{r}\cdot\xi} \kappa(\xi)i\xi\widehat{\phi}(\xi)\,d\xi\right) = 4\pi\rho(\mathbf{r})\,. \tag{15.29}$$

Taking the Fourier transform (15.27) of (15.29) provides the simple relation

$$\widehat{\phi}(\xi) = \frac{4\pi\widehat{\rho}(\xi)}{|\xi|^2\kappa(\xi)} \tag{15.30}$$

which can be used to compute ϕ (and thus **e**) from ρ.

The expression (15.28) can be simplified in certain limits. We have

$$\frac{1}{(2\pi)^3} \int_{\mathbb{R}^3} e^{-i\mathbf{r}\cdot\xi} \kappa(\xi)\widehat{\mathbf{e}}(\xi)\,d\xi \approx \epsilon_j \mathbf{e}(\mathbf{r}) \tag{15.31}$$

where $j = 1$ when **e** is very smooth and $j = 0$ when **e** consists of only high frequencies. However, for general fields **e**, it is not possible to approximate the Fourier integral in this way. Thus, we cannot think of (15.28) as a partial differential equation, except approximately in special cases.

When a nondielectric material (or one with a much smaller dielectric effect) is immersed in a dielectric (e.g., a protein in water), it might be plausible to approximate (15.31) with $j = 1$ in the dielectric, switching to $j = 0$ at the interface of the nondielectric material (which introduces high frequencies due to the abrupt change in material). This leads to the standard Poisson equation with a spatially varying permittivity ϵ_j that jumps from $j = 1$ in the dielectric to $j = 0$ in the nondielectric material; this is often used to model macromolecular systems in solvent [329].

However, it is not clear what to do when very small nondielectric objects, such as nanotubes [447], are introduced into a dielectric. The scale of a nanotube is so small that there would be almost no ϵ_0 region in such models, so that any predictions of electrostatics would be essentially the same as if there were pure dielectric. It is possible to introduce a spatially varying permittivity that changes more smoothly between the two extremes [447], but this does not capture accurately the behavior of the wave number dependence.

The characteristic scale λ represents a correlation length relating the way changes in the dielectric influence each other spatially. When the dielectric molecules are constrained, for example, at a material boundary, the characteristic scale λ increases. This is because the dielectric molecules become ordered near a wall, and thus, changes propagate further than in bulk. It is also clear that these changes may be anisotropic, with changes parallel to the wall more affected than perpendicular to the wall. Such changes near an interface could cause λ to increase effectively to infinity at the surface of the bounding material. Thus, it might be reasonable to view the kernel κ in this case as continuous across material boundaries.

The equation (15.28) involves a pseudo-differential operator [126]. Due to the special form of $\kappa(\nu)$, it is possible to write (15.28) as a fourth-order elliptic partial differential operator for the potential ϕ:

$$\nabla \cdot \left(\left(\epsilon_1 - \epsilon_0 \lambda^2 \Delta \right) \nabla \phi \right) = \left(1 - \lambda^2 \Delta \right) \rho \qquad (15.32)$$

provided that λ is constant [227, 228, 514]. However, if λ is a function of $\mathbf{r}$, this is no longer valid. Also, the limit $\lambda \to \infty$ is harder to interpret in this setting.

15.2.6 Nonlocal Relationship Between p and e

The model (15.25) has been used by different people [85, 144, 280, 305] to account for the frequency dependence of the (zero temporal frequency) dielectric relationship. It is often expressed as a nonlocal dependence of the polarization on the electric field and written in the form

$$\mathbf{p}(\mathbf{r}) = \int K(\mathbf{r}, \mathbf{r}') \mathbf{e}(\mathbf{r}') \, d\mathbf{r}' \qquad (15.33)$$

where the averaging kernel K satisfies

$$K(\mathbf{r}, \mathbf{r}') = K(\mathbf{r} - \mathbf{r}') = \int e^{i\mathbf{k} \cdot (\mathbf{r} - \mathbf{r}')} \kappa(\mathbf{k}) \, d\mathbf{k} \qquad (15.34)$$

with the expression κ representing the Debye-like frequency dependence (15.24). Taking Fourier transforms, we see that the "nonlocal" model (15.24) is the same as (15.26). However, it is not possible to represent the mollifier K as an ordinary function. Clearly,

$$K = \delta + \frac{\epsilon_1 - \epsilon_0}{\epsilon_0} \widetilde{K}, \qquad (15.35)$$

where

$$\widetilde{K}(\mathbf{r}) = \int e^{i\mathbf{k} \cdot \mathbf{r}} \tilde{\kappa}(\mathbf{k}) \, d\mathbf{k} \qquad (15.36)$$

and $\tilde{\kappa}$ is defined by

$$\tilde{\kappa}(\mathbf{k}) = \frac{1}{1 + |\mathbf{k}|^2 \lambda^2}. \qquad (15.37)$$

We easily identify $\tilde{\kappa}$ as the Fourier transform of the fundamental solution of the Laplace operator $1 - \lambda^2 \Delta$, so that

$$\widetilde{K}(\mathbf{r}) = \frac{\lambda e^{-|\mathbf{r}|/\lambda}}{4\pi |\mathbf{r}|}. \qquad (15.38)$$

Although this expression appears singular, we realize it is less singular than the Dirac δ-function, which simply evaluates a function at a point instead of averaging. The exponential decay insures that the averaging is fairly local in nature. The kernel for the nonlocal expression for the polarization can be written formally (in the sense of distributions [443]) as

$$K(\mathbf{r}) = \delta + \frac{\lambda(\epsilon_1 - \epsilon_0)e^{-|\mathbf{r}|/\lambda}}{4\pi\epsilon_0|\mathbf{r}|}. \tag{15.39}$$

More precisely, we have

$$\epsilon_0\mathbf{p}(\mathbf{r}) = \epsilon_0\mathbf{e}(\mathbf{r}) + \frac{\lambda(\epsilon_1 - \epsilon_0)}{4\pi}\int \frac{e^{-|\mathbf{r}-\mathbf{r}'|/\lambda}}{|\mathbf{r} - \mathbf{r}'|}\mathbf{e}(\mathbf{r}')\,d\mathbf{r}'. \tag{15.40}$$

15.3 Computational Techniques

Recently, computational techniques have been developed to allow efficient implementation of nonlocal dielectric models [531, 532]. These now allow experimentation with nonlocal models to explore the differences one obtains compared with other dielectric models [448], at a cost that is a modest factor times conventional local dielectric models. Moreover, there are now models that incorporate both ionic effects and nonlocal dielectric effects [533, 534]. We draw the following material from [448], with the authors' permission.

Discontinuous (local) dielectric models are widely used to approximate the electrostatic field surrounding proteins, small molecules, and other systems involving a dielectric material [278]. In some cases, one uses the dielectric solvent model together with a quantum mechanics model [491]. More commonly, the discontinuous dielectric model is used as an implicit solvent model in molecular dynamics simulations [33, 39, 242, 425]. Recently, nonlocal models [42, 43, 530–532, 534] have been of interest. Here, we compare the predictions of the two models of the electrostatic potential around solutes of different sizes. Here, we choose the solute to be a water molecule in order to compare with bulk water properties. In [448], computations are presented with the solute being a small protein, BPTI. In both cases, we see dramatic differences between the predictions of the two models near the solute.

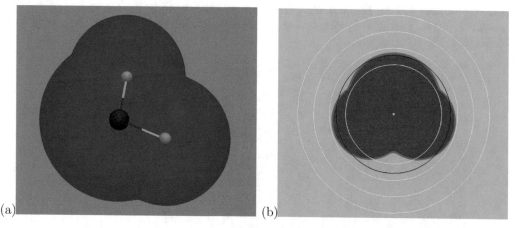

(a)   (b)

Fig. 15.1 Domain where the dielectric discontinuity appears. (a) Depiction of a water molecule surrounded by a region of low dielectric constant due to the polarization of the electron distribution. (b) The red and cyan domains depict the solute (i.e., the central water molecule) and the solvent domain, respectively. The red dot at the center indicates the centroid of the water molecule. The contours are at distances of 1.5Å, 1.8Å, 2.0Å, 2.5Å, and 3Å. All of these are in yellow except 1.8Å, which is in black.

15.3.1 Local Dielectric Model

Recall that Debye's postulate can be written as

$$\nabla \cdot (\epsilon \nabla \phi) = \rho. \tag{15.41}$$

The local dielectric model interprets (15.41) as an equation in three-dimensional space with ϵ being a discontinuous, piecewise constant function. Figure 15.1(a) depicts the region of low dielectric in purple. Note that the charge center of the water molecule is within the triangle spanned by the oxygen and hydrogen centers, and thus, the charge center is not at the center of this low dielectric domain. This allows one to understand asymmetries in hydration of ions [350] due to charge differences.

In the example depicted in Figure 15.1(b), the value in the green region corresponds to the experimentally determined value for bulk water. The value inside the solute (indicated in red) is a small factor of the vacuum value, ϵ_0. Here, we take the solute value to be $2\epsilon_0$. We use the approximation for bulk water given by (15.21). Thus, $\epsilon = 80\epsilon_0$ for $T \approx 20.3$ degrees centigrade, and we will use this value in our simulations.

15.3.2 Nonlocal Dielectric Model

Nonlocal models [42, 43, 531, 532] interpret (15.41) as an equation with ϵ being a nonlocal operator. Such a system is in general very costly to invert. Fortunately, it is possible to convert certain such systems into a system of two partial differential equations, the complexity of which is no greater than twice that of the local dielectric model [531, 532]. Our objective here is to compare the solutions of the local and nonlocal models. We do so for a simple model system.

The basis for the nonlocal model is the fact that the dielectric molecules, water in this case, have a length scale comparable to lengths of importance near the solute. In particular, near the solute (within a distance corresponding to a small factor times the diameter of a water molecule), the electrostatic potential ϕ goes from a positive high to a negative low in a distance comparable to the length of a water molecule (see Figure 15.2(a)). Thus, waters near the solute have difficulty reorienting to align with the electrostatic field as required by the Debye ansatz (15.8). In particular, the layer of water nearest the solute is referred to as a solvation layer, and it is known to have qualitatively different behavior than bulk water [76, 90, 224, 288, 378, 466].

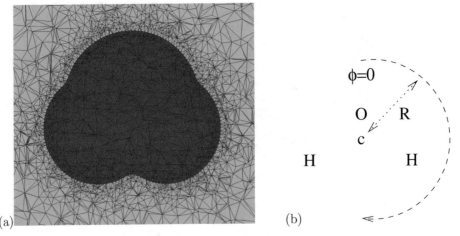

Fig. 15.2 (a) Mesh with 62407 vertices used for the domain depicted in Figure 15.1. (b) Coordinates for plots in subsequent figures.

15.3.3 A Model System

There are many small molecules that could be considered for comparing dielectric models, but the simplest may be the water molecule itself. First of all, it is small, and secondly, it is highly polar, so that it can be expected to generate an interesting electrostatic potential. But perhaps most important is the fact that the water molecule provides an ultimate test since an implicit solvent model for it should support predictions consistent with properties of bulk water.

Since a water molecule is easily solvated, the nonlocal correlation distance λ is set to be 3.5 Å. However, computations with larger correlation distances were similar, with even more dramatic differences between the nonlocal and local models. In [42], optimized values of λ were determined for ionic solvation. As usual, the dielectric constants are set to be 2 and 80 for the solute and solvent domains, respectively.

To model a water molecule, we take the coordinates of the oxygen and two hydrogens to be as follows: O (0.000 0.391 0.000), H1 (−0.758 −0.196 0.000), H2 (0.758 −0.196 0.000). Thus, they are all on the XY-plane, the geometric centroid is the origin $(0, 0, 0)$, the distance between oxygen and each hydrogen atoms is 0.9584Å, and the angle between two hydrogens is 104.45 degrees. The charges are -2 on the oxygen and $+1$ on each hydrogen. The domain where the dielectric discontinuity appears is depicted in Figure 15.1. The red and cyan domains in Figure 15.1 depict the solute (i.e., the central water molecule) and the solvent domain, respectively. The red dot at the center indicates the centroid of the water molecule.

In Figure 15.2(a), a typical mesh with around 140000 vertices is depicted. The actual calculations were done on a mesh with 237799 vertices.

Due to the difficulty of comprehending the potentials in three dimensions, we have chosen special coordinates, as shown in Figure 15.2(b). That is, we consider the potential on circles in the plane of the water molecule at different distances from the centroid of the triangle formed by the oxygen and two hydrogens of the water molecule.

15.3.4 Model Differences

In Figure 15.3, we compare the two potentials for the two models at distance of $R = 2.5$Å and $R = 3$Å from the water centroid, with $\lambda = 3.5$. In [448], computations for other values of λ are also presented. We see that the two models make predictions that differ by an order of magnitude. As expected, the nonlocal model does not screen the water charges nearly as much as the discontinuous model. Typical water distances would be near 3Å, so there would be essentially no dielectric matter within a sphere of radius $R = 2$Å from the water centroid. The only dielectric effect at this scale would be due to the polarization of the electron cloud, and this is the same effect that occurs within the solute domain. The nonlocal model captures this behavior since the effective dielectric constant has been reduced due to the high-frequency fluctuation of the electrostatic field.

At a distance of $R = 3$Å from the water centroid, the difference between the two models, as shown in Figure 15.3(b), is reduced to a smaller factor. Indeed, as we move farther and farther from the water centroid, the model differences become less acute. For example, at a distance of $R = 3.5$Å, the differences (not shown) between the two models differ by less than a factor of two. The fact that the two models tend to the same prediction at larger distances is a good consistency check for the numerical implementation of both models.

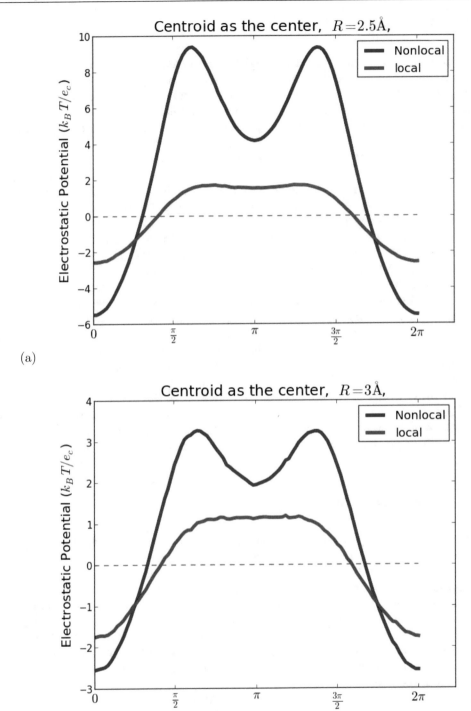

(a)

(b)

Fig. 15.3 Potentials for the nonlocal and discontinuous (local) dielectric models at a distance of (a) $R = 2.5$Å and (b) $R = 3$Å from the water centroid, with $\lambda = 3.5$.

15.3.5 Model Sensitivity

A major difference between the local (discontinuous) dielectric model and the nonlocal model has to do with its sensitivity to cavity definition. To study this, we have varied the cavity by shrinking it by 10%. Recall that the red and cyan domains in Figure 15.1 depict the solute (i.e., the central water molecule) and the solvent domain, respectively, for both original and shrunk cases. The thin layer colored in blue indicates the difference of the solute domains between the original and shrunk cases. Thus, it is treated as a part of the solute in the original case but the solvent in the shrunk case.

In Figure 15.4, we show the predictions for the two models for the original solute domain and for the perturbed solute domain, at a distance of $R = 1.8$Å from the water centroid. We see that the local (discontinuous) dielectric model gives a radically different prediction due to the perturbation, whereas the nonlocal model is not changed very much. The perturbations in the local (discontinuous) dielectric model are an order of magnitude larger than for the nonlocal model.

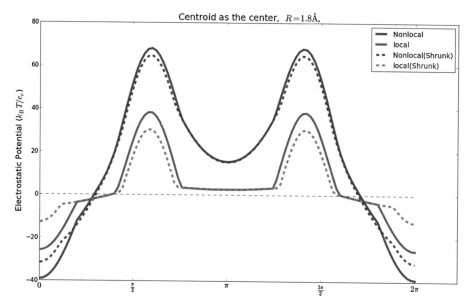

Fig. 15.4 The two potentials for the two models at a distance of $R = 1.8$Å from the water centroid, for both the standard domain for the water molecule and one shrunk by 10%. The horizontal axis is the angle ϕ defined in Figure 15.2(b). The vertical axis the electrostatic potential in units $k_B T/\epsilon_c$.

Another interesting feature of Figure 15.4 is the distinct discontinuity of the slope of the potential curve as the angle varies through the point corresponding to the cavity boundary. For the nonlocal model, a small discontinuity can still be observed, but it is much more dramatic for the local model.

The reason that the nonlocal model is less sensitive is that it naturally adjusts the effective dielectric constant when the frequency of the oscillation increases, as occurs near the solute. Thus, a much smaller effective perturbation occurs as a result of the change in the definition of the solute cavity.

On the other hand, at greater distances from the water centroid, these perturbations are less pronounced, as depicted in Figure 15.3 at distances of $R = 2.5$Å and $R = 3$Å from the water centroid. The perturbations in the local (discontinuous) dielectric model are now comparable with those in the nonlocal model. In fact, the differences are a bit larger for the nonlocal model, but the absolute predictions are also much larger, so the relative perturbation is smaller.

15.3.6 Conclusions About Continuum Dielectric Models

Recent algorithmic advances [531, 532] have made it feasible to compare nonlocal [43] and local [33, 39, 425] dielectric models. We see that the predictions of the two models can differ substantially near a solute, although the differences are diminished away from the solute. It should now be possible to see to what extent nonlocal solvation models yield different predictions in complex systems in which dielectric behavior plays a role, such as quantum mechanics models [491] and molecular dynamics simulations [33, 39, 425].

15.4 Computing the Energy

Estimating the electrostatic energy of a protein–water–ion system is not easy. At a quantum mechanical level of description, the energy is an eigenvalue that emerges as a global quantity of the system. In molecular mechanics, the partial charges of individual atom sites are approximated, accounting for polarization effects (most individual atoms would have net charge zero in their native state). In this setting, the electrostatic (enthalpic) energy is given by

$$\sum_{i \in I_\rho} \sum_{j \neq i \in I_\rho} q_i q_j |\mathbf{r}_j - \mathbf{r}_i|^{-1}, \qquad (15.42)$$

where the q_i's denote the partial charges assigned at locations $\mathbf{r}_i$. Note that care is taken in the double sum to insure that $j \neq i$ to avoid what is known as **self-energy**. There is an energetic contribution for an individual atom at a quantum mechanical level of description, but putting $i = j$ in the expression in (15.42) would give an infinite contribution. Likewise, the other terms for $j \neq i$ represent only approximations to the quantum mechanics assessment of the energy.

To approximate the quantum-level description better, the electron distributions could be represented as continuous distributions. Ultimately, even the positions of the atom nuclei are known only in a probabilistic sense. So we can imagine a fully continuous charge description of this type, but the scales of representation would be vastly different. The nuclei are more localized than the electrons by several orders of magnitude. Thus, it is problematic to use this approach.

Going in the other direction, the pattern of charges in the molecular mechanics description is often modeled as being continuous. This is useful computationally, but it has some pitfalls as we now discuss.

When the potential ϕ is continuous, the energy associated with the system modeled by (15.14) is given by the expression

$$\int \phi \rho \, dx = \sum_{i \in I_\rho} q_i \langle \delta_{\mathbf{r}_i}, \phi \rangle, \qquad (15.43)$$

where we have used distribution theory (the angle brackets) to allow interpretation of the point charges rigorously as delta functions:

$$\langle \delta_{\mathbf{r}}, v \rangle = v(\mathbf{r}) \tag{15.44}$$

for any test function v. Unfortunately, we know that, with any of the models considered (nonlocal or nonlinear), the potential ϕ will have a singularity like $|\mathbf{r} - \mathbf{r}_i|^{-1}$ for $\mathbf{r}$ near $\mathbf{r}_i$ when ρ involves true delta functions. Thus, the expression (15.43) includes a sum of infinities.

15.4.1 Analytical Approximations

We can imagine approximating the singularities by a number of techniques. For simplicity, let us assume we are in a vacuum so that the exact potential ϕ is given by (15.2). For example, we might replace the δ functions by an integral over a the surface of the sphere of radius η, suitably scaled. That is, we define

$$\langle \delta_{\eta,\mathbf{r}}, v \rangle = \frac{1}{4\pi\eta^2} \oint_{S_{\eta,\mathbf{r}}} v \, ds, \tag{15.45}$$

where $S_{\eta,\mathbf{r}}$ denotes the sphere of radius η around $\mathbf{r}$. Although this is still a singular distribution, in the sense that it cannot be represented by an integrable function on $\mathbb{R}^3$, the corresponding potential is now continuous. The potential is the same away from the singularity, but it is constant (equal to $1/\eta$) inside a ball of radius η:

$$\phi_{\eta,\mathbf{r}}(\mathbf{x}) = \begin{cases} |\mathbf{x} - \mathbf{r}|^{-1} & |\mathbf{x} - \mathbf{r}| \geq \eta \\ \eta^{-1} & |\mathbf{x} - \mathbf{r}| \leq \eta. \end{cases} \tag{15.46}$$

Now at least the pairing

$$\langle \delta_{\eta,\mathbf{r}}, \phi_{\eta,\mathbf{r}} \rangle = \eta^{-1} \tag{15.47}$$

is finite. This is often called the Born (electrostatic) approximation [224].

Making the systematic change of $\delta_{\eta,\mathbf{r}}$ for δ_{η} in (15.43) changes the singular terms to

$$\sum_{i \in I_\rho} q_i \langle \delta_{\eta,\mathbf{r}_i}, \phi_\eta \rangle = \sum_{i \in I_\rho} \sum_{j \neq i \in I_\rho} q_i^2 \eta^{-1} + \sum_{i \in I_\rho} \sum_{j \neq i \in I_\rho} \frac{q_i q_j}{4\pi\eta^2} \oint_{S_{\eta,\mathbf{r}_i}} |\mathbf{r}_j - \mathbf{r}|^{-1} \, ds, \tag{15.48}$$

where we recall that $S_{\eta,\mathbf{r}_i}$ denotes the sphere of radius η around $\mathbf{r}_i$, assuming that the minimum distance between $\mathbf{r}_i$ and $\mathbf{r}_j$ is bigger than η. The latter sum converges when $\eta \to 0$:

$$\lim_{\eta \to 0} \sum_{i \in I_\rho} \sum_{j \neq i \in I_\rho} \frac{q_i q_j}{4\pi\eta^2} \oint_{S_{\eta,\mathbf{r}_i}} |\mathbf{r}_j - \mathbf{r}|^{-1} \, ds = \sum_{i \in I_\rho} \sum_{j \neq i \in I_\rho} q_i q_j |\mathbf{r}_j - \mathbf{r}_i|^{-1}, \tag{15.49}$$

and this limiting expression is the correct electrostatic energy in this case. The other term

$$\eta^{-1} \sum_{i \in I_\rho} q_i^2 \tag{15.50}$$

is clearly a spurious self-energy term, and it diverges as $\eta \to 0$. Note that there is no chance of cancellation due to having charges of opposite signs. All the contributions in (15.50) are of one sign.

On the positive side, we see that the expression

$$\int \phi_\eta \rho_\eta \, dx - \eta^{-1} \sum_{i \in I_\rho} q_i^2 \tag{15.51}$$

tends to the right limit as $\eta \to 0$.

15.4.2 Smoothing the Delta Function

It might be a concern that the approximation of the Dirac delta function in Section 15.4.1 is still too singular. Another approach to interpret the energy is to approximate the Dirac delta functions by smooth functions. This has some physical basis, since the electrons in an atom are distributed over a region on the order of an Ångstrom in size, a scale we would actually resolve in these models. On the other hand, as discussed earlier, the nuclear charge is concentrated on a much smaller scale. Thus, let us imagine that we replace δ systematically by δ_η for some fixed $\eta > 0$. To be precise, we take

$$\delta_\eta(\mathbf{r}) = \eta^{-3} \chi(\eta^{-1} \mathbf{r}) \tag{15.52}$$

for all points $\mathbf{r}$, where χ is a smooth function that vanishes for $|\mathbf{r}| \geq 1$ and satisfies $\int \chi(\mathbf{r}) \, d\mathbf{r} = 1$. Then, for sufficiently smooth functions ψ, we have

$$\lim_{\eta \to 0} \int \psi(\mathbf{r}) \delta_\eta(\mathbf{r} - \mathbf{r}_i) \, d\mathbf{r} = \psi(\mathbf{r}_i). \tag{15.53}$$

Define ϕ_η to be the solution of (15.14) with $\rho = \rho_\eta$ defined by

$$\rho_\eta(\mathbf{r}) = \sum_{i \in I_\rho} q_i \delta_\eta(\mathbf{r} - \mathbf{r}_i). \tag{15.54}$$

Thus, the approximate energy expression

$$\int \phi_\eta \rho_\eta \, dx = \sum_{i \in I_\rho} q_i \int \phi_\eta(\mathbf{r}) \delta_\eta(\mathbf{r} - \mathbf{r}_i) \, d\mathbf{r} = \sum_{i \in I_\rho} q_i \int \phi_\eta(\mathbf{r} + \mathbf{r}_i) \delta_\eta(\mathbf{r}) \, d\mathbf{r} \tag{15.55}$$

appears to be a reasonable replacement for (15.43).

For any constant M,

$$\int \phi_\eta \rho_\eta \, dx - M \sum_{i \in I_\rho} q_i^2 = \sum_{i \in I_\rho} q_i \int \left(\phi_\eta(\mathbf{r} + \mathbf{r}_i) - M q_i \right) \delta_\eta(\mathbf{r}) \, d\mathbf{r}. \tag{15.56}$$

It is plausible that

$$\phi_\eta(\mathbf{r} + \mathbf{r}_i) \approx \frac{c q_i}{\eta} \quad \forall |\mathbf{r}| \leq \eta \tag{15.57}$$

for some constant c that would depend only on the choice of ψ. Choosing $M = c/\eta$, we conclude that

$$\int \phi_\eta \rho_\eta \, dx - \frac{c}{\eta} \sum_{i \in I_\rho} q_i^2 \qquad (15.58)$$

should remain bounded as $\eta \to 0$.

15.4.3 Numerical Computations

When equations like (15.14) are solved numerically, we obtain an approximation ϕ^h, where $h > 0$ is the spatial discretization parameter. In the case of finite element approximation [60], the expression

$$\int \phi^h \rho \, dx = \sum_{i \in I_\rho} q_i \left\langle \delta(\mathbf{r} - \mathbf{r}_i), \phi^h \right\rangle = \sum_{i \in I_\rho} q_i \, \phi^h(\mathbf{r}_i) \qquad (15.59)$$

is perfectly well defined [451]. However, as h tends to zero, the values $\phi^h(\mathbf{r}_i)$ will diverge [452]. In particular, we expect [452] that

$$\phi^h(\mathbf{r}) \approx \frac{c_i q_i}{h_i}, \quad \forall |\mathbf{r} - \mathbf{r}_i| \le h_i, \qquad (15.60)$$

where h_i is the local mesh size near $\mathbf{r}_i$ and c_i is a constant depending on the local mesh behavior and where the singularity falls within a mesh element. Thus,

$$\int \phi^h \rho \, dx = \sum_{i \in I_\rho} q_i \, \phi^h(\mathbf{r}_i) \approx \sum_{i \in I_\rho} \frac{c_i q_i^2}{h_i} \qquad (15.61)$$

which we recognize as some sort of self-energy term. We can deal with this self-energy term in a rigorous way and show how to remove it as follows.

The system (15.14) is linear, so we can write

$$\phi^h = \sum_{i \in I_\rho} q_i \, \phi_i^h, \qquad (15.62)$$

where ϕ_i^h is the discrete (finite difference or finite element) solution of (15.14) with $\rho = \delta(\mathbf{r} - \mathbf{r}_i)$. The self-energy term that we want to eliminate is precisely

$$\sum_{i \in I_\rho} \langle q_i \, \phi_i^h, q_i \delta(\mathbf{r} - \mathbf{r}_i) \rangle = \sum_{i \in I_\rho} q_i^2 \phi_i^h(\mathbf{r}_i). \qquad (15.63)$$

Thus, the desired energy is (15.59) minus (15.63):

$$E^h = \sum_{i \in I_\rho} q_i \, \phi^h(\mathbf{r}_i) - \sum_{i \in I_\rho} q_i^2 \phi_i^h(\mathbf{r}_i) = \sum_{i \in I_\rho} \sum_{j \in I_\rho} q_i \, q_j \phi_j^h(\mathbf{r}_i) - \sum_{i \in I_\rho} q_i^2 \phi_i^h(\mathbf{r}_i)$$

$$= \sum_{i \in I_\rho} \sum_{i \neq j \in I_\rho} q_i \, q_j \phi_j^h(\mathbf{r}_i). \qquad (15.64)$$

Computation of (15.64) has exactly the right terms, avoiding the self-interaction terms. However, (15.64) appears to require the solution of lots of separate problems, so in general this would be very expensive. But, in one common case, computing energy differences, it simplifies.

15.4.4 Computing Energy Differences

Suppose that we want to compare the energy difference $E_1 - E_2$ for two different systems of charges indexed by I_1 and I_2, respectively. That is, we have fixed charge distributions ρ_i given by

$$\rho_k = \sum_{i \in I_k} q_i \delta(\mathbf{r} - \mathbf{r}_i) \tag{15.65}$$

for $k = 1, 2$. Let us assume for simplicity that $I^1 = I^2 \cup \Delta I$ for some set of indices ΔI. In particular, computing pKa's (Section 4.5) often involves sets that differ only in one element.

Let ϕ^k denote the solution of the system (15.14) with $\rho = \rho_k$ for $k = 1, 2$. Here, we have temporarily dropped the mesh parameter h since there are too many indices. Then, using the energy definition (15.64), we have

$$\Delta E = E^1 - E^2 = \sum_{i \in I_1} q_i\, \phi^1(\mathbf{r}_i) - \sum_{i \in I_1} q_i^2\, \phi_i^h(\mathbf{r}_i) - \sum_{i \in I_2} q_i\, \phi^2(\mathbf{r}_i) + \sum_{i \in I_2} q_i^2\, \phi_i^h(\mathbf{r}_i)$$

$$= \sum_{i \in I_1} q_i\, \phi^1(\mathbf{r}_i) - \sum_{i \in I_2} q_i\, \phi^2(\mathbf{r}_i) - \sum_{i \in \Delta I} q_i^2\, \phi_i^h(\mathbf{r}_i) \tag{15.66}$$

$$= \sum_{i \in \Delta I} q_i\, \phi^1(\mathbf{r}_i) + \sum_{i \in I_2} q_i \left(\phi^1(\mathbf{r}_i) - \phi^2(\mathbf{r}_i) \right) - \sum_{i \in \Delta I} q_i^2\, \phi_i^h(\mathbf{r}_i).$$

We can write

$$\Delta \rho = \rho_1 - \rho_2 = \sum_{i \in \Delta I} q_i\, \delta(\mathbf{r} - \mathbf{r}_i), \tag{15.67}$$

and by linearity of the system (15.14), we can think of $\Delta \phi = \phi^1 - \phi^2$ as the solution of (15.14) with $\rho = \Delta \rho$:

$$\Delta \phi = \phi^1 - \phi^2 = \sum_{j \in \Delta I} q_j\, \phi_j^h. \tag{15.68}$$

Thus, the energy difference can be expressed as

$$\Delta E = \sum_{i \in \Delta I} q_i\, \phi^1(\mathbf{r}_i) + \sum_{i \in I_2} q_i \sum_{j \in \Delta I} q_j\, \phi_j^h(\mathbf{r}_i) - \sum_{i \in \Delta I} q_i^2\, \phi_i^h(\mathbf{r}_i). \tag{15.69}$$

Therefore, we are required only to compute ϕ^1 and ϕ_i^h for $i \in \Delta I$.

15.4.5 Thermodynamic Smoothing

Another physical interpretation of smoothing the delta function is available. One drawback of the η-smoothing in Section 15.4.2 is that it is motivated mainly by the distribution of the electrons. The proton nucleus would still look like a point charge in this picture. However, it is useful to ask what this model is supposed to be representing. Unless we are at absolute zero, all of the atoms are moving due to thermal agitation and thus so are the nuclei as well. Thus, the uncertainty in position of the nuclei in thermal motion is far greater than it is due only to quantum effects.

PDB files list **temperature factors** for each atom, also known as **B-factors**, which indicate the observed thermal motion (or maybe just the blurriness) of each atom. These smoothed distributions of charge could potentially be used to provide a more physically appropriate smoothing of the charge delta functions. The dielectric model is after all an ensemble average over time, and if the temperature factors accurately represent the mobility of atoms, then this could be factored into the model in a potentially useful way.

The B-factors are different for each atom, and thus, we would have an approximation that varied from location to location. However, now the charge would be more legitimately distributed (and continuous), so the need for correction related to self-charge terms would be eliminated. On the other hand, the assessment of the electrostatic contribution to enthalpy would now be potentially temperature dependent.

15.4.6 Potential Splitting

In [532], a splitting for the potential is introduced, of the form

$$\phi(\mathbf{r}) = \Phi(\mathbf{r}) + V_0(\mathbf{r}) = \Phi(\mathbf{r}) + \frac{1}{4\pi\epsilon_0} \sum_{i\in I_\rho} \frac{q_i}{|\mathbf{r} - \mathbf{r}_i|}, \tag{15.70}$$

where the q_i's denote fixed charges assigned at locations $\mathbf{r}_i$ and Φ is a smooth function associated with other charges, e.g., due to the solvent and representing the dielectric effect. Let us denote

$$\rho_0 = \sum_{i\in I_\rho} q_i \delta_{\mathbf{r}_i} = \epsilon_0 \Delta V_0, \tag{15.71}$$

representing the fixed charges. Then, $\Delta\phi = \epsilon_0^{-1}\rho_0 + \Delta\Phi$, and we can compute the energy as

$$\frac{1}{4\pi\epsilon_0} \sum_{i\in I_\rho} \sum_{i\neq j\in I_\rho} \frac{q_i q_j}{|\mathbf{r}_j - \mathbf{r}_i|} + \frac{1}{\epsilon_0}\langle\rho_0, \Phi\rangle + \langle\Delta\Phi, V_0\rangle + \langle\Delta\Phi, \Phi\rangle$$

$$= \sum_{i\in I_\rho} \sum_{i\neq j\in I_\rho} \frac{q_i q_j}{|\mathbf{r}_j - \mathbf{r}_i|} + \frac{2}{\epsilon_0}\langle\rho_0, \Phi\rangle - \langle\nabla\Phi, \nabla\Phi\rangle, \tag{15.72}$$

since $\langle\Delta\Phi, V_0\rangle = \langle\Phi, \Delta V_0\rangle = \epsilon_0^{-1}\langle\Phi, \rho_0\rangle$.

15.5 Nonlinear Dielectric Models

Large electric gradients near the surface could cause saturation of the dielectric effect, leading to nonlinear models of the dielectric coefficient [24, 25, 199, 252, 276, 281, 430, 431, 480].

Nonlinear models arise because the polarization field $\nabla\phi_\gamma$ in the relationship (15.8) cannot continue to increase indefinitely for $\nabla\phi$ arbitrarily large. The field $\nabla\phi_\gamma$ is due to the fortuitous orientation of the solvent charge groups (e.g., water molecules), but once perfect alignment is achieved, no more improvement can result. (There is an additional effect due to the polarization of the electron distribution of the individual solvent charge groups, but this also will saturate with sufficiently large field strength.) We can write this saturation property mathematically by taking a limit in (15.8), viz.

$$\lim_{|\nabla\phi|\to\infty} (1 - \varepsilon)\nabla\phi = \lim_{|\nabla\phi|\to\infty} \nabla\phi_\gamma = C, \tag{15.73}$$

for some constant C depending only on the direction of $\nabla\phi$. Let us write ε as a function of $\nabla\phi$ by introducing a vector $\xi = \nabla\phi$. Then, (15.73) becomes

$$\lim_{|\xi|\to\infty} (1 - \varepsilon(\xi))\xi = C. \tag{15.74}$$

This constant can potentially depend on the direction of approach to infinity for a simple lattice, but in general, it will be isotropic. The value of C can be estimated by considering the case of perfect alignment of water molecules. The dipole of water is 1.85 Debye, and the (maximum) density of water corresponds to about one water molecule per 30 Ångstroms cubed (Å^3), corresponding to a box of side 3.1 Ångstroms. A more intuitive measure of dipole strength is q_e–Å, where q_e is the charge of an electron. Thus, the dipole of water is 0.386 q_e–Å. Thus, the maximum strength of the water dipole in bulk is about $0.04 q_e$–Å^{-2}.

One simple model that satisfies (15.74) is

$$\varepsilon(\xi) = \varepsilon_0 + \frac{\varepsilon_1}{1 + \lambda|\xi|} \tag{15.75}$$

for some constants ε_0, ε_1, and λ. This model (15.75) is easy to analyze [445], but we do not intend to suggest it as a serious model. Rather, it should be viewed as a prototype nonlinear dielectric model, one that can be easily analyzed and implemented in existing codes for the purposes of software verification.

The Langevin–Debye model [110, 430, 431] defines

$$\varepsilon(\xi) = c_0 + c_1 \frac{L(c_2|\xi|)}{|\xi|}, \tag{15.76}$$

where c_0, c_1, and c_2 are physical constants and L is the Langevin function

$$L(x) = \coth(x) - \frac{1}{x} = \frac{e^x + e^{-x}}{e^x - e^{-x}} - \frac{1}{x} \approx \tfrac{1}{3}x - \tfrac{1}{45}x^3 + \cdots \tag{15.77}$$

Other nonlinear coefficient models are reviewed in [252].

Both the nonlocal and nonlinear models of the dielectric response have the effect of representing frequency dependence of the dielectric effect. The gradient $|\nabla\phi(x)|$ provides a proxy for frequency content, although it will not reflect accurately high-frequency, low-power electric fields. Thus, a combination of nonlocal and nonlinear models for the dielectric response might need to be used in general to capture the full effect.

15.6 Exercises

Exercise 15.1. Plot the Langevin function (15.77) on the interval $[0, 30]$. Compare the expression (15.76) with the function (15.75). More precisely, compare $3L(x)/x$ with $1/(1+x)$ on various intervals.

Exercise 15.2. Consider the Langevin function L defined in (15.77), and define

$$M(x) = \frac{1}{3\sqrt{1 + (x/3)^2}}. \tag{15.78}$$

Prove that the ratio $r(x) = (L(x)/x)/M(x)$ satisfies $0.93171 \leq r(x) \leq 1$ for $x \geq 0$.

Exercise 15.3. Assume that the $\mathbf{r}_i$ and q_i for $i \in I_\gamma$ can be enumerated as $i = (j, k)$, where j is the index for the group and k is the index within each group, with $\mathbf{r}_i := \mathbf{r}_j - \mathbf{r}_{jk}$ and $q_i := q_{jk}$ where the jth group of charges q_{jk} sums to zero for all j:

$$\sum_k q_{jk} = 0. \tag{15.79}$$

Show that the expression for the charge density can be simplified as

$$\gamma(\mathbf{r}) = \sum_j \sum_k q_{jk} \delta(\mathbf{r} - \mathbf{r}_j - \mathbf{r}_{jk}) \tag{15.80}$$

Exercise 15.4. The expression δ in (15.1) can be interpreted in several ways. As a first abstraction, we can take it to be the Dirac delta function, which provides a rigorous model of a point charge [443]. In [230], a mollification of the Dirac delta function is introduced, which makes it possible to reason classically about expressions involving δ. This is a very useful device, and it can also be given a physical interpretation. We can think of δ representing the actual charge cloud that would be seen at a quantum scale. With this interpretation, there is an assumption being made, namely that the local charge distribution can be represented by a single function $\delta(\mathbf{r})$, independent of the charge q and independent of the atom in question. This is of course not exact, but it gives a physical interpretation to the mollifier used in [230]. A closer approximation might be obtained by letting q be fractional, with positions $\mathbf{r}_{jk}$ chosen to improve the representation [330].

Exercise 15.5. Let us suppose that the charge groups are homogeneous in the sense that

$$\mathbf{r}_{jk} = \mathcal{R}(\theta_j)\rho_k \tag{15.81}$$

for fixed vectors ρ_k and for some angle $\theta_j \in S_2$ (where S_2 denotes the unit 2-sphere), and further that $q_{jk} = q_k$ independent of j. This would be the case for water, for example [330]. Then,

$$\sum_k q_{jk} \delta(\mathbf{r} - \mathbf{r}_j - \mathbf{r}_{jk}) = \sum_k q_k \delta(\mathbf{r} - \mathbf{r}_j - \mathcal{R}(\theta_j)\rho_k) \tag{15.82}$$

$$= \mathcal{F}(\theta_j, \mathbf{r} - \mathbf{r}_j)$$

where $\mathcal{F}$ is defined by

$$\mathcal{F}(\theta, \mathbf{r}) = \sum_k q_k \delta(\mathbf{r} - \mathcal{R}(\theta)\rho_k). \tag{15.83}$$

Now suppose that δ is rotationally invariant. Then,

$$\mathcal{F}(\theta, \mathcal{R}(\theta)\mathbf{r}) = \sum_k q_k \delta(\mathcal{R}(\theta)\mathbf{r} - \mathcal{R}(\theta)\rho_k)$$

$$= \sum_k q_k \delta(\mathbf{r} - \rho_k) \tag{15.84}$$

$$= \nabla \cdot \mathcal{W}(\mathbf{r}),$$

with $\mathcal{W}(\mathbf{r}) = \nabla \psi(\mathbf{r})$ where ψ solves a Poisson equation of the form

$$\Delta \psi = \sum_k q_k \delta(\mathbf{r} - \rho_k). \tag{15.85}$$

If δ is the Dirac δ-function, then $\mathcal{W}$ is a generalized multipole expression

$$\mathcal{W}(\mathbf{r}) = -\sum_k q_k \frac{\mathbf{r} - \rho_k}{|\mathbf{r} - \rho_k|^3}. \tag{15.86}$$

Exercise 15.6. Then,

$$\begin{aligned}
\sum_k q_{jk} \delta(\mathbf{r} - \mathbf{r}_j - \mathbf{r}_{jk}) &= \mathcal{F}(\theta_j, \mathbf{r} - \mathbf{r}_j) \\
&= \nabla \cdot \mathcal{W}(\mathcal{R}(\theta_j)^t (\mathbf{r} - \mathbf{r}_j)).
\end{aligned} \tag{15.87}$$

Therefore, we have an exact representation of the dielectric field γ defined in (15.80), viz.,

$$\gamma(\mathbf{r}) = \sum_j \nabla \cdot \mathcal{W}(\mathcal{R}(\theta_j)^t (\mathbf{r} - \mathbf{r}_j)). \tag{15.88}$$

Exercise 15.7. Suppose that $\nabla \cdot (\epsilon \mathbf{e}) = \rho$ where ρ is a continuous function in a neighborhood of $\partial \Omega$ and ϵ is a piecewise constant function with values ϵ_0 inside Ω and ϵ_1 outside Ω. Let e_1 be the limit of $\mathbf{n} \cdot \mathbf{e}$ from the outside of Ω, and let e_0 be the limit of $\mathbf{n} \cdot \mathbf{e}$ from the inside of Ω, where $\mathbf{n}$ is the unit normal to $\partial \Omega$. Prove that $\epsilon_0 e_0 = \epsilon_1 e_1$ on $\partial \Omega$. Show that this does not depend on whether $\mathbf{n}$ is the inner or outer normal.

Exercise 15.8. An alternate approach to understanding the non-Debye dielectric is the following. Recast the Debye ansatz as solving (see (15.7))

$$\mathbf{d}(x) = \varepsilon(x) \mathbf{e}(x)$$

for some symmetric matrix-valued function ε. That is, for each x, we are given $\mathbf{d}$ and $\mathbf{e}$, and we want to find a matrix ε such that $\mathbf{d} = \varepsilon \mathbf{e}$. Prove that this is possible provided $\mathbf{e} \neq 0$. (Hint: write $\varepsilon = U^T \Lambda U$, and this becomes $D = \Lambda E$ where $D = U \mathbf{d}$, $E = U \mathbf{e}$. Choose a rotation matrix U so that $E_i > 0$ for all i. Then, $\Lambda_{ii} = D_i / E_i$.)

Exercise 15.9. Consider the problem in Exercise 15.8. Show that the symmetric matrix ε can be chosen so that

$$\|\varepsilon\| = \sqrt{3} \frac{\|D\|}{\|\mathbf{e}\|} = \sqrt{3} \frac{\|\mathbf{d}\|}{\|\mathbf{e}\|}.$$

(Hint: For example, choose U in the Hint for Exercise 15.8 so that $E = (\|\mathbf{e}\|/\sqrt{3})(1, 1, 1)$. Thus, $\Lambda_{ii} = (\sqrt{3}/\|\mathbf{e}\|) D_i$. Write $\lambda = (\Lambda_{11}, \Lambda_{22}, \Lambda_{33})$.)

Chapter 16
Wrapping Technology

Harold Abraham Scheraga was born in 1921 and is a major figure in structural biology who has contributed decisively to the study of the hydrogen bond and the hydrophobic effect. He was also an early contributor to the concept of wrapping.

The dehydron provides a marker for drug–target interaction, and it has guided the redesign of drugs to improve specificity and affinity. Wrapping technology analyzes interface dehydrons in a target–ligand complex to predict their change in strength and stability induced by changes in nonpolar microenvironment upon target–ligand binding. Differences in target dehydron patterns are exploited to redesign drugs to control selectivity and improve affinity.

16.1 Rational Drug Discovery

Rational drug discovery is rapidly evolving [13, 234, 274, 365, 502], and wrapping technology is one technique that has been used to redesign existing drugs to improve selectivity and affinity [148, 168, 170]. It is based on the observation that the strength and stability of dehydrons can be modulated by an external agent of appropriate design. More precisely, dehydrons can be further desolvated by the attachment of a drug that contributes to dehydration, as depicted in Figure 16.1. Through retrospective studies [169], it appears that dehydrons have played a significant role in drug function due to their prominence in protein–ligand interactions [116, 326]. Wrapping technology exploits differences in wrapping among proteins, especially those with similar structures, to enhance selectivity and affinity of drug ligands.

Inhibitor design is often guided by structural descriptors of protein binding sites, such as accessibility [94, 302], curvature [37], and hydrophobicity [83, 363]. However, out of 814 protein-inhibitor PDB complexes studied in [148], at least 488 of them had binding cavity hydrophobicity not significantly higher than the rest of the surface [148]. While intermolecular electrostatic bonds are obvious candidates to promote protein–ligand association and specificity [535], we have seen that dehydrons often play an equally significant role, as is the case for antibody binding (Chapter 13).

Specificity of association is an essential aspect of biological proteins and a critical goal of drug discovery. But many drug targets, such as protein kinases [95], share similar properties which foster promiscuity unless differentiating features are targeted. For example, dehydrons are often not conserved in families of paralog proteins [162]. By contrast, it appears that

© Springer International Publishing AG 2017
L.R. Scott and A. Fernández, *A Mathematical Approach to Protein Biophysics*,
Biological and Medical Physics, Biomedical Engineering,
DOI 10.1007/978-3-319-66032-5_16

surface nonpolar moieties are a highly conserved feature of protein interfaces [326]. Thus, side effects resulting from off-target–ligand binding may be minimized by selectively targeting dehydrons.

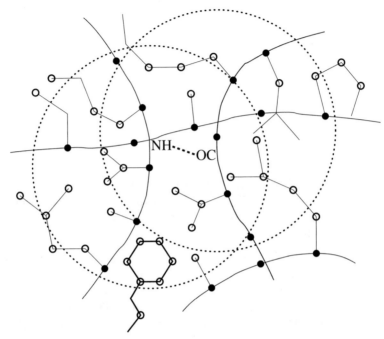

Fig. 16.1 Depiction of a drug enhancing the wrapping in the desolvation domain of a hydrogen bond, cf. Figure 8.2. The desolvation domain is depicted as the union of two spheres. The solid black dots represent nonpolar carbonaceous groups on the sidechains indicated by broken lines. The drug ligand provides an additional nonpolar carbonaceous group (big black hexagon with a tail) in the desolvation domain.

donor	acceptor	dNO	dHO	$\angle_{DHA}$	$\angle_{DAB}$	$\angle_{HAB}$	tw	Aw	Bw	W	N	α
ARG A 87	ALA A 28	2.75	1.83	152.05	169.96	162.04	19	15	4	3	4	33
ARG B 87	ALA B 28	2.76	1.80	160.92	167.76	166.29	17	4	13	3	11	38
ASN A 88	ASP A 29	3.43	2.45	166.16	151.77	150.02	13	13	0	6	1	9
ASN B 88	ASP B 29	3.30	2.42	147.25	148.48	142.12	14	1	13	5	7	18
GLY A 49	GLY A 52	3.03	2.13	148.84	124.10	117.21	14	10	4	4	5	35
GLY B 49	GLY B 52	2.90	2.00	148.39	164.25	174.50	16	5	11	5	12	31
GLY A 51	ILE B 50	2.93	2.08	141.22	171.07	162.14	18	9	9	3	5	15
GLY A 52	ILE B 50	3.40	2.50	150.39	130.99	123.38	15	11	4	4	5	15
PHE B 99	PRO A 1	3.12	2.20	152.57	164.21	158.65	15	7	8	3	0	0

Table 16.1 Dehydrons in the PDB file 2BPX which depicts the HIV-1 protease complexed with Indinavir (designated MK1 in the PDB file). Key: dXO denotes the distance from X(=N or H) to O in the hydrogen bond, tw is the total number of wrappers, xw is the number of wrappers from chain x (=A or B), W is the number of waters in the desolvation shell (radius 6.5 Å), N is the number of nonpolar carbons in MK1 in the desolvation shell (radius 6.5 Å), α is the number of all atoms in MK1 in the desolvation shell. The angles are as described in Section 6.4.

Over three-fourths of the protein-inhibitor PDB complexes studied in [148] have nonpolar groups in the inhibitors within the desolvation domain of protein dehydrons. That is, inhibitors are typically dehydron wrappers. This situation is illustrated in Figure 16.2 which shows the inhibitor Indinavir (Crixivan) bound to the functionally dimeric HIV-1 protease

(PDB file 2BPX) (cf. [148, Figure 1]). Six intramolecular dehydrons in the protease, the backbone hydrogen bonds Ala28-Arg87, Asp29-Asn88, and Gly49-Gly52 in both chains A and B (marked as green lines in [148, Figure 1] and depicted in Figure 16.2 by showing the indicated backbone atoms as spheres), frame the cavity associated with substrate binding. These dehydrons help to explain the mechanism of substrate binding and clarify the gating mechanism of the flap [148].

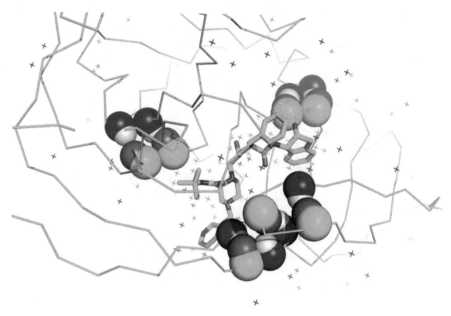

Fig. 16.2 HIV-1 protease in complex with inhibitor Indinavir that acts as dehydron wrapper, cf. Table 16.1. The protease is shown as lines between C_α atoms, Indinavir is depicted as sticks, and the spheres denote the dehydron constituents (donor and acceptor atoms).

The dehydrons shown in Figure 16.2 appear in three clusters at distributed sites which form a distinctive pattern around the nonpolar regions of the ligand. Clusters of dehydrons are common, since decreased wrapping around one hydrogen bond is likely to contribute to decreased wrapping nearby. Indinavir contributes several desolvating groups (Table 16.1) to wrap the dehydrons in the enzymatic cavity. The association of Indinavir deactivates the dehydrons that promote water removal and charge de-screening required to facilitate the enzymatic nucleophilic attack [169]. Additional intermolecular dehydrons in the protease are also significantly desolvated by nonpolar groups in Indinavir, contributing to its binding affinity. In total, eight dehydrons are identified in Table 16.1 that benefit from the water removal induced by Indinavir binding.

We see in Table 16.1 the extensive intrusion of Indinavir (designated MK1 in the PDB file) in the desolvation spheres around the most under-wrapped hydrogen bonds in the PDB file 2BPX. Shown in Table 16.1 are all hydrogen bonds whose total number of wrappers is less than 20 using a desolvation shell radius of 6.5 Å around the CA carbons. Also indicated in Table 16.1 is the breakdown of the wrappers by chain; the maximum number of wrappers for the intramolecular hydrogen bonds from one chain is 15, so these are extremely under-wrapped before the dimer is formed. The intermolecular hydrogen bonds between the two chains draw wrappers from the two chains even more strongly, and two of these have significant contributions to wrapping from MK1. Interestingly, there are no other hydrogen bonds in

2BPX with more than eight atoms from MK1 in the corresponding desolvation domain, or more than four nonpolar carbons from MK1.

In addition to the extensive dehydron network involved in the binding of Indinavir to the HIV-1 protease, the PDB file 2BPX reveals conventional intermolecular polar interactions between Indinavir and the HIV-1 protease. In Table 16.2, all of the polar atoms (nitrogens and oxygens) in MK1 are listed, together with the hydrogen bonds they make. Of the nine polar atoms, only four make hydrogen bonds with the HIV-1 protease; two with mainchain acceptors and two with sidechains. Three others are hydrogen bonded to waters, and the remaining two do not appear to be making strong polar bonds. Thus, the inventory of enthalpic contributions by intermolecular polar interactions between Indinavir and the HIV-1 protease is smaller than the enthalpic contribution of the dehydron network depicted in Table 16.1.

bond	MK1 atom	distance	heavy atom
none	N1		
water	N2(H)	3.06	HOH B 332
water?	N3	3.59	HOH A 308
MCHB	N4(H)	3.02	O GLY B 27
SCHB	N5	3.63	NH1 ARG B 8
water	O1	2.90	HOH A 308
SCHB	O2(H)	2.71	OD2 ASP A 25
water	O3	2.97	HOH A 308
MCHB	O4(H)	2.90	N ASP B 29

Table 16.2 Polar contacts between Indinavir (MK1) and the HIV-1 protease in PDB file 2BPX. Key: "bond" indicates the type of hydrogen bond (HB) made (MC is mainchain, SC is sidechain). The last column gives the heavy atom participating in the hydrogen bond. The interaction between N3 and HOH A 308 is likely less significant than the ones between HOH A 308 and O1 and O2, which are much closer.

16.2 Reviewing Imatinib

The cancer drug Imatinib (a.k.a., Gleevec, STI-571) was the first cancer drug used to target specific proteins involved in the cancer process, tyrosine kinases, and the success of Imatinib has had a substantial impact on drug design [450]. Janet Rowley, Brian Druker, and Nicholas Lydon received the 2012 Japan Prize "for their contribution to the development of a new therapeutic drug targeting cancer-specific molecules" according to the prize announcement, the drug in question being Imatinib.

Imatinib, shown in Figure 16.3, was originally developed to target the Bcr-Abl protein in the treatment of chronic myeloid leukemia. However, its success led to attempts to use it to target other tyrosine kinases. The C-KIT kinase is a therapeutic target for treating gastrointestinal stromal tumors (GIST), and it is inhibited by Imatinib at the nanomolar level.

The PDB file 1T46 details the binding of Imatinib to the tyrosine kinase c-Kit. As listed in Table 16.3, Imatinib makes only four polar contacts with c-Kit in 1T46, and two of the other polar sites on Imatinib make apparent hydrogen bonds with water. Of the former, the potential hydrogen bond between N21 and OE1 GLU 640 does not have an ideal geometry, and GLU 640 makes a salt bridge with LYS 623. Similarly, the geometry of the hydrogen bond between O29 and the mainchain donor on Asp 810 is not ideal. The sites N8 and N51 do not make obvious polar contacts with either.

Curiously, Imatinib has minimal interactions with dehydrons in c-Kit. There are 16 hydrogen bonds in 1T46 wrapped by fewer than 21 nonpolar carbons within dehydration domains consisting of spheres of radius 6.5 Å around the two CA atoms for the donor and acceptor. None of these desolvation domains contain any atoms from Imatinib (MK1 in 1T46). Thus, unlike the impact of Indinavir on the dehydrons in the HIV-1 protease, Imatinib appears to bind without wrapping its target in an obvious way, if at all.

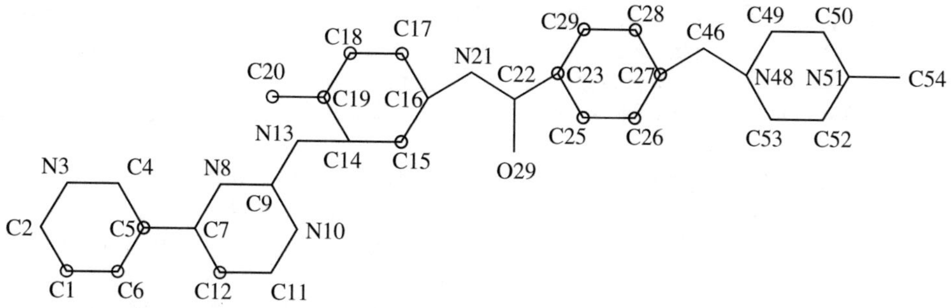

Fig. 16.3 PDB atom names for Imatinib (STI in PDB file 1T46).

Unfortunately, there are several deficiencies of Imatinib, and some attempts have been made to improve the ligand. We review two of these, one which responds to a type of Imatinib resistance, and the other minimizes side effects.

bond	MK1 atom	distance	heavy atom
MCHB	N3	2.85	N CYS 673
SCHB	N13(H)	2.95	OG1 THR 670
SCHB	N21(H)	2.86	OE1 GLU 640
MCHB	O29	3.08	N ASP 810
water	N10	2.82	O HOH 1105
water	N48	3.17	O HOH 1108

Table 16.3 Polar contacts between Imatinib (STI) and c-Kit in PDB file 1T46. Key: bond indicates the type of hydrogen bond (HB) made (MC is mainchain, SC is sidechain). The last column gives the heavy atom participating in the hydrogen bond. The hydrogen bond between N3 and HOH A 308 is unlikely due since it competes with the more likely partners O1 and O2.

16.2.1 Imatinib Resistance

In [170], the authors identified the possible dehydron formed by Phe 811 and Ala 814 in 1T46 as a possible target for a modification to Imatinib. The dehydration domain consisting of the spheres of radius 6.5 Å around the two CA atoms of Phe 811 (acceptor) and Ala 814 (donor) contains 21 nonpolar carbons, 10 water molecules, and 9 atoms from Imatinib (code STI in 1T46). Of the latter, the polar carbon C11, which is covalently bonded to N10 in Imatinib, was identified for enhancement by methylation, which would further enhance wrapping of this bond.

The Imatinib atoms in the desolvation domain of Ala 814–Phe 811 are all within the 6.5 Å sphere around Phe 811 CA. The distance from Phe 811 CA to C11 is 4.92 Å, and the closest nonpolar carbon is C12 (6.08 Å). If we had chosen the desolvation radius to be 6 Å, then there would have been no nonpolar Imatinib atoms in the desolvation domain. The carbon C15 (distance 6.42 Å) is also nonpolar, but the carbon C22 (distance 6.40 Å) is polar. The polar atom O29 (6.15 Å) is also nearby, and the distance from Phe 811 CA to N10 is 5.22 Å.

Moreover N10 is 2.82 Å from the water HOH 1105, indicating a hydrogen bond, and this water appears to be making a polar bond with the end of the sidechain Lys 623. This water is also only 3.45 Å from Phe 811 CA.

Thus, the Imatinib nonpolar atoms are peripheral to the desolvation domain of Ala 814–Phe 811, and the methylation at C11, the closest heavy atom to Phe 811 CA, could thus be expected to enhance this hydrogen bond. According to MolProbity, the distance from Phe 811 CA to the hydrogen attached to C11 is 3.97 Å, so the location of the new methyl group would be in a very favorable position for desolvation. Thus, the addition of a methyl group at C11 could be anticipated to add a new type of binding mechanism (wrapping) of Imatinib to c-Kit.

The authors of [170] tested a modification, called WBZ_7, to Imatinib created by methylation at the C11 position. WBZ_7 was tested and shown to have a slightly higher affinity than Imatinib for native c-Kit. Thus, the strategy of adding a wrapping mechanism to Imatinib was successful, despite the danger that enlarging the ligand might cause a steric hindrance that would reduce affinity. In fact, affinity was improved, and since a novel mechanism of binding was added, it seems plausible that the binding of WBZ_7 would be more robust to perturbations in the target, c-Kit.

It was known that just such a modification to c-Kit, the mutation D816V (Asp 816 is converted to Val 816), confers resistance to Imatinib. The affinity of Imatinib to the mutant D816V is reduced by three orders of magnitude compared with the affinity to the wild-type c-Kit kinase. Fortunately, WBZ_7 was found to bind strongly to the D816 mutant of c-Kit. It is interesting to speculate why this might occur.

We know that WBZ_7 was designed to bind more robustly to c-Kit, by adding a binding mode based on desolvating a dehydron, a binding mode missing from Imatinib binding to c-Kit. If a mutation in a target causes a significant structural change in the target, this may move polar contacts enough to cause a significant decrease in the binding enthalpy between the drug and target. The mutation D816V of c-Kit has a significant impact on wrapping of the hydrogen bond formed by Phe 811 and Ala 814, as well as others. The residue Asp 816 (that is, its CB atom) is found in the desolvation domain of several hydrogen bonds in 1T46, as indicated in Table 16.4. Thus, the mutation to Val 816 will likely have a significant effect on the strength and stability of some of the hydrogen bonds in mutant c-Kit, possibly leading to substantial structure changes. There are no Imatinib atoms in the desolvation domains of any of these hydrogen bonds.

donor	acceptor	d_{NO}	d_{HO}	$\angle DHA$	$\angle DAB$	$\angle HAB$	tw	ω
GLY 598	GLY 601	2.72	1.74	166.2	164.9	159.9	26	5
ARG 815	GLY 812	3.25	2.28	163.1	126.6	129.6	17	16
ASN 819	ASP 816	3.09	2.23	142.5	124.4	117.0	11	6
TYR 823	ASP 820	2.92	1.95	164.8	125.3	120.2	20	5
ALA 814	PHE 811	2.81	1.84	163.5	132.9	134.3	21	10

Table 16.4 Hydrogen bonds in the PDB file 1T46 which contain the CB atom from Asp 816 in their desolvation domains consisting of spheres of radius 6.5 Å around the donor and acceptor CA atoms. The PDB file 1T46 depicts the tyrosine kinase c-Kit complexed with Imatinib (designated STI in the PDB file). Key: d_{XO} denotes the distance from X(=N or H) to O in the hydrogen bond, tw is the total number of wrappers in a desolvation domain, and ω is the number of waters in the desolvation domain. None of these desolvation domains contain any of the Imatinib (STI) atoms. For comparison, the same data are given for the Ala 814–Phe 811 hydrogen bond, whose desolvation domain contains 9 STI atoms, of which 2 are nonpolar. The angles are as described in Section 6.4.

The mutation D816V also could have significant effects on interactions in addition to the mainchain hydrogen bonds listed in Table 16.4. The oxygen OD1 in ASP 816 in 1T46 is 3.23 Å from the water HOH 1060, indicating a likely hydrogen bond. The mutation D816V would likely remove this water from the environment. The oxygen OD2 in ASP 816 is 2.96 Å from ND2 in ASN A 819 (and according to MolProbity, 1.99 Å from HD22 in ASN 819), indicating a likely sidechain hydrogen bond. This bond could not be supported by Val 816, further contributing to possible structural change.

Thus it is reasonable to assume that the mutation D816V changes structural features of c-Kit that make binding by Imatinib harder. Since Imatinib appears to bind solely due to polar contacts, small structural changes can easily disrupt them. By contrast, wrapping a dehydron is much more robust to small structural changes, since the desolvation effects of nonpolar groups are less directional than polar contacts. The binding of WBZ_7 to the D816V mutant may continue to be driven in part by the F811–A814 dehydron.

16.2.2 Avoiding Side Effects

Another modification to Imatinib, called WBZ_4, was studied [171] in an attempt to retain binding to c-Kit but avoid binding to other tyrosine kinases. WBZ_4 is derived from Imatinib by methylation at the C2 position in STI (Figure 16.3). The carbon atom C2 in STI is found in a 6.5 Å desolvation domain around three hydrogen bonds listed in Table 16.5. Although these hydrogen bonds are not extremely underwrapped, it does appear that there are several waters in their vicinity. Molecular dynamics studies [171] showed that the mainchain hydrogen bond between C673 and G676 is substantially enhanced by the methylation at the C2 position in STI, including removal of intervening waters. In Figure 16.4, this hydrogen bond is depicted together with a nearby water (PDB file 1T46).

From Table 16.5, we see that the hydrogen bond between C673 and G676 has reduced wrapping and a substantial number of neighboring waters. The position (according to Mol-Probity) of the hydrogen H21 attached to C2 in STI is favorable for placing a methyl group. H21 is about 3 Å from the center of the hydrogen bond between C673 and G676, and it is about 3 Å from HOH 1187 in 1T46. The nearest atom to H21 that could cause a steric clash is CE2 in Tyr 672, at a distance of 3.05 Å. The distance from C2 in STI to CE2 in Tyr 672 is 3.54 Å.

Unlike the WBZ_7 variant of Imatinib, WBZ_4 was designed not to bind to the Abl kinases, due to their impact on heart disease [171]. Through a series of tests, WBZ_4 was shown [171] to bind selectively to c-Kit in preference to the Abl kinases.

donor	acceptor	d_{NO}	d_{HO}	$\angle_{DHA}$	$\angle_{DAB}$	$\angle_{HAB}$	tw	ω	N	A
GLY A 676	CYS A 673	2.89	1.93	162.22	132.28	126.23	23	13	3	10
LEU A 800	GLY A 676	2.83	1.93	147.44	166.36	155.79	25	9	3	9
PHE A 681	ASP A 677	3.02	2.06	159.71	157.60	151.02	31	9	2	6

Table 16.5 Hydrogen bonds in the PDB file 1T46 which contain the C2 atom from Imatinib (STI) in their desolvation domains consisting of spheres of radius 6.5 Å around the donor and acceptor CA atoms. Key: d_{XO} denotes the distance from X(=N or H) to O in the hydrogen bond, tw is the number of wrappers excluding ones from Imatinib, ω is the number of waters in the desolvation domain, N is the number of nonpolar carbons in MK1 in the desolvation domain, A is the number of all atoms in MK1 in the desolvation shell. The angles are as described in Section 6.4.

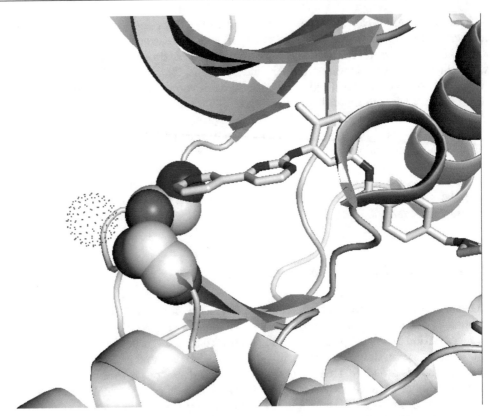

Fig. 16.4 Depiction of the hydrogen bond C673-G676 in 1T46, with the corresponding backbone heavy atoms shown as spheres. A nearby water is (HOH 1141) is shown as magenta dots. Imatinib (STI) is shown using the sticks mode in PyMol. The position of the anticancer drug Imatinib closest to this dehydron and to the water was methylated to create WBZ_4.

16.3 Three-body Interactions

Certain drug lead optimization steps, in which a nonpolar group in the drug is introduced next to a polar group in the target, are often are considered to be "counterintuitive" [51, 419, 420, 508]. For example, "some unintuitive results, such as waters near protein backbone being significantly less favorable, on average, than those around polar sidechains" are found in [51]. They also observe "that high-energy hydration sites often exist near protein motifs typically characterized as hydrophilic, such as backbone amide groups."

Examining two stereoisomers (PDB files 3fmh and 3fmk) binding the P38 kinase, the authors state [508] that the "basis for the favorability of interactions of the tetrazole with the backbone amide of Ala111 and Asp112 may be somewhat counterintuitive...." But the tetrazole in one of the stereoisomers is wrapping dehydrons 112–115 and 111–158 from P38 (one of them).

Often decisions regarding lead optimization are based on free-energy calculations of target–ligand affinities [508] and computations of the reversible work required to transfer water molecules from the protein–water interface to bulk solvent [51]. *One of the most "counterintuitive" techniques is precisely what wrapping technology does: modifying the lead scaffold by incorporating a nonpolar moiety (a methyl, etc.) to displace water from the vicinity of a backbone amide in a structured region of the protein target* [51, 419, 420]. By vicinal, we mean a water molecule whose oxygen is within 4 Å of a protein atom. The water vicinal to the backbone amide in a structured region is known to be the "hottest" in terms of its free-energy content relative to bulk water (on average +1.95 kcal per mole) [51, Table I], confirming that dehydrons are suitable targets for drug optimization. The identification of dehydrons as promoters of "hot" water [458] represents a convergence of predictions by independent technologies.

Water vicinal to dehydrons is frustrated in its hydrogen bonding possibilities as it binds to the carbonyl due to confinement [158]. This is consistent with its lower entropy as computed by WaterMap [51, 420]. However, the displacement of water from the vicinity of the dehydron stabilizes and strengthens the hydrogen bond. This is a three-body effect not represented in pairwise interactions in typical molecular mechanics models such as used in these papers. In this case, the three bodies are the amide, carbonyl, and the introduced nonpolar moiety in the ligand.

What is nonintuitive is that such a nonpolar group can interact favorably with two polar groups. But as the water molecule is removed, the dielectric environment is modified to enhance the electrostatic interaction between the amide and carbonyl [142, 159, 174]. To capture such an effect with conventional pairwise potentials, the partial charges of the amide and carbonyl would need to be made dependent on the environment. The removal of the water molecule enhances the preformed interaction of the dehydron due to a nanoscale modulation of the dielectric.

Pairwise potentials, using fixed partial charges, fail to represent the electrostatic energy of hydrogen bonds even in a fixed context [277]. When water is removed, the change in context may require even more complicated models, but it may be possible to augment force fields to capture such context-dependent behavior [112, 142, 253]. In any case, a three-body model (of a nonpolar moiety approaching a polar pair) can use experimental data to improve the accuracy of current models.

In [178], this overlooked effect is reviewed by comparing the design of the compound WBZ_4 with its parental compound Imatinib [171]. As shown in Figure 16.4, Imatinib does not displace a water vicinal to its C2 position (Figure 16.3) and the dehydron C673-G676 in c-Kit (PDB file 1T46). By contrast, WBZ_4, which is methylated at position C2, presumably removes such water molecules. WaterMap captures this dehydron wrapping effect only partially. They report in [420, Figure 5 caption] a free-energy difference at the hydration site of ≈ 0.9 kcal/mole when comparing WBZ_4 primary target c-Kit with Imatinib's primary target ABL. This estimation misses the three-body effect resulting from the proximity of the extra methyl group in WBZ_4 to the dehydron C673-G676 in c-Kit (PDB file 1T46). Using the experimental data in [171, Table 1], it is shown in [178] that the WaterMap estimate of the free-energy difference is low by about 1.02 kcal/mole. As noted in (9.8), it has been experimentally determined that water removal from a preformed dehydron lowers free energy by about 0.93 kcal/mole [173], a value comparable to the entropic effect captured by WaterMap but complementary to it. This example demonstrates that the quantitative assessment of the binding free energy can be improved by incorporating a simple model for the strengthening and stabilization of the preformed target structure due to the removal of interfacial water upon drug binding.

16.4 Scope of Wrapping Technology

The examples given so far of wrapping technology are just the tip of an iceberg. Many other possibilities exist that have not been fully explored. We give one example to point the way. The protein insulin-like growth factor 1 receptor kinase (IGF1R) is a target for breast cancer suppression [136, 489]. On the other hand, IGF1R is homologous to the insulin receptor protein (IR). Thus undesirable cross-reactivities between the two kinases can be expected, and IGF1R would need to be selectively targeted.

The possibility of multiple binding partners for a given protein inhibitor can arise in paralog proteins because they share similar domain structures [358]. Toxic side effects can result unless nonconserved features are specifically targeted. Remarkably, dehydrons are differentiating features among proteins with nearly identical structures. Figure 16.5 illustrates the closely aligned structures of two paralogs, insulin receptor (IR, PDB file 3EKK) and the insulin-like growth factor 1 receptor kinase (IGF1R, PDB file 3D94) [471]. The backbone structures of the proteins align closely, but the dehydrons are surprisingly different, due to the differences in packing determined by the composition of the individual sidechains. This is detailed in Table 16.6 which provides a hydrogen bond alignment of the two structures. Note that the residue alignment is particularly simple in this case, as the numbering in the two PDB files differs by 27 in all cases. In some cases, hydrogen bonds are found in the different structures with different amounts of wrapping. In other cases, there is no aligned hydrogen bond, suggesting that the corresponding region is unstructured. For example, the least well-wrapped hydrogen bond (ALA A 1036 — ASN A 1033) in 3EKK has no analog in 3D94. On the other hand, both corresponding residues are conserved (ALA A 1009 and ASN A 1006) and by relaxing the constraints on hydrogen bond parameters, we find the putative dehydron ALA A 1009 — ASN A 1006 with 13 wrappers.

3D94 donor — acceptor	W	3EKK donor — acceptor	W
GLU A 963 — ASP A 960	18	GLU A 990 — ASP A 987	19
not found		ASN A 1033 — PHE A 1007	18
GLY A 978 — GLY A 981	17	GLY A 1005 — GLY A 1008	14
not found		GLU A 1022 — ILE A 1019	14
not found		ASP A 1017 — ALA A 1023	13
not found		ALA A 1036 — ASN A 1033	9
VAL A 1113 — GLY A 1055	18	VAL A 1140 — GLY A 1082	20
SER A 1063 — SER A 1059	17	SER A 1090 — SER A 1086	20
ARG A 1065 — ARG A 1062	14	not found	
GLY A 1122 — ASN A 1110	17	GLY A 1149 — ASN A 1137	20
GLY A 1140 — ARG A 1137	15	not found	
ASP A 1156 — GLU A 1152	14	ASP A 1183 — GLU A 1179	13
GLU A 1197 — ARG A 1193	14	ASP A 1224 — LYS A 1220	17
GLU A 1213 — ASP A 1209	21	ASP A 1240 — GLU A 1236	18
MET A 1225 — ASN A 1222	18	MET A 1252 — ASN A 1249	16
not found		ARG A 1253 — PRO A 1250	16
SER A 1235 — GLU A 1231	18	not found	
GLU A 1246 — GLY A 1243	14	GLU A 1273 — SER A 1270	13
not found		ASN A 1282 — SER A 1279	15

Table 16.6 Dehydron alignment for the PDB files 3D94 (IR) and 3EKK (IGF1R). The column "W" lists the number of wrappers. The desolvation domain radius chosen was 6.5 Å.

Different modifications of Imatinib have been designed using the wrapping technology reviewed here, and they have been tested experimentally with remarkable success [148, 168, 170, 171]. A common view is that "identification of structure features that allow differentiation between effect and side effect profiles of medicinal agents is currently rate limiting in drug discovery" [186]. Wrapping technology has had significant success in improving specificity and affinity [148, 168, 170] precisely because it exploits such features, namely, dehydrons.

Experimentally based approaches to rational drug design [186, 234, 365, 502] seek to reveal "what" is significant for drug discovery. However, wrapping technology exploits "why" ligands

Fig. 16.5 Aligned backbones for two paralog kinases, IR (PDB file 3D94) and IGF1R (PDB file 3EKK) represented by virtual bonds joining α-carbons.

interact with proteins [148, 168, 170]. Thus, we counter the view that "computational solutions that precisely link molecular structure to broad biological response are currently not possible" [186].

Typical analysis of protein–ligand binding involves pairwise, intermolecular interactions [219] combined with steric considerations [94]. Surprisingly, the modulation of the strength and stability of intramolecular hydrogen bonds by hydrophobic groups on ligands (drugs) plays a significant role in many target–ligand complexes. Such ligands do not utilize pairwise intermolecular interactions but rather promote water exclusion near dehydrons upon association.

We have focused here on an analysis of static information related to protein–ligand binding, yet "the need to account for the dynamic behavior of a receptor has long been recognized as a complicating factor in computational drug design" [77]. Not only does induced fit play a role [140, 219], but structural flexibility can also lead to unwanted promiscuity [89, 140]. Also see [471, Chapter 11].

Kinase targets pose a major challenge to structure-based drug design because their flexible loopy regions framing the ATP pocket and the transphosphoesterification environment are likely to adopt an induced fit, generally unpredictable, upon association with ligands. But it has been possible to verify [140] that WBZ_4 induces a dehydron in JNK-1 in a predictable manner. This anti-JNK activity is currently being exploited [421] to position WBZ_4 as an anti-ovarian cancer drug.

The relationship between disorder and defective wrapping [162, 396] helps to explain the order-upon-binding scenario and may be turned into a design concept whereby the putative ligand becomes a wrapper of the packing defects that the loopy region forms as it visits specific conformations. Thus, the activation loop of C-Kit kinase was stabilized in the activated, induced-fit conformation by redesigning the drug inhibitor Imatinib to wrap better a dehydron in the activation loop [170].

Wrapping technology represents a new paradigm for rational design of drugs that exclude water near solvent accessible hydrogen bonds on the surface of the protein target. This represents a clear departure from the standard approach which focuses on the possibility of promoting pairwise interactions between ligand and target. Not only does wrapping technology utilize a novel modality crucial to the functioning of large classes of ligands, it also guides modifications to existing ligands as opposed to a combinatorial search paradigm. A major obstacle to targeting kinases is their high level of structural similarity, but it is overcome by wrapping technology which exploits packing differences. Wrapping technology identifies potential ligands with high selectivity, utilizing packing defects as selectivity filters and switches that can allow the targeting of specific proteins.

16.5 Exercises

Exercise 16.1. In Chapter 7, different approaches to defining interfaces were sketched. In some cases, attempts have been made to differentiate different types of interfaces based on function [81, 257]. Determine the numbers of intramolecular dehydrons in various interfaces and compare this with the corresponding density of intermolecular hydrogen bonds in Section 7.1.

Exercise 16.2. Analyze further the data in Table 16.6 to account for the missing hydrogen bonds (marked "not found") in the alignment. Follow the model given for the hydrogen bond ALA A 1036 — ASN A 1033 in 3EKK in the text.

Chapter 17
Epilogue and Future Perspectives

John Alexander Reina Newlands (1837–1898) proposed an early version of the periodic table of elements, coining the term 'octave' to describe the periodicity that he observed, a precursor to what we now know at the octet rule. Although initially ridiculed for his work published in 1865, he received the Davy Medal in 1887.

As we stated at the beginning, this book is intended to initiate a new approach to protein biophysics, not to summarize the main results of a mature field. This chapter is in the position where one would expect an Epilogue. Instead, we offer a prologue outlining where research might go in the future. We begin by summarizing key aspects of the book to provide a succinct background for what we propose for the future. We then suggest some new directions that may, or may not, be succeptible to mathematical understanding.

17.1 Recapitulation of Key Results

We have introduced some new ideas that help to explain how proteins behave in a digital way. We have focused on wrapping of mainchain hydrogen bonds as a motif that explains multiple molecular effects. Our interpretation of the genetic code in Section 8.7 shows that wrapping is less constrained than polarity with respect to single-letter codon mutations. Thus, it is not surprising that wrapping plays such a dominant role in functional innovation.

We have shown that dehydrons play a central role in protein-ligand associations. They exert an intermolecular force that enables and promotes such associations. Binding via dehydrons requires a less stringent geometric match between protein and ligand, unlike intermolecular hydrogen bonds, and yet they are sufficiently stereo-specific to determine binding locations on the protein.

In addition, we have seen that wrapping can affect the electronic structure of the peptide bond, effecting rotational flexibility. The biological implications of this fact have not been fully explored.

The effects of wrapping extend to all electrostatic interactions, but we have purposely focused primarily on its effect on mainchain hydrogen bonds, where wrapping enhances

© Springer International Publishing AG 2017
L.R. Scott and A. Fernández, *A Mathematical Approach to Protein Biophysics*,
Biological and Medical Physics, Biomedical Engineering,
DOI 10.1007/978-3-319-66032-5_17

strength and stability. The situation can be diametrically different for other electrostatic interactions; wrapping is likely destabilizing for salt-bridges [223].

Wrapping typically affects the dielectric behavior of water, and thus, we have reviewed in detail the mathematics behind the concept of dielectric. We have also reviewed classical electrostatic interactions in detail, in particular, dipole–dipole interactions, as well as more recently discovered ones such as cation-π. We have examined dipole–dipole interactions in the depth required to correctly assess them.

The effects of wrapping span multiple scales, ranging from the subatomic to the molecular, and via signaling, to the cellular scale. There is surely more to learn about the effects of wrapping in general and the dehydron in particular. But we hope that this book will initiate a new, more quantitative phase in protein biophysics, and that this in turn will provide a more rigorous basis for drug design and the understanding of molecular-based diseases in the future. With this in mind, we highlight some directions of research that are in need of further mathematical development.

17.2 Dehydrons Deprotonate

It has recently been discovered that dehydrons functionalize water locally by turning water molecules into proton acceptors, thereby enabling certain (enzymatic) reactions in protein catalysis [154, 156–158]. This means that dehydrons can be regarded as chemical entities. Many of the mechanisms of biological chemistry will need to be re-examined in light of this effect.

The simple explanation for this effect is, in part, that dehydrons make nearby water more mobile [458], as indicated in Figures 2.4 and 2.5. This increased mobility makes the water more available than normal as an agent for deprotonation. The rate at which water can deprotonate is related to pH; the reason that HO^- and H_3O^+ coexist with H_2O is that water molecules steal protons from each other. But in addition, what makes the increased collisions effective is the fact that the dehydrons frustrate the vicinal water so that there is an unused electron pair on the oxygen due to a preferred orientation of the water [158]. In summary, there are three complementary effects of dehydrons on vicinal water: increased mobility, hydrogen bond frustration, and restricted orientation. This combination bestows chemical functionality upon water near dehydrons.

But a deeper understanding of deprotonation requires a quantum mechanics treatment [156–158] called QM/MM [435]. The idea of QM/MM is to calculate classical molecular dynamics (MM) trajectories with the potential energy based in part on a quantum-mechanical representation for the presumed reactants. The sampling strategy for the trajectories relies on Jarzynski's Theorem [246]. For more details, see [159, Chapter 7 and Appendix A].

17.3 Protein Targets with Unknown Structure

Structure-based design is covered by a recent book [204] and is widely used in drug design. When structure is known, the computational task is greatly simplified, but not eliminated. The concept of dehydron, as introduced so far in this book, is based on protein structure. Structure analysis can be used to identify hot spots [152]. High-throughput docking and hit optimization guided by molecular dynamics [536] can be verified via structure examination [497]. Machine learning can be combined with structure-based protein design [270].

In the absence of structure, the critical issue is to obtain some proxy for structural information. This can be done using software to predict protein structure [84]. Unfortunately, such techniques identify many spurious decoys [429]. On the other hand, it is not necessary to obtain the full structure to predict defects in structure (dehydrons) [396] based on sequence alone. Once the positions in sequence where dehydrons are predicted, they can be used to identify hot spots [152] and interaction sites [471, Chapter 5].

It is known that structural defects (dehydrons) can be approximately determined via an examination of disorder data [396, 537]. There are now many techniques that can estimate protein disorder from sequence alone [544].

The efficacy of the technology utilizing disorder scores to estimate the location of dehydrons is illustrated by a recently patented invention to treat heart failure through disruption of the myosin-MyBP-C interface [357].

MyBP-C refers to a multi-domain myosin-binding protein with unreported structure that is a central regulator of cardiac contraction [97, 405, 424]. MyBP-C molecules constitute molecular brakes modulating the displacement of myosin motors, with a brake-release mechanism hinging on site-specific phosphorylation. By sequence-based inference of dehydron-rich regions in MyBP-C, patented drug leads were identified to cure heart failure [357]. This process is explained more fully in [177], and it makes clear that having software that directly predicts dehydrons from sequence is both feasible and potentially useful.

17.4 Dynamical Analysis of Dehydrons

Predicting the induced folding of a disordered part of a protein upon ligand binding is as daunting as solving the general protein folding problem. However, when the disordered region comprises only a few residues, and the region aligns with a dehydron in a paralogous protein, it may be possible to induce a dehydron when a ligand brings sufficient wrapping groups to the region. An illustration of this dynamic aspect of drug ligand design is the positioning of WBZ_4 as an anti-JNK agent, leading to the current development of an anti-ovarian cancer agent [421] as described in Section 16.4. Molecular dynamics simulations augmented by including three-body effects described in Section 16.3 will be essential to capture such induced structuring.

17.5 Non-Debye Dielectric Response

Recall the definition of the polarization field $\mathbf{p}$ from Section 15.2. We define the non-Debye [154, 155] component as

$$\zeta = \epsilon_0 \mathbf{p} \cdot \mathbf{e}^\perp, \tag{17.1}$$

where $\mathbf{e}^\perp$ is defined in (15.10). The non-Debye component can be large near protein surfaces with complicated nano-scale detail [154, 155]. Moreover, there is a fundamental asymmetry in the dielectric response [39, 42] to ions based on their charges. That is, anions appear to solvate more easily than cations [158]. Thus, the dielectric behavior can depend on the local solute properties.

Similar to the Debye ansatz, it has been proposed [153] that $\zeta \mathbf{e}^\perp = \omega \nabla \phi$ where ϕ is related to the frustration of interfacial water at each location, and ω is a physical constant determined by experiment. This ansatz lead to the conclusion that the non-Debye electrostatic energy may be stored as interfacial tension. This suggests that the relationship between water frustration and non-Debye dielectric response needs to be further explored. See [159, Chapter 2] for more details. In particular, nonlocal effects related to the non-Debye dielectric response need to be assessed.

17.6 Thermodynamics and Statistical Mechanics

We have taken a naïve approach to utilizing the potential energy $U(x)$ of an atomic system, where $x \in \mathbb{R}^{3n}$ denotes the coordinates of the system, in particular we always sought the configuration of minimum potential energy to characterize behavior. The correct measure of energy in a thermalized system is the **free energy**. Statistical mechanics provides a way to include thermal fluctuations in energetic models in a way that satisfies the laws of thermodynamics [132, 139, 205, 266, 338, 374, 381, 389, 398, 404, 426, 427, 442, 484, 485, 492]. The total energy of the system is a combination of the potential energy and the kinetic energy. The enthalpy of a system H can be expressed as the expected or average potential energy

$$H = \frac{1}{Z} \int_\Omega U(\mathbf{r}_1, \ldots, \mathbf{r}_n) e^{-\beta U(\mathbf{r}_1, \ldots, \mathbf{r}_n)} \, d\mathbf{r}_1 \cdots d\mathbf{r}_n, \tag{17.2}$$

where T is temperature, $\beta = 1/k_B T$, k_B is Boltzmann's constant, and Z is a dimensionless number called the **partition function** for the system:

$$Z = \int_\Omega e^{-\beta U(\mathbf{r}_1, \ldots, \mathbf{r}_n)} \, d\mathbf{r}_1 \cdots d\mathbf{r}_n. \tag{17.3}$$

Here $\Omega \subset \mathbb{R}^{3n}$ denotes the state space, that is, the set of allowable positions $\mathbf{r}_i$. Thus, Z is simply a normalization factor so that the quantity

$$P(\mathbf{r}_1, \ldots, \mathbf{r}_n) = \frac{1}{Z} e^{-\beta U(\mathbf{r}_1, \ldots, \mathbf{r}_n)} \tag{17.4}$$

forms a probability distribution over the state space. The probability distribution P in (17.3) is called the **Boltzmann distribution**. It says that lower energy states (smaller values of U) are more likely. The interpretation is that in a thermalized system all states are possible, with probability P; the lowest energy U has more influence but is not the only contributor. As $T \to 0$, P tends to a discrete distribution with point masses at the states with lowest energy configurations $U_{\min}$.

The quantity $G = -\frac{1}{\beta} \log Z$ is the Helmholtz free energy of the system. In considering two possible system configurations, the one with lower free energy G is the one that will occur physically. Two things make it difficult to estimate G accurately. One is the lack of accuracy of $U(x)$ for a given x that we have encountered in many situations, such as the potential energy of a hydrogen bond. The other has to do with the high dimensionality of the integral (17.3), so that sampling the state space Ω becomes a major difficulty [501]. However, in some cases [450] for a low-dimensional system, it is possible to compute such integrals via standard numerical techniques to high accuracy. Although it is still challenging to estimate $U(x)$ accurately, such simple approaches are able to shed light on configurations where a simple analysis based on the potential energy minimum alone would be misleading [450]. In any case, thermodynamic arguments are essential for a rigorous analysis of protein systems [150]. Of course, this makes everything much more complicated, but it also introduces a larger set of engaging mathematical challenges.

Chapter 18
Units

Larkin Burneal Scott (1917–1990) developed analog and digital computational enhancements to analytical instruments at the Perkin-Elmer Corporation, such as the baseline corrector for the DSC-II.

It is helpful to pick the right set of units in order to reason easily about a physical subject. In different contexts, different units are appropriate. Small boat enthusiasts will recognize the need to determine whether depth on a chart is labeled in feet or fathoms. It is common in the United States coastal waters to label the depths in feet where mostly small boats will be expected to be found. But commercial vessels might prefer to think in fathoms (a fathom is six feet) since their depth requirements will be some number of fathoms (and thus a much larger number of feet). The phrase "mark twain" was used on riverboats for which twelve feet of water provided safe passage.

A mistake about units could be disastrous. In boating, thinking that a depth given in feet is really fathoms could lead to a grounding of the boat. Perhaps it is reasonable to expect that such gross errors would never occur, but a commercial airline was once forced to land at a converted airfield in Gimli, Manitoba because the weight of the fuel was computed in pounds instead of kilograms. As a result, the plane lost all power, including most of its electrical power needed for running the plane, at a high altitude. The aircraft and occupants survived the incident, but the former became known as the Gimli Glider.

In astronomy, we may measure distances in light-years. But this is the wrong unit for our discussion. Just like the choice between fathoms for commercial vessels and feet for small pleasure boats, we need to find the right size for our mental models.

18.1 Basic Units *vs.* Derived Units

basic units	length, time, mass, charge, temperature
derived units	energy, viscosity, kinematic viscosity, permittivity, speed of light

Table 18.1 An example of basic *versus* derived units.

© Springer International Publishing AG 2017
L.R. Scott and A. Fernández, *A Mathematical Approach to Protein Biophysics*,
Biological and Medical Physics, Biomedical Engineering,
DOI 10.1007/978-3-319-66032-5_18

We encounter units for many things: length, time, mass, charge, viscosity, energy, and so forth. There are only so many of these that are independent. Once we choose a set of units, others must be derived from them. In Table 18.1, we give a simple example of a particular choice of basic and derived units.

As an example, energy (E) is measured in units of mass (m) times velocity (v)-squared, and velocity has units length (ℓ) over time (t):

$$E = mv^2 = m(\ell/t)^2. \tag{18.1}$$

However, there is no canonical definition of which units are basic and which are derived.

We might want certain units to have a prescribed value, and thus, we take those units to be basic. For example, we might want the permittivity of free space to be one, as we discuss in Section 18.2.3.

18.1.1 SI Units

basic units	meter, second, kilogram, ...
derived units	joule (energy), newton (force), ...

Table 18.2 The SI system of basic and derived units.

One standard set of units is the SI system. Mass is measured in (kilo)grams, distance in meters, time in seconds. There are other basic units as well, but let us stop here as it allows us to define the standard units of energy and force.

The SI standard unit of energy is the joule. Energy has units mass times velocity-squared, as we know from Einstein's famous relation. A joule is a newton-meter, the work related to applying the force of a newton for a distance of a meter. A newton is one kilogram-meter/second^2, so a joule as one kilogram-(meter/second)^2.

These quantities are familiar macroscopic measures. A kilogram is the weight of a good book, and a meter per second is understandable as a walking speed: 3.6 kilometers per hour. Thus, a joule is the energy required to get a book up to walking speed. In many cases, the older unit **calorie** is used, which differs from the joule by a small factor: one calorie is 4.1868 joules.

18.2 Biochemical Units

There are natural units associated with biochemical phenomena which relate more to the nanoscale. For example, the frequently used unit for energy is **kcal/mole**. This of course refers to one-thousand calories per mole of particles, or per 6.022×10^{23} particles, which is Avogadro's number. That is a big number, but we can squash it down with the right word: It is 0.6022 yotta-particles (**yotta** is a prefix which means 10^{24}, just as kilo means 10^3 or nano means 10^{-9}). The kcal is 4.1868 kilojoules, or 3.9683 Btu (British thermal units, a unit used in describing the power of both residential and commercial heating and cooling systems).

Another unit used at the molecular level is the electron-volt. This has the value $1.6021766 \times 10^{-19}$ joules, or 23.045 kcal/mole. For example, the ionization energy of hydrogen is 13.53 eV, or 311.79 kcal/mole.

18.2.1 Molecular Length and Mass Units

At the molecular scale, the typical units of mass (e.g., the gram) are much too large to be meaningful. Moreover, the units such as the meter and gram are based on macro-scale quantities. The Ångstrom is just a shorthand notation for 10^{-10} meters, and it is the commonly used distance in molecular discussions. A more appropriate length scale for the atomic scale [479] might be the Bohr radius $a_0 = 0.529189$Å, which is based on properties of the electron distribution for the hydrogen atom, but at least we see that it is on the same order as an Ångstrom. Similarly, a more natural mass unit would be based on an atomic mass, e.g., the **dalton** (or **Da**) which is essentially the mass of the hydrogen atom. More precisely, it is one-twelfth of the mass of carbon twelve. The dalton is almost identical to the previous standard known as the **atomic mass unit** (or **amu**). The dalton mass unit $= 1.66053 \times 10^{-27}$ kilograms.

18.2.2 Molecular Time Units

A natural timescale for biochemistry is the femtosecond (10^{-15} second) range. This is the temporal scale to observe the dynamics of molecules above the quantum level. For example, time-stepping schemes for molecular dynamics simulation are often a few femtoseconds, although some systems (e.g., liquid argon) appear to be stable for timesteps up to 100 femtoseconds. The **svedberg** is a time unit equal to 100 femtoseconds (10^{-13} second).

This svedberg provides a timescale that resolves molecular motion, but does not over resolve it: It is a scale at which to see details evolving the way a mechanical system would evolve in our everyday experience. We perceive things happening in a fraction of a second and are aware of motions that take place over many seconds. Runners and other athletes are timed to hundredths of a second, so we can think of that as a timestep for our perception. Thus, our typical perception of motion covers 10^4 or 10^5 of our perceptual timesteps. By this reckoning, there are about 2×10^{11} timesteps in a typical human lifetime. Biological events, such as protein folding, take up to 10^{11} svedbergs, and even more. Note that a typical human height is about 2×10^{10} Ångstroms.

There is a natural length scale associated with any temporal scale when electromagnetic waves will be of interest. Just like the light-year, it is natural to consider the distance light travels in the natural time unit here, the svedberg, about 2.9979×10^{-5} meters, or 30 micrometers. This may seem odd. You might have expected a spatial unit on the order of an atomic unit such as the Ångstrom, but this is 0.03 millimeters, a scale we can almost resolve with a magnifying glass. This means that light is still very fast at these molecular scales. We hesitate to give this length a name, but it is clearly a light-svedberg.

If we pick the svedberg as time unit and the Ångstrom as spatial unit, then the natural velocity scale is the Ångstrom per svedberg, which is equal to $10^{-10+13} = 1000$ meters per second, about three times the speed of sound in air at sea level. With the dalton as mass unit, the natural energy unit in these units is one dalton-(Ångstrom per svedberg)^2. One dalton-(Å/svedberg)$^2 = 1.66 \times 10^{-21}$ joules $= 0.239$ kcal/mole. Thus, the chosen units of mass, length and time lead to a nearly unit value for the commonly used unit of energy, kcal/mole.

basic units	Ångstrom, svedberg (10^{-13}s), dalton (mass)
derived units	(energy) dalton-(Å/svedberg)^2 = 0.239 kcal/mole

Table 18.3 Some units relevant for biochemistry.

18.2.3 Charge Units

The natural unit of charge for protein chemistry is the charge of the electron, q_e. When we look at macromolecules, we can resolve individual units and their charges. The coulomb is an aggregate charge constant defined so that $q_e = 1.602 \times 10^{-19}$ C. That is, C = 6.242 $\times$ $10^{18}q_e$. The actual definition of a coulomb is the charge associated with an ampere flowing for a second. Thus, a hundred amp-hour battery has 360,000 coulombs of charge, or about $2.25 \times 10^{24}q_e$, which corresponds to 3.7 moles of electrons.

Permittivity has units charge-squared per energy-length:

$$\text{permittivity} = \frac{\text{charge}^2}{\text{energy-length}} = \frac{\text{charge}^2\text{time}^2}{\text{mass-length}^3}. \tag{18.2}$$

Thus, it is possible to have the permittivity, charge, and energy be one in any units by varying the spatial (length) unit:

$$\text{length} = \frac{\text{charge}^2}{\text{energy-permittivity}}. \tag{18.3}$$

However, it would not be possible to specify length, permittivity, energy, and charge independently.

The permittivity of free space ε_0 is 8.8542×10^{-12}F m^{-1} (farads per meter). A farad is a coulomb-squared per newton-meter. That is, we also have

$$\begin{aligned}
\varepsilon_0 &= 8.8542 \times 10^{-12}\text{C}^2\text{N}^{-1}\text{m}^{-2} \\
&= 3.450 \times 10^{26}q_e^2\text{N}^{-1}\text{m}^{-2} = 3.450 \times 10^{26}q_e^2\text{J}^{-1}\text{m}^{-1} \\
&= 1.444 \times 10^{27}q_e^2\text{cal}^{-1}\text{m}^{-1} = 1.444 \times 10^{30}q_e^2\text{kcal}^{-1}\text{m}^{-1} \\
&= 2.40 \times 10^6 q_e^2(\text{kcal/mole})^{-1}\text{m}^{-1} = 2.40 \times q_e^2(\text{kcal/mole})^{-1}\mu\text{m}^{-1} \\
&= 0.72q_e^2(\text{kcal/mole})^{-1}\text{lfs}^{-1},
\end{aligned} \tag{18.4}$$

where "lfs" stands for light-femtosecond, the distance traveled by light (in a vacuum) in a femtosecond. Thus, we see that in the units in which energy is measured in kcal/mol, charge is measured in units of the charge of the electron, q_e, and length is the light-femtosecond (lfs), we find the permittivity of free space to be on the order of unity. It is noteworthy that Debye [110] used units so that $\varepsilon_0 = 1$, together with energy measured in kcal/mol and charge measured in units of q_e. This means that the implied spatial unit is 1.39 lfs, or about 417 nanometers, or just under half a micron, a length related to the Debye screening length in water [225]. If this is the chosen spatial unit, then $\varepsilon_0 = 1$ in these units. For reference, very large viruses [552] are between one- and two-tenths of a micron in diameter.

We can compute permittivity in a more common length scale as

$$\varepsilon_0 = 0.00024 q_e^2 (\text{kcal/mole})^{-1} \overset{\circ}{A}{}^{-1}, \tag{18.5}$$

but we see it is now quite small.

basic units	kcal/mole (energy), q_e (charge), ε_0 (permittivity)
derived units	(length) 417 nanometers

Table 18.4 The units implied by Debye's assumptions.

18.2.4 Conversion Constants

Boltzmann's constant, $k_B = 1.380 \times 10^{-23}$ joules per degree Kelvin, relates energy to temperature. This seems really small, so let us convert it to the "kcal/mole" energy unit. We get

$$
\begin{aligned}
k_B &= 1.380 \times 10^{-23} \text{J/K} \\
&= \frac{1.380 \times 6.022}{4.1868} \text{cal/mole-K} \\
&= 1.984 \text{cal/mole-K}.
\end{aligned}
\tag{18.6}
$$

For example, at a temperature of T=303K, we have $k_B T = 0.601$ kcal/mole.

If temperature is in degrees Kelvin, velocities are measured in Ångstroms per picosecond (around 224 miles per hour), and masses in daltons, then $k_B \approx 0.831$.

Planck's constant,

$$h = 6.626068 \times 10^{-34} \, \text{m}^2\text{-kg/s} = 39.90165 \, \text{Ångstroms}^2\text{-dalton/picosecond}, \tag{18.7}$$

has units energy-time, which is a unit of **action**. The other Planck constant $\hbar = h/2\pi$ is then $\hbar = 6.35055$ Ångstroms^2-dalton/picosecond.

The ratio of Planck's constant to Boltzmann's constant has an interesting interpretation. It is $h/k_B = 4.80 \times 10^{-11}$ seconds per degree Kelvin, or 48 picoseconds per degree Kelvin.

18.3 Quantum Chemistry Units

basic units	ε_0 (permittivity), a_0 (length), m_e (mass), 10^{-3} svedberg (time)
derived units	hartree $= \hbar^2$ (energy)

Table 18.5 Quantum chemistry units.

For one particle of mass m moving in a force field having potential V, it can be written

$$\iota\hbar\frac{\partial\psi}{\partial t} = \mathcal{H}\psi \tag{18.8}$$

where $\hbar$ is Planck's constant divided by 2π, ι is the imaginary unit and

$$\mathcal{H}\phi(\mathbf{r}) := -\frac{\hbar^2}{2m}\nabla^2\phi(\mathbf{r}) + V(\mathbf{r})\phi(\mathbf{r}). \qquad (18.9)$$

The Schrödinger equation (18.8) has three terms which must have the same units in order to be dimensionally correct. If we divide (18.8) by $\hbar$, then the diffusion term is multiplied by the constant $\hbar/2m$. Fortunately, $\hbar/m$ has units of length-square over time, as required. In the Schrödinger equation (18.8), we have implicitly assumed that the permittivity of free space $\varepsilon_0 = 1/4\pi$. We can do this, as noted above, but we need to choose the right spatial and energy units to make it all work out. Unfortunately, if the energy unit is kcal/mole, the natural scale for biochemistry, then the spatial unit is quite large, four orders of magnitude larger than the typical scale of interest.

A more typical choice of spatial unit at the quantum scale [479] would be to use the Bohr radius $a_0 = 0.529189\text{Å}$. This scale differs only by a factor of about two from what we have been considering so far. But the natural unit of mass is the mass of the electron $m_e = 9.10938 \times 10^{-31}$ kg $= 5.48579 \times 10^{-4}$ dalton. In these units, Planck's constant is

$$h = 39.90165 \,\text{Angstroms}^2\text{-dalton/picosecond} = 137.45 \, a_0^2\text{-}m_e/\text{femtosecond}. \qquad (18.10)$$

If we also adopt the hartree[1] [457] E_h for the unit of energy, and we adopt the mass of the electron m_e as the unit of mass, then things are better. By definition, the hartree E_h is

$$E_h = m_e c^2 \alpha^2 = 4.356 \times 10^{-18}\text{joules} = 1.040 \times 10^{-21}\text{kcal} = 626.5\,\text{kcal/mole}, \qquad (18.11)$$

(cf. Table 3.2) where $\alpha \approx .007$ is a dimensionless number known as the **fine structure constant**, cf. Exercise 18.7. Moreover, we also have the coefficient of the potential in (18.8) equal to E_h; that is, $e^2/(4\pi\varepsilon_0 a_0) = E_h$.

The time-derivative term in (18.8) is multiplied by $\hbar$, which fortunately has units of energy times time. Planck's constant $h = 6.626068 \times 10^{-34}$ joule-seconds $= 1.521 \times 10^{-16}$ hartree-seconds $= 0.1521$ hartree-femtoseconds. Dividing by 2π, we find that Planck's constant

$$\hbar = 0.02421 \text{ hartree-femtoseconds}. \qquad (18.12)$$

That is, if we take the time unit to be femtoseconds, then the coefficient of the time-derivative term is $= 0.02421$, or about one over forty. This is a small term. It implies that changes can happen on the scale of a few tens of attoseconds, whereas on the scale of a few femtoseconds (the typical time step of molecular dynamics simulations), the time-derivative term in (18.8) can plausibly be ignored, or rather time-averaged. To cast this in terms of the units suggested for biochemistry, the natural timescale for quantum chemistry is about 10^{-3} smaller, about a milli-svedberg.

In thinking of the Schrödinger equation in classical terms as describing the probability of an electron's position as it flies around the nucleus, it is interesting to think about the timescale for such a motion. At the speed of light, it takes an attosecond to go 3 Ångstroms. The timescale of the Schrödinger equation is 24 attoseconds, and in this time anything moving at the speed of light would go 72 Ångstroms. If the Schrödinger equation represents the average behavior of electrons moving around the nucleus at anything approaching the speed of light, then they can make many circuits in this basic time unit of the Schrödinger equation. So it is plausible that it represents such an average of dynamic behavior.

---

[1] Douglas Hartree (1897–1958) pioneered approximation methods for quantum chemistry calculations.

It is possible to have $\hbar = 1$ if we pick the right time units. From (18.12), we see that we need the time unit to be $t = 0.02421$ femtoseconds $= 24.21$ attoseconds. At the speed of light, you would go 72.58 Ångstroms in 24.21 attoseconds.

18.4 Laboratory Units

In a laboratory, it would be confusing to use units appropriate at the molecular scale. A typical mass unit would be a milligram (mg), and a typical volume unit would be a milliliter (mL). Inside cells, protein concentrations can exceed 100 mg per mL.

The terms "millimolar" and "micromolar" are used to express the fractional concentration of one substance in another, e.g., water. A mole of water has 6.022×10^{23} particles, which is Avogadro's number, and this many atoms of oxygen, and twice this number of hydrogens, weighs 18.0153 grams. At 4 degrees Centigrade, where water has its maximum density, one gram of water occupies one cubic centimeter, or one milliliter. Thus a mole of water occupies 0.0180153 liters (at $4°$ C), so a liter of water has 55.508 moles of water, near 4 degrees C. Thus, we say that the molar concentration of water is 55.508 moles per liter.

The molar concentration of a mixture is more complicated. By definition, a solute (e.g., salt) has a molar concentration of xM (moles per liter) if there are x moles of the solute in a liter of solute and solvent (e.g., water). This mixture is easy to create: You prepare x moles of the solute and add solvent until you fill a liter container. You figure out what x moles weighs based on its atomic weight. But it is not known a priori how much solvent you will need to add, since the way they mix is not predictable in advance.

A one-molar solution of salt (Na–Cl) in water would require one mole of salt, or 58.4425 grams, based on the combined atomic weights of sodium and chlorine. Correspondingly, a 200mM solution would require only 11.6885 grams. At $4°$ C, a liter of water weights about a kilogram. So it is reasonable to assume that a 200mM solution of Na–CL and water would be mostly water. The number of atoms in a 200mM solution is 0.2 times Avogodro's number; The number of atoms of water in a liter is 55.508 times Avogodro's number. If we ignore the number of waters that would flow over the one-liter container when the salt is added, the ratio of water molecules to salt molecules would be 277.54. For a one-molar solution (1M), the number would be about five times smaller, nearer to 55 waters per Na–Cl pair. Accounting for the displacement of water by salt, these numbers would be slightly smaller.

18.5 Mathematical Units

There is a natural set of units that might be called mathematical units. They are based on the observation that many named constants are really just conversion factors. For example, Boltzmann's constant converts temperature to energy. Thus with the right temperature scale, Boltzmann's constant is one (cf. Exercise 18.4). Similarly, Planck's constant has units energy times time, and it will be one with the right relationship between energy and time. This places a constraint on the relationship between mass, length, and time. A natural mass unit is the dalton, since it is roughly the mass of the smallest atom. With the dalton as the mass unit, the largest masses in the Schrödinger equation are of order one, although the smallest (i.e., the electrons) have a tiny mass in this unit. It is natural to take the speed of light to be one, so this sets a relationship between length and time.

If we divide Planck's constant by the speed of light, we get $\hbar/c = 0.212 \times 10^{-15}$ dalton-meters. If we want $\hbar = 1$ and $c = 1$ [380], then we need to have the length unit to be 0.212×10^{-15} meters $= 0.212$ femtometers. The diameter of a proton is approximately one femtometer.

If we divide Planck's constant by the speed of light-squared, we get $\hbar/c^2 = 0.7066 \times 10^{-24}$ dalton-second. If we want $\hbar = 1$ and $c = 1$, then we need to have the time unit to be 0.7066×10^{-24} seconds $= 0.7066$ yoctoseconds. If these independent calculations are correct, we would find that the speed of light is about 0.3 femtometers per yoctosecond. A femtometer per yoctosecond is 10^9 meters per second, so we have agreement.

To summarize, if we take length to be measured in multiples of L $= 0.212$ femtometers, time to be measured in multiples of t $= 0.7066$ yoctoseconds, and mass in daltons, then $c = \hbar = 1$. See Exercise 18.5 for the similar case where the unit of mass is the mass of the electron. As noted above, a joule in these units is 6.7006×10^9 dalton-$(L/t)^2$. Similarly, in these units $k_B = 9.2468 \times 10^{-14} K^{-1}$.

basic units	speed of light, Planck's constant, Boltzmann's constant
derived units	time unit = 0.7 yoctoseconds, length unit = 0.2 femtometers

Table 18.6 Some units to simplify mathematical equations.

18.6 Evolutionary Units

There are also other timescales of interest in biology, and geology. The **molecular clock** refers to the time it takes for a single point mutation to occur in DNA. This is measured by comparing divergent genomes of related species, for which the time of divergence is estimated from the fossil record. The unit of measure for sequence divergence is **percentage divergence**, which refers to the fraction of times each individual sequence entry is expected to have been modified in a given time unit. Typically, this is a very small number, so the time unit is often taken to be large. Thus, a 2% sequence divergence per million years means that the probability of mutation of each individual sequence entry is only 0.02 in a million years. However, in a hundred million years, we would expect each entry to be modified twice.

Estimates for molecular clocks vary on the order of one percent divergence per 10^6 years, although turtle mitochondrial DNA mutation tends to be slower [22], possibly due to their slower Internet connections. Fortunately, this is a slow scale from a human perspective. However, over geologic time, it is significant.

It is interesting to note that using typical estimates of the age of the earth [4], there has not been enough time for this type of mutation to cause a complete change to a typical chromosome. Time is usually measured in units of Mya (millions of years ago), or bya (billions of years ago); the latter unit is often replaced by the shorter Ga (giga-annum). The age of the earth is estimated to be at least 4.5 Ga [4].

We can put these two pieces of data together to estimate how many times a complete genome might have been modified. If the variation is occurring at a rate of one percent per 10^6 years, then in 10^8 years it could become completely modified. But given the estimated age of the earth, no genome would be expected to have been fully modified more than about forty five times via single point mutations.

From the point of view of a dynamical system, a genome that has been modified 45 times might well have reached a stable equilibrium (if there is one). On the other hand, more recently evolved species would not yet be at a point where many cycles have taken place, so their genomes might be still far from an equilibrium point. Species that are older that 10^6 years would have experienced substantial modification (more than one base-pair in a hundred modified). But species much younger than 10^8 years might exhibit little limiting behavior, still undergoing substantial modification due to random mutations.

18.7 Other Physical Properties

We now consider some other physical properties that are measured in much the same way that basic units are. Some of them could themselves be treated as basic units, such as viscosity. Others deal with more complex issues, such as the pH scale which describes a mixture of materials.

18.7.1 The pH Scale

At a pH of k, there are 10^{-k} moles of hydronium ions (and hydroxyl ions) per liter of water. As observed in Section 18.4, a liter of water at 4 degrees Centigrade has 55.508 moles of water. Thus, the ratio of hydronium ions to water molecules at a pH of k is roughly one hydronium ion per $5.5508 \times 10^{k+1}$ water molecules. Humans seem happiest at pH seven, which corresponds to a ratio of approximately one hydronium ion per half billion water molecules. However, the pH in cells can be much lower.

18.7.2 Polarity and Polarization

The Debye is the standard unit for dipole moment and is 3.338×10^{-30} coulomb-meters. A more useful unit would be a q_e-Ångstrom, where q_e is the charge of an electron, and this turns out to be about 4.8 Debye. Recall that a coulomb is $6.242 \times 10^{18} q_e$. Thus, a Debye is 0.2084 q_e-Ångstrom. The dipole moment of water ranges from about 1.9 Debye to 3.5 Debye depending on the environment [100, 210].

Polarization is the effect of an external field to change the strength of a dipole. An interesting feature is that the polarization coefficient has units of volume (i.e., length-cubed). Thus, there is a natural motif that can be used to illustrate the polarizability of an object: the volume of its representation. For example, if we are representing atoms as spheres, the volume of the sphere could be taken to be its polarization coefficient.

Polarization is a tensor, and it need not be isotropic. However, in many cases, a scalar approximation is appropriate. The polarizability of water is $\alpha \approx 1.2 \text{Å}^3$.

18.7.3 Water Density

Water is a molecule with a complex shape, but it is possible to estimate the volume that an individual molecule occupies. A mole of water, which consists of 6.022141×10^{23} water molecules, weighs 18.0153 grams. Thus, a gram of water has 0.33427×10^{23} water molecules.

At 4 degrees Centigrade, where water has its maximum density, one gram of water occupies one cubic centimeter, or 10^{24}Å^3. Therefore, 0.33427×10^{23} water molecules occupy 10^{24}Å^3, at 4° C; one water molecule thus occupies $29.916 = 10/0.33427$Å^3. This corresponds to a cube of just over 3.1 Ångtroms on a side. It is interesting to compare this distance with a typical O–O distance (3.0Å, cf. Table 6.1).

The density of water is the number of water molecules per unit volume. If a single water molecule occupies 29.92Å^3, this corresponds to 0.0334 water molecules per Å^3, or a density of 0.0334Å^{-3}. Near protein surfaces, the density can be significantly higher, and there is a correlation between the average dipole orientation of nearby waters and the local density [226].

18.7.4 Fluid Viscosity and Diffusion

Fluids display an aggregate behavior known as **viscosity**. Fluid dynamicists [44] call the viscosity μ and physicists [293] call it η. The units of the coefficient of viscosity (often called **dynamic viscosity**) are mass per length-time. A standard unit of viscosity is the poise,[2] which is one gram per centimeter-second. One poise is 0.1 pascal-second, where a pascal is a unit of pressure or stress. One pascal is one newton per meter-squared, where we recall that a newton (one kilogram-meter/second-squared) is a measure of force.

The viscosity of water at 293 degrees Kelvin (20 degrees Centigrade) is about one centipoise, or about 0.001 pascal-second. The viscosity of olive oil is about 80 times larger, so the ratio of viscosities of olive oil and water is roughly the ratio the dielectric of water and vacuum. The viscosity of air is 0.0018 centipoise, a factor of over five-hundred smaller.

18.7.5 Kinematic Viscosity

Another scaling factor is significant in fluid flow, namely the fluid density. The ratio of viscosity (or dynamic viscosity) and density is called **kinematic viscosity**, usually labeled ν. This has units length-squared per time, since density has units of mass per length-cubed. Thus, kinematic viscosity has the same units as a spatial diffusion constant. The stoke is one centimeter-squared per second. The kinematic viscosity of water is about one millimeter-squared per second, or one centistoke, whereas the kinematic viscosity of air is roughly two times *larger*. That is, air is more viscous than water! The viscosity of fluids varies significantly with temperature, but we have provided values at roughly the same temperature (293 K) for comparison.

Viscous drag is the effective force of viscosity in opposing motion. It provides a retarding force in the direction opposite to the motion. The drag coefficient has the units of force divided by velocity, or mass per time unit.

18.8 Exercises

Exercise 18.1. Using the Bohr radius $a_0 = 0.5291772$Å as the basic unit of length, called a bohr, the svedberg as time unit, and the amu as mass unit, compute the unit of energy, one amu-(bohr/svedberg)^2 in terms of the unit kcal/mole.

---

[2] The unit of viscosity is named for Jean Louis Marie Poiseuille (1799–1869) who, together with Gotthilf Heinrich Ludwig Hagen (1797–1884) established the basic properties of viscous flow in simple geometries.

Exercise 18.2. Determine a basic unit of length L such that, with the svedberg as time unit, and the dalton as mass unit, then the unit of energy of one dalton-$(L/\text{svedberg})^2$ is exactly one kcal/mole. Compare L with the van der Waals radius of different atoms (which are closest?).

Exercise 18.3. Determine a three-dimensional volume which can be used to tile space and fits a water molecule better than a cubic box. Use this volume to estimate the density of water.

Exercise 18.4. Suppose we take the dalton as the mass unit and that we choose space and time units so that the speed of light and Planck's constant are both one. What is the temperature scale that makes Boltzmann's constant equal to one?

Exercise 18.5. Suppose we take the mass of the electron as the mass unit, and that we want units so that the speed of light and Planck's constant are both one. What are the corresponding time and length scales?

Exercise 18.6. Suppose temperature is in degrees Kelvin and mass is in daltons. Determine a velocity scale such that $k_B = 1$.

Exercise 18.7. The fine structure constant is

$$\alpha = \frac{q_e^2}{2\,h\,\varepsilon_0\,c}, \qquad (18.13)$$

where m_e and q_e are the mass and charge of the electron, respectively, ε_0 is the permittivity of free space, h is Planck's constant, and c is the speed of light in a vacuum. Prove that α is dimensionless. Determine other combinations of various physical constants that are also dimensionless.

Exercise 18.8. The Rydberg constant R_∞ is

$$R_\infty = \frac{m_e\,q_e^4}{8\,\varepsilon_0^2\,h^3\,c}, \qquad (18.14)$$

where m_e and q_e are the mass and charge of the electron, respectively, ε_0 is the permittivity of free space, h is Planck's constant, and c is the speed of light in a vacuum. Prove that the hartree $E_h = 2R$.

Exercise 18.9. We have two equations for the Hartree E_h, namely, $E_h = m_e c^2 \alpha^2$ and $q_e^2/(4\pi\varepsilon_0 a_0) = E_h$. Show that these are compatible, that is, $4\pi\varepsilon_0 = q_e^2/a_0 m_e c^2 \alpha^2$.

Exercise 18.10. The fine structure constant is $\alpha \approx 1/137.035999074$. Suppose that we want the product $c\alpha = 1$ in our units, instead of $c = 1$ [282]. What is the corresponding new length scale L with the other choices of units in Section 18.5 being the same?

Chapter 19
Notes

William Gardiner Pritchard (1942–1994) was an experimental fluid dynamicist who worked closely with theoretical and computational mathematicians to explore the domains of validity of fluid models.

The following notes are intended to complement the discussion in the indicated chapters.

Chapter 2

We are indebted to the noted photographer Ron Scott for the suggestion of the grain of sand in an oyster to explain epidiorthotric effects. Regarding epidiorthotric forces in psychosocial contexts, we have in mind something like a sense of insecurity that often drives highly talented people. Groucho Marx expressed this via the conundrum that he would never join a club that would stoop so low as to have him as a member. A prolific inventor, Larkin Burneal Scott (1917–1991), expressed it in a jest as follows: "I don't have an inferiority complex; I *am* inferior."

Chapter 3

The origin of the octet rule is the periodic table, initiated in 1869 by Dmitri Ivanovitch Mendeleev. It is useful to reflect on the history of the development of the understanding of different types of electrostatic forces and structural features of proteins. In Table 19.1, we present these together with approximate dates of emergence and representative citations. Dates should be considered only approximate and not a definitive statement about priority. Similarly, the references presented are intended only to give a sense of some central contributions.

Force/structure name	emergence date(s)	selected references
van der Waals	1873	[499]
Keesom dipole	1915	[322]
covalent bond	1916–19	[294, 310]
Debye induction	1920	[322]
hydrogen bond	1920–33–44	[297]
London dispersion	1930	[322]
alpha/beta structure	1951	[384]
hydrophobic force	1954–present	[118]
cation-π interaction	1996	[123]
dehydron	2003	[161, 172, 174]

Table 19.1 A brief history of the development of understanding of the principal bonds that are significant for protein structure and interaction.

© Springer International Publishing AG 2017
L.R. Scott and A. Fernández, *A Mathematical Approach to Protein Biophysics*,
Biological and Medical Physics, Biomedical Engineering,
DOI 10.1007/978-3-319-66032-5_19

Chapter 20
Glossary

The following definitions are intended only as an informal description. Other sources should be consulted for definitive meanings. Our objective is simply to provide a rapid, if approximate, way for the reader to return to the main part of the book without needing to resort to another text or an online resource. The definitions often depend on other definitions; terms that are used in one definition and are defined separately are indicated in a separate font *like this*. Many terms are defined fully in the text, and the index provides pointers to these explications. If a term is not found here, consult the index next.

An **aliphatic chain** is a group of atoms in which the electron distribution is localized around each atom, e.g., the noncyclic chains of CH_n groups in Leucine and Isoleucine.

Allostery is derived from the Greek meaning "other shape." An **allosteric** effect [354] is one that is induced at one part of a molecule by an effect (e.g., ligand binding) at another.

An **amino acid** is a molecule that forms the basis of the sidechain of a protein.

An **amide group** is the $N\text{–}H$ pair in a peptide bond as shown in Figure 14.1.

Something is **amphiphilic** if it is a combination of hydrophobic and hydrophilic parts.

An **antigen** is the entity to which antibodies bind. In general, these should be entities that are foreign such as bacteria.

An **aromatic chain** is a system in which the electron distribution is distributed around many atoms, e.g., the cyclic chains of CH_n groups in phenylalanine.

A protein **antagonist** blocks the intended binding of a ligand to a receptor.

A **beta-hairpin** (or β-hairpin) is a secondary structural unit consisting of two β strands with a turn between them so that the β strands align to some degree.

A **beta strand** (or β strand) is a secondary structural unit consisting of one component of a β sheet, that is, a sequence of residues that have the same alternating backbone angle θ defined in (5.3) required for a β sheet but do not have a defined partner with which to form hydrogen bonds.

The **backbone** is the name for the sequence of C_α carbons in a protein chain, that is, the lower left and upper right C's in Figure 14.1.

A **bidentate** interaction involves two separate interactions between heavy atoms, such as can occur between terminal atoms on Arg and either Glu or Asp.

A **capsid** is the outer coat of a virus, typically a protein complex.

A **carbonyl group** is the $C = O$ pair in a peptide bond as shown in Figure 14.1.

C. elegans, or more completely **Caenorhabditis elegans**, is a worm.

© Springer International Publishing AG 2017
L.R. Scott and A. Fernández, *A Mathematical Approach to Protein Biophysics*,
Biological and Medical Physics, Biomedical Engineering,
DOI 10.1007/978-3-319-66032-5_20

A **chain** is an individual protein in a *protein complex*

A **coil** (a.k.a., loop) in a protein structure is a sequence of residues without alpha helix or beta-sheet structure. It can be quite long (several tens of residues) and is generally believed to be without a predetermined geometry, i.e., it can easily change its shape.

The **conformation** of a protein is the three-dimensional shape that it adopts. A change in conformation means that a new shape is adopted.

A **covalent** bond is an electrostatic bond in which the electrons of different atoms become intertwined and can no longer be identified as belonging to a distinct atom.

A **crystal** is a lattice of objects, such as proteins, that can form under certain conditions. The repeated (periodic) structure in particular allows them to be imaged using X-rays.

A **dimer** is an object made of two *monomers*, typically the same or very similar.

A protein **domain** is the basic unit of *tertiary structure*. A single protein can consist of a single domain or many domains. A domain is an autonomously foldable portion of a protein chain.

DNA is the acronym for deoxyribonucleic acid.

E. coli, or more completely **Escherichia coli**, is a bacterium commonly found in food.

Electron density refers to the fact that electrons cannot be located exactly, but rather a probabilistic description is used in quantum mechanics to describe where they spend a fraction of their time.

Endocytosis is a process of cell surface folding that ingests a substance.

The term **epidiorthotric** was introduced in Section 2.2.3 to refer to an effect that occurs as the result of a repair of a defect.

An **epitope** is a small region of a protein involved in a binding event, such as the part of an *antigen* where an antibody binds.

A protein **fold** is the topological description of a *tertiary structure*.

Homo sapien (sometimes written **H. sapien**) is the formal biological name for a human being.

Something is **hydrophobic** if it repels water.

Something is **hydrophilic** if it attracts water.

Hydrophobic packing refers to the placement of carbonaceous groups in the vicinity.

A **hydroxyl** group is the OH group at the ends of the sidechains of serine, threonine, and tyrosine.

A **ligand** is anything that binds to something.

A **loop** in a protein structure is an alternate designation for a *coil*.

A **missense mutation** is gene codon mutation that is not synonymous and not a *nonsense mutation*.

A **moiety** is a portion of a whole, usually with a defined property or structure.

A **monodentate** interaction involves only separate interactions between heavy atoms, as opposed to *bidentate* interactions. Interactions between terminal atoms on Arg and either Glu or Asp can be either monodentate or bidentate.

A **monomer** is a single unit, e.g., a peptide, that can join with one or more other monomers of the same type to form a larger complex. The term can be used for something as small as a single molecule or as large as a protein.

A **motif** is a characteristic feature.

A **multimer** is something formed from small units, often called *monomers*, cf. *polymer*.

Mus musculus is the formal biological name for the common house mouse.

A **noncovalent** bond, or interaction, is an electrostatic interaction in which the electrons of the atoms in the bond remain sufficiently apart to remain identified with their atoms, even though they may be strongly correlated.

A **nonsense mutation** is gene codon mutation to a stop codon.

A **packing defect** is a deficiency in wrapping, that is, a lack of adequate carbonaceous groups in the vicinity.

A **partial charge** is a model to account for the fact that electrons may be unevenly distributed in a molecule.

A **peptide** is the basic unit of a protein.

Something is **polar** if it has a positive *partial charge* on one side and a negative *partial charge* on the other.

A **polymer** is something formed from small units, often called *monomers*, cf. *multimer*.

To **polymerize** is to form a larger system from small units, e.g., a chain such as a protein.

A **polypeptide** is the result of polymerizing peptides.

A graph has **power-law distribution** if the number of vertices with degree k is roughly $k^{-\gamma}$ for a fixed γ.

The **primary structure** of a protein is the sequence of its amino acids.

A **protein complex** is a collection of two or more proteins that are bound together.

The **quaternary structure** of a protein system is the three-dimensional arrangement of the different protein *chains* of the system.

An amino acid **residue** is part of the amino acid that remains when it is cleaved to form a sidechain on a protein.

RNA is the acronym for ribonucleic acid.

A graph is **scale free** if obeys a *power-law distribution*, that is, if the number of vertices with degree k is roughly $k^{-\gamma}$ for a fixed γ.

A **small world** graph is one in which most of the vertices have low degree, such as a graph with a *power-law distribution*.

The **secondary structure** of a protein is the set of alpha helices, beta sheets, turns, and loops.

A **sidechain** is another name for *residue*.

The **solvent accessible surface** of a protein is the surface obtained by rolling a ball of a fixed diameter (usually related to the size of a water molecule) around a protein. Such a surface need not be connected.

A protein is **soluble** if it can form a stable and functional form in water.

A **somatic mutation** is one that occurs after birth, and is neither inherited or passed on to offspring, such as occur during antibody maturation.

A **steric** effect is one that involves the shape of an object (the word steric derives from the Greek word for "shape"; also see the explanation of *allostery*).

Protein **structure** is hierarchical, involving *primary, secondary, tertiary*, and, for a *protein complex, quaternary* structure.

A **subunit** of a *protein complex* is one of the proteins in the collection.

The **tertiary structure** of a protein is the three-dimensional shape of the fully folded proteins.

A **tetramer** is an object made of four *monomers*, typically the same or very similar.

The **three-letter code** for RNA and DNA is the sequence of three letters that code for a particular amino acid.

A **trimer** is an object made of three *monomers*, typically the same or very similar.

A **turn** in a protein structure is a short sequence (typically four) of residues without alpha helix or beta-sheet structure [518].

UWHB is the acronym for underwrapped hydrogen bond, a.k.a. dehydron.

Vicinal means "in the vicinity" or nearby.

A **widget** is an object that can be used in larger systems in a generic way.

References

1. S.R. Accordino, J.A. Rodriguez Fris, G.A. Appignanesi, A. Fernández, A unifying motif of intermolecular cooperativity in protein associations. Eur. Phys. J. E **35**(7), 1–7 (2012)
2. H. Adams, K.D.M. Harris, G.A. Hembury, C.A. Hunter, D. Livingstone, J.F. McCabe, How strong is a π-facial hydrogen bond? Chem. Commun. 2531–2532 (1996)
3. N. Agmon, Tetrahedral displacement: the molecular mechanism behind the Debye relaxation in water. J. Phys. Chem. **100**(1), 1072–1080 (1996)
4. C.J. Allègre, G. Manhès, C. Göpel, The age of the earth. Geochimica et Cosmochimica Acta **59**(8), 1445–1456 (1995)
5. M.P. Allen, D.J. Tildesley, *Computer Simulation of Liquids* (Oxford Science Publications, Clarendon Press, Oxford, 1989)
6. I. Anapolitanos, *On van der Waals forces*. Ph.D. thesis, University of Toronto, 2011
7. D.E. Anderson, J.H. Hurley, H. Nicholson, W.A. Baase, B.W. Matthews, Hydrophobic core repacking and aromatic-aromatic interaction in the thermostable mutant of T4 lysozyme Ser 117 →. Protein Sci. **2**(8), 1285–1290 (1993)
8. D.E. Anderson, W.J. Becktel, F.W. Dahlquist, pH-induced denaturation of proteins: a single salt bridge contributes 3–5 kcal/mol to the free energy of folding of T4 lysozyme. Biochemistry **29**(9), 2403–2408 (1990)
9. A. Andreeva, D. Howorth, J.-M. Chandonia, S.E. Brenner, T.J.P. Hubbard, C. Chothia, A.G. Murzin, Data growth and its impact on the SCOP database: new developments. Nucl. Acids Res. **36**, D419–425 (2008)
10. J. Applequist, J.R. Carl, K.-K. Fung, Atom dipole interaction model for molecular polarizability. Application to polyatomic molecules and determination of atom polarizabilities. J. Am. Chem. Soc. **94**(9), 2952–2960 (1972)
11. D.H. Ardell, G. Sella, No accident: genetic codes freeze in error-correcting patterns of the standard genetic code. Philos. Trans. R. Soc. Lond. Ser. B: Biol. Sci. **357**(1427), 1625 (2002)
12. I.T. Arkin, H. Xu, M.O. Jensen, E. Arbely, E.R. Bennett, K.J. Bowers, E. Chow, R.O. Dror, M.P. Eastwood, R. Flitman-Tene, B.A. Gregersen, J.L. Klepeis, I. Kolossvary, Y. Shan, D.E. Shaw, Mechanism of Na+/H+ antiporting. Science **317**(5839), 799–803 (2007)
13. M.R. Arkin, J.A. Wells, Small-molecule inhibitors of protein-protein interactions: progressing towards the dream. Nat. Rev.-Drug Discov. **3**(4), 301–317 (2004)
14. J. Arnórsdóttir, A.R. Sigtryggsdóttir, S.H. Thorbjarnardóttir, M.M. Kristjánsson, Effect of proline substitutions on stability and kinetic properties of a cold adapted subtilase. J. Biochem. **145**(3), 325–329 (2009)
15. P.J. Artymiuk, C.C.F. Blake, Refinement of human lysozyme at 1.5 Å resolution analysis of non-bonded and hydrogen-bond interactions. J. Mol. Biol. **152**(4), 737–762 (1981)
16. K. Asai, S. Hayamizu, K. Handa, Prediction of protein secondary structure by the hidden Markov model. Comput. Appl. Biosci. **9**(2), 141–146 (1993)
17. F. Avbelj, J. Moult, Role of electrostatic screening in determining protein main chain conformational preferences. Biochemistry **34**(3), 755–764 (1995)

© Springer International Publishing AG 2017
L.R. Scott and A. Fernández, *A Mathematical Approach to Protein Biophysics*,
Biological and Medical Physics, Biomedical Engineering,
DOI 10.1007/978-3-319-66032-5

18. F. Avbelj, Amino acid conformational preferences and solvation of polar backbone atoms in peptides and proteins. J. Mol. Biol. **300**(5), 1335–1359 (2000)

19. F. Avbelj, R.L. Baldwin, Role of backbone solvation in determining thermodynamic β propensities of the amino acids. Proc. Natl. Acad. Sci. USA **99**(3), 1309–1313 (2002)

20. F. Avbelj, S.G. Grdadolnik, J. Grdadolnik, R.L. Baldwin, Intrinsic backbone preferences are fully present in blocked amino acids. Proc. Natl. Acad. Sci. USA **103**(5), 1272–1277 (2006)

21. F. Avbelj, P. Luo, R.L. Baldwin, Energetics of the interaction between water and the helical peptide group and its role in determining helix propensities. Proc. Natl. Acad. Sci. USA **97**, 10786–10791 (2000)

22. J.C. Avise, B.W. Bowen, T. Lamb, A.B. Meylan, E. Bermingham, Mitochondrial DNA evolution at a turtle's pace: evidence for low genetic variability and reduced microevolutionary rate in the Testudines. Mol. Biol. Evol. **9**(3), 457–473 (1992)

23. R. Azriel, E. Gazit, Analysis of the minimal amyloid-forming fragment of the islet amyloid polypeptide. An experimental support for the key role of the phenylalanine residue in amyloid formation. J. Biol. Chem. **276**(36), 34156–34161 (2001)

24. C. Azuara, E. Lindahl, P. Koehl, H. Orland, M. Delarue, PDB_Hydro: incorporating dipolar solvents with variable density in the Poisson-Boltzmann treatment of macromolecule electrostatics. Nucleic Acids Res. **34**(Web Server issue), W38 (2006)

25. C. Azuara, H. Orland, M. Bon, P. Koehl, M. Delarue, Incorporating dipolar spriolvents with variable density in Poisson-Boltzmann electrostatics. Biophys. J. **95**(12), 5587–5605 (2008)

26. M. Bachmayr, Adaptive low-rank wavelet methods and applications to two-electron Schrödinger equations. Ph.D. thesis, Universitätsbibliothek, 2012

27. M. Bachmayr, W. Dahmen, Adaptive near-optimal rank tensor approximation for high-dimensional operator equations. Found. Comput. Math. 1–60 (2013)

28. R.P. Bahadur, P. Chakrabarti, F. Rodier, J. Janin, Dissecting subunit interfaces in homodimeric proteins. Proteins: Struct. Funct. Genet. **53**, 708–719 (2003)

29. R.P. Bahadur, P. Chakrabarti, F. Rodier, J. Janin, A dissection of specific and non-specific protein-protein interfaces. J. Mol. Biol. **336**, 943–955 (2004)

30. Y. Bai, S.W. Englander, Hydrogen bond strength and β-sheet propensities: the role of a side chain blocking effect. Proteins-Struct. Funct. Genet. **18**(3), 262–266 (1994)

31. Y. Bai, J.S. Milne, L. Mayne, S.W. Englander, Primary structure effects on peptide group hydrogen exchange. Proteins-Struct. Funct. Genet. **17**, 75–86 (1993)

32. E.N. Baker, R.E. Hubbard, Hydrogen bonding in globular proteins. Prog. Biophys. Mol. Biol. **44**, 97–179 (1984)

33. N.A. Baker, Improving implicit solvent simulations: a poisson-centric view. Curr. Opin. Struct. Biol. **15**(2), 137–143 (2005)

34. R.L. Baldwin, In search of the energetic role of peptide hydrogen bonds. J. Biol. Chem. **278**(20), 17581–17588 (2003)

35. R.L. Baldwin, Energetics of protein folding. J. Mol. Biol. **371**(2), 283–301 (2007)

36. R.L. Baldwin, Protein folding: making a network of hydrophobic clusters. Science **295**, 1657–1658 (2002)

37. Y.-E.A. Ban, H. Edelsbrunner, J. Rudolph, Interface surfaces for protein-protein complexes, in *Proceedings of the 8th Annual International Conference on Research in Computational Molecular Biology (RECOMB)* (2004), pp. 205–212

38. A.L. Barabasi, *Linked: The New Science of Networks* (Perseus, New York, 2002)

39. J.P. Bardhan, M.G. Knepley, Communication: modeling charge-sign asymmetric solvation free energies with nonlinear boundary conditions. J. Chem. Phys. **141**(13), 131103 (2014)

40. K. Bartik, C. Redfield, C.M. Dobson, Measurement of the individual pK_a values of acidic residues of hen and turkey lysozymes by two-dimensional ^{1}H NMR. Biophys. J. **66**, 1180–1184 (1994)

41. A.P. Bartók, M.J. Gillan, F.R. Manby, G. Csányi, Machine-learning approach for one-and two-body corrections to density functional theory: applications to molecular and condensed water. Phys. Rev. B **88**(5), 054104 (2013)

42. M.V. Basilevsky, D.F. Parsons, An advanced continuum medium model for treating solvation effects: nonlocal electrostatics with a cavity. J. Chem. Phys. **105**(9), 3734–3746 (1996)

43. V. Basilevsky, G.N. Chuev, Nonlocal solvation theories, in *Continuum Solvation Models in Chemical Physics: From Theory to Applications*, ed. by B. Mennucci, R. Cammi (Wiley, New York, 2008)

44. G.K. Batchelor, *An Introduction to Fluid Dynamics* (Cambridge University Press, Cambridge, 2000)

45. P.A. Bates, P. Dokurno, P.S. Freemont, M.J.E. Sternberg, Conformational analysis of the first observed non-proline cis-peptide bond occurring within the complementarity determining region (CDR) of an antibody. J. Mol. Biol. **284**(3), 549–555 (1998)

46. E.R. Batista, S.S. Xantheas, H. Jonsson, Multipole moments of water molecules in clusters and ice Ih from first principles calculations. J. Chem. Phys. **111**(13), 6011–6015 (1999)

47. E. Bellacchio, K.L. McFarlane, A. Rompel, J.H. Robblee, R.M. Cinco, V.K. Yachandra, Counting the number of disulfides and thiol groups in proteins and a novel approach for determining the local pK_a for cysteine groups in proteins in vivo. J. Synchrotron Radiat. **8**(3), 1056–1058 (2001)

48. A.Y. Ben-Naim, *Hydrophobic Interactions* (Springer, Berlin, 1980)

49. R. Berisio, V.S. Lamzin, F. Sica, K.S. Wilson, A. Zagari, L. Mazzarella, Protein titration in the crystal state. J. Mol. Biol. **292**, 845–854 (1999)

50. D. Berleant, M. White, E. Pierce, E. Tudoreanu, A. Boeszoermenyi, Y. Shtridelman, J.C. Macosko, The genetic code-more than just a table. Cell Biochem. Biophys. **55**(2), 107–116 (2009)

51. T. Beuming, Y. Che, R. Abel, B. Kim, V. Shanmugasundaram, W. Sherman, Thermodynamic analysis of water molecules at the surface of proteins and applications to binding site prediction and characterization. Proteins: Struct. Funct. Bioinform. **80**(3), 871–883 (2012)

52. C.C. Bigelow, On the average hydrophobicity of proteins and the relation between it and protein structure. J. Theor. Biol. **16**(2), 187–211 (1967)

53. C. Biot, E. Buisine, M. Rooman, Free-energy calculations of proteinligand cationπ and aminoπ interactions: from vacuum to proteinlike environments. J. Am. Chem. Soc. **125**(46), 13988–13994 (2003)

54. M. Bittelli, M. Flury, K. Roth, Use of dielectric spectroscopy to estimate ice content in frozen porous media. Water Resour. Res. **40**, W04212 (2004)

55. A.A. Bogan, K.S. Thorn, Anatomy of hot spots in protein interfaces. J. Mol. Biol. **280**, 1–9 (1998)

56. A. Bondi, van der Waals volumes and radii. J. Phys. Chem. **68**(3), 441–451 (1964)

57. P.A. Bopp, A.A. Kornyshev, G. Sutmann, Static nonlocal dielectric function of liquid water. Phys. Rev. Lett. **76**, 1280–1283 (1996)

58. D. Borwein, J.M. Borwein, K.F. Taylor, Convergence of lattice sums and Madelung's constant. J. Math. Phys. **26**(11), 2999–3009 (1985)

59. C. Branden, J. Tooze, *Introduction to Protein Structure*, 2nd edn. (Garland Pub., New York, 1991)

60. S.C. Brenner, L.R. Scott, *The Mathematical Theory of Finite Element Methods*, 3rd edn. (Springer, Berlin, 2008)

61. E. Breslow, V. Mombouyran, R. Deeb, C. Zheng, J.P. Rose, B.C. Wang, R.H. Haschemeyer, Structural basis of neurophysin hormone specificity: geometry, polarity, and polarizability in aromatic ring interactions. Protein Sci. **8**(4), 820–831 (1999)

62. D.J. Brooks, J.R. Fresco, A.M. Lesk, M. Singh, Evolution of amino acid frequencies in proteins over deep time: inferred order of introduction of amino acids into the genetic code. Mol. Biol. Evol. **19**, 1645–1655 (2002)

63. William Fuller Brown, Jr. Dielectrics, in *Handbuch der Physik*, vol. 14 (Springer, Berlin, 1956), pp. 1–154

64. R.G. Bryant, The dynamics of water-protein interactions. Annu. Rev. Biophys. Biomol. Struct. **25**, 29–53 (1996)

65. M. Bryliński, L. Konieczny, I. Roterman, Ligation site in proteins recognized in silico. Bioinformation **1**(4), 127–129 (2006)

66. M. Bryliński, K. Prymula, W. Jurkowski, M. Kochańczyk, E. Stawowczyk, L. Konieczny, I. Roterman, Prediction of functional sites based on the fuzzy oil drop model. PLoS Comput. Biol. **3**(5), e94 (2007)

67. A.D. Buckingham, P.W. Fowler, Do electrostatic interactions predict structures of van der Waals molecules? J. Chem. Phys. **79**(12), 6426–6428 (1983)

68. A.D. Buckingham, P.W. Fowler, A model for the geometries of van der Waals complexes. Can. J. Chem. **63**, 2018–2025 (1985)

69. S.K. Burley, G.A. Petsko, Amino-aromatic interactions in proteins. FEBS Lett. **203**(2), 139–143 (1986)

70. S.K. Burley, G.A. Petsko, Electrostatic interactions in aromatic oligopeptides contribute to protein stability. Trends Biotechnol. **7**(12), 354–359 (1989)

71. A. Buzzell, Action of urea on tobacco mosaic virus II. The bonds between protein subunits. Biophys. J. **2**(2P1), 223–233 (1962)

72. Christopher Bystroff, Vesteinn Thorsson, David Baker, HMMSTR: a hidden Markov model for local sequence-structure correlations in proteins. J. Mol. Biol. **301**(1), 173–190 (2000)

73. O.M. Cabarcos, C.J. Weinheimer, J.M. Lisy, Competitive solvation of K^+ by benzene and water: cation-π interactions and π-hydrogen bonds. J. Chem. Phys. **108**(13), 5151–5154 (1998)

74. D.R. Caffrey, S. Somaroo, J.D. Hughes, J. Mintseris, E.S. Huang, Are protein-protein interfaces more conserved in sequence than the rest of the protein surface? Protein Sci. **13**(1), 190–202 (2004)

75. J.W. Caldwell, P.A. Kollman, Cation-π interactions: nonadditive effects are critical in their accurate representation. J. Am. Chem. Soc. **117**(14), 4177–4178 (1995)

76. C. Caleman, J.S. Hub, P.J. van Maaren, D. van der Spoel, Atomistic simulation of ion solvation in water explains surface preference of halides. Proc. Natl. Acad. Sci. **108**(17), 6838–6842 (2011)

77. H.A. Carlson, J.A. McCammon, Accommodating protein flexibility in computational drug design. Mol. Pharmacol. **57**(2), 213–218 (2000)

78. F.J. Carver, C.A. Hunter, D.J. Livingstone, J.F. McCabe, E.M. Seward, Substituent effects on edge-to-face aromatic interactions. Chem.-A. Eur. J. **8**(13), 2847–2859 (2002)

79. A. Cauerhff, F.A. Goldbaum, B.C. Braden, Structural mechanism for affinity maturation of an anti-lysozyme antibody. Proc. Natl. Acad. Sci. USA **101**(10), 3539–3544 (2004)

80. P. Chakrabarti, R. Bhattacharyya, Geometry of nonbonded interactions involving planar groups in proteins. Prog. Biophys. Mol. Biol. **95**(1–3), 83–137 (2007)

81. P. Chakrabarti, J. Janin, Dissecting protein-protein recognition sites. Proteins: Struct. Funct. Genet. **47**, 334–343 (2002)

82. J. Chalupsky, J. Vondrasek, V. Spirko, Quasiplanarity of the peptide bond. J. Phys. Chem. A **112**(4), 693–699 (2008)

83. D. Chandler, Interfaces and the driving force of hydrophobic assembly. Nature **437**(7059), 640–647 (2005)

84. J. Chen, X. Zhang, A. Fernández, Molecular basis for specificity in the druggable kinome: sequence-based analysis. Bioinformatics **23**(5), 563–572 (2007)

85. D.A. Cherepanov, Force oscillations and dielectric overscreening of interfacial water. Phys. Rev. Lett. **93**, 266104 (2004)

86. H. Choi, H. Kang, H. Park, New angle-dependent potential energy function for backbonebackbone hydrogen bond in proteinprotein interactions. J. Comput. Chem. **31**(5), 897–903 (2010)

87. L.T. Chong, Y. Duan, L. Wang, I. Massova, P.A. Kollman, Molecular dynamics and free-energy calculations applied to affinity maturation in antibody 48G7. Proc. Natl. Acad. Sci. **96**(25), 14330–14335 (1999)

88. T.C. Choy, Van der Waals interaction of the hydrogen molecule: an exact implicit energy density functional. Phys. Rev. A **62**, 012506 (2000)

89. J.E. Chrencik, J. Orans, L.B. Moore, Y. Xue, L. Peng, J.L. Collins, G.B. Wisely, M.H. Lambert, S.A. Kliewer, M.R. Redinbo, Structural disorder in the complex of human pregnane X receptor and the macrolide antibiotic rifampicin. Mol. Endocrinol. **19**(5), 1125–1134 (2005)

90. G.N. Chuev, M.V. Basilevsky, Molecular models of solvation in polar liquids. Rus. Chem. Rev. **72**, 735–757 (2003)

91. T. Clackson, J.A. Wells, A hot spot of binding energy in a hormone-receptor interface. Science **267**, 383–386 (1995)

92. A.J. Cohen, P. Mori-Sánchez, W. Yang, Insights into current limitations of density functional theory. Science **321**(5890), 792–794 (2008)

93. M. Cohen, D. Reichmann, H. Neuvirth, G. Schreiber, Similar chemistry, but different bond preferences in inter versus intra-protein interactions. Proteins: Struct. Funct. Bioinform. **72**(2), 741–753 (2008)

94. M.S. Cohen, C. Zhang, K.M. Shokat, J. Taunton, Structural bioinformatics-based design of selective, irreversible kinase inhibitors. Science **308**(5726), 1318–1321 (2005)

95. P. Cohen, Protein kinases-the major drug targets of the twenty-first century? Nat. Rev.-Drug Discov. **1**(4), 309–315 (2002)

96. C. Colovos, T.O. Yeates, Verification of protein structures: patterns of nonbonded atomic interactions. Protein Sci. **2**(9), 1511–1519 (1993)

97. B.A. Colson, T. Bekyarova, D.P. Fitzsimons, T.C. Irving, R.L. Moss, Radial displacement of myosin cross-bridges in mouse myocardium due to ablation of myosin binding protein-C. J. Mol. Biol. **367**(1), 36–41 (2007)

98. P.R. Connelly, R.A. Aldape, F.J. Bruzzese, S.P. Chambers, M.J. Fitzgibbon, M.A. Fleming, S. Itoh, D.J. Livingston, M.A. Navia, J.A. Thomson et al., Enthalpy of hydrogen bond formation in a protein-ligand binding reaction. Proc. Natl. Acad. Sci. USA **91**(5), 1964–1968 (1994)

99. S. Costantini, G. Colonna, A.M. Facchiano, ESBRI: a web server for evaluating salt bridges in proteins. Bioinformation **3**(3), 137 (2008)

100. F.-X. Coudert, R. Vuilleumiear, A. Boutin, Dipole moment, hydrogen bonding and IR spectrum of confined water. ChemPhysChem **7**(12), 2464–2467 (2006)

101. C.J. Cramer, *Essentials of Computational Chemistry*, 2nd edn. (Wiley, New York, 2004)

102. E. Thomas, *Creighton, Proteins: Structures and Molecular Properties* (W. H. Freeman, New York, 1993)

103. P.B. Crowley, A. Golovin, Cation-π interactions in protein-protein interfaces. Proteins: Struct. Funct. Bioinform. **59**, 231–239 (2005)

104. L.E. Cybulski, M. Martín, M.C. Mansilla, A. Fernández, D. De Mendoza, Membrane thickness cue for cold sensing in a bacterium. Curr. Biol. **20**(17), 1539–1544 (2010)

105. B.I. Dahiyat, D. Benjamin Gordon, S.L. Mayo, Automated design of the surface positions of protein helices. Protein Sci. **6**(6), 1333–1337 (1997)

106. M. Davies, C. Toseland, D. Moss, D. Flower, Benchmarking pK_a prediction. BMC Biochem. **7**(1), 18 (2006)

107. A. De Simone, G.G. Dodson, C.S. Verma, A. Zagari, F. Fraternali, Prion and water: tight and dynamical hydration sites have a key role in structural stability. Proc. Natl. Acad. Sci. USA **102**, 7535–7540 (2005)

108. A. De Simone, R. Spadaccini, P.A. Temussi, F. Fraternali, Toward the understanding of MNEI sweetness from hydration map surfaces. Biophys. J. **90**(9), 3052–3061 (2006)

109. A. De Simone, A. Zagari, P. Derreumaux, Structural and hydration properties of the partially unfolded states of the prion protein. Biophys. J. **93**(4), 1284–1292 (2007)

110. P. Debye, *Polar Molecules* (Dover, New York, 1945)

111. W.L. DeLano, M.H. Ultsch, A.M. de Vos, J.A. Wells, Convergent solutions to binding at a protein-protein interface. Science **287**, 1279–1283 (2000)

112. O. Demerdash, E.-H. Yap, T. Head-Gordon, Advanced potential energy surfaces for condensed phase simulation. Annu. Rev. Phys. Chem. **65**, 149–174 (2014)

113. O.N.A. Demerdash, J.C. Mitchell, Using physical potentials and learned models to distinguish native binding interfaces from de novo designed interfaces that do not bind. Proteins: Struct. Funct. Bioinform. **81**(11), 1919–1930 (2013)

114. G.D. Demetri, Structural reengineering of imatinib to decrease cardiac risk in cancer therapy. J. Clin. Investig. **117**(12), 3650–3653 (2007)

115. A.H. DePace, A. Santoso, P. Hillner, J.S. Weissman, A critical role for amino-terminal glutamine/asparagine repeats in the formation and propagation of a yeast prion. Cell **93**(7), 1241–1252 (1998)

116. C. Deremble, R. Lavery, Macromolecular recognition. Curr. Opinion Struct. Biol. **15**(2), 171–175 (2005)

117. F. Despa, R.S. Berry, The origin of long range attraction between hydrophobes in water. Biophys. J. **92**, 373–378 (2007)

118. F. Despa, A. Fernández, R.S. Berry, Dielectric modulation of biological water. Phys. Rev. Lett. **93**, 269901 (2004)

119. M. Di Giulio, On the origin of the genetic code. J. Theor. Biol. **187**(4), 573–581 (1997)

120. C.M. Dobson, Protein folding and misfolding. Nature **426**, 884–890 (2003)

121. M. Dobson, M. Luskin, Iterative solution of the quasicontinuum equilibrium equations with continuation. J. Sci. Comput. **37**(1), 19–41 (2008)

122. J.E. Donald, D.W. Kulp, W.F. DeGrado, Salt bridges: geometrically specific, designable interactions. Proteins: Struct. Funct. Bioinform. **79**(3), 898–915 (2011)

123. D.A. Dougherty, Cation-pi interactions in chemistry and biology: a new view of benzene, Phe, Tyr, and Trp. Science **271**(5246), 163–168 (1996)

124. J. Doyle, Beyond the sperical cow. Nature **411**, 151–152 (2001)

125. Y. Duan, C. Wu, S. Chowdhury, M.C. Lee, G. Xiong, W. Zhang, R. Yang, P. Cieplak, R. Luo, T. Lee et al., A point-charge force field for molecular mechanics simulations of proteins based on condensed-phase quantum mechanical calculations. J. Comput. Chem. **24**(16), 1999–2012 (2003)

126. J.J. Duistermaat, *Fourier Integral Operators* (Springer, Berlin, 1996)

127. J. Dunbar, H.P. Yennawar, S. Banerjee, J. Luo, G.K. Farber, The effect of denaturants on protein structure. Protein Sci. **6**, 1727–1733 (1997)

128. W.E.W. Ren, E. Vanden-Eijnden, A general strategy for designing seamless multiscale methods. JCP **1**(1), 1–1 (2010)

129. S.R. Eddy, Profile hidden Markov models. Bioinformatics **14**(9), 755–763 (1998)

130. D. Eisenberg, R.M. Weiss, T.C. Terwilliger, The hydrophobic moment detects periodicity in protein hydrophobicity. Proc. Natl. Acad. Sci. USA **81**(1), 140–144 (1984)

131. A. Elgsaeter, B.T. Stokke, A. Mikkelsen, D. Branton, The molecular basis of erythrocyte shape. Science **234**, 1217–1223 (1986)

132. J.L. Ericksen, *Introduction to the Thermodynamics of Solids* (Springer, Berlin, 1998)

133. L. Esposito, A. De Simone, A. Zagari, L. Vitagliano, Unveiling the omega/psi correlation in high resolution protein structures. Acta Crystallogr. Sect. A **61**(a1), C482 (2005)

134. L. Esposito, A. De Simone, A. Zagari, L. Vitagliano, Correlation between ω and ψ dihedral angles in protein structures. J. Mol. Biol. **347**(3), 483–487 (2005)

135. E. Eyal, S. Gerzon, V. Potapov, M. Edelman, V. Sobolev, The limit of accuracy of protein modeling: influence of crystal packing on protein structure. J. Mol. Biol. **351**(2), 431–442 (2005)

136. D.H. Fagan, D. Yee, Crosstalk between IGF1R and estrogen receptor signaling in breast cancer. J. Mammary Gland Biol. Neoplasia **13**(4), 423–429 (2008)

137. M. Fandrich, M.A. Fletcher, C.M. Dobson, Amyloid fibrils from muscle myoglobin. Nature **410**(6825), 165–166 (2001)
138. E. Fatuzzo, W.J. Merz, *Ferroelectricity* (North-Holland, Amsterdam, 1967)
139. E. Fermi, *Thermodynamics* (Dover, New York, 1956)
140. A. Fernández, S. Bazán, J. Chen, Taming the induced folding of drug-targeted kinases. Trends Pharmacol. Sci. **30**(2), 66–71 (2009)
141. A. Fernández, A. Sanguino, Z. Peng, E. Ozturk, J. Chen, A. Crespo, S. Wulf, A. Shavrin, C. Qin, J. Ma et al., An anticancer C-Kit kinase inhibitor is reengineered to make it more active and less cardiotoxic. J. Clin. Investig. **117**(12), 4044 (2007)
142. A. Fernández, Conformation-dependent environments in folding proteins. J. Chem. Phys. **114**, 2489–2502 (2001)
143. A. Fernández, Cooperative walks in a cubic lattice: protein folding as a many-body problem. J. Chem. Phys. **115**, 7293–7297 (2001)
144. A. Fernández, Intramolecular modulation of electric fields in folding proteins. Phys. Lett. A **299**, 217–220 (2002)
145. A. Fernández, Buffering the entropic cost of hydrophobic collapse in folding proteins. J. Chem. Phys. **121**, 11501–11502 (2004)
146. A. Fernández, Keeping dry and crossing membranes. Nat. Biotechnol. **22**, 1081–1084 (2004)
147. A. Fernández, Direct nanoscale dehydration of hydrogen bonds. J. Phys. D: Appl. Phys. **38**, 2928–2932 (2005)
148. A. Fernández, Incomplete protein packing as a selectivity filter in drug design. Structure **13**, 1829–1836 (2005)
149. A. Fernández, What factor drives the fibrillogenic association of beta-sheets? FEBS Lett. **579**, 6635–6640 (2005)
150. A. Fernández, Nanoscale thermodynamics of biological interfacial tension. Proc. R. Soc. A **299**, 1–10 (2010)
151. A. Fernández, *Transformative Concepts for Drug Design* (Springer, Berlin, 2010)
152. A. Fernández, Epistructural tension promotes protein associations. Phys. Rev. Lett. **108**(18), 188102 (2012)
153. A. Fernández, The principle of minimal episteric distortion of the water matrix and its steering role in protein folding. JChemPhys **139**(8), 085101 (2013)
154. A. Fernández, Communication: chemical functionality of interfacial water enveloping nanoscale structural defects in proteins. JChemPhys **140**(22), 221102 (2014)
155. A. Fernández, Water promotes the sealing of nanoscale packing defects in folding proteins. J. Phys.: Condens. Matter **26**(20), 202101 (2014)
156. A. Fernández, Packing defects functionalize soluble proteins. FEBS Lett. **589**(9), 967–973 (2015)
157. A. Fernández, Quantum theory of interfacial tension quantitatively predicts spontaneous charging of nonpolar aqueous interfaces. Phys. Lett. A (2015)
158. A. Fernández, Acidbase chemistry of frustrated water at protein interfaces. FEBS Lett. **590**(2), 215–223 (2016)
159. A. Fernández, *Physics at the Biomolecular Interface* (Springer, Berlin, 2016)
160. A. Fernández, R.S. Berry, Extent of hydrogen-bond protection in folded proteins: a constraint on packing architectures. Biophys. J. **83**(5), 2475–2481 (2002)
161. A. Fernández, R.S. Berry, Proteins with H-bond packing defects are highly interactive with lipid bilayers: implications for amyloidogenesis. Proc. Natl. Acad. Sci. USA **100**, 2391–2396 (2003)
162. A. Fernández, R.S. Berry, Molecular dimension explored in evolution to promote proteomic complexity. Proc. Natl. Acad. Sci. USA **101**, 13460–13465 (2004)
163. A. Fernández, R.S. Berry, Golden rule for buttressing vulnerable soluble proteins. J. Proteome Res. **9**(5), 2643–2648 (2010)
164. A. Fernández, M. Boland, Solvent environment conducive to protein aggregation. FEBS Lett. **529**, 298–303 (2002)
165. A. Fernández, J. Chen, A. Crespo, Solvent-exposed backbone loosens the hydration shell of soluble folded proteins. J. Chem. Phys. **126**(24), 245103 (2007)
166. A. Fernández, A. Crespo, S. Maddipati, L.R. Scott, Bottom-up engineering of peptide cell translocators based on environmentally modulated quadrupole switches. ACS Nano **2**, 61–68 (2008)
167. A. Fernández, J. Kardos, L.R. Scott, Y. Goto, R.S. Berry, Structural defects and the diagnosis of amyloidogenic propensity. Proc. Natl. Acad. Sci. USA **100**(11), 6446–6451 (2003)
168. A. Fernández, S. Maddipati, A priori inference of cross reactivity for drug-targeted kinases. J. Med. Chem. **49**(11), 3092–3100 (2006)

169. A. Fernández, K. Rogale Plazonic, L.R. Scott, H.A. Scheraga, Inhibitor design by wrapping packing defects in HIV-1 proteins. Proc. Natl. Acad. Sci. USA **101**, 11640–11645 (2004)

170. A. Fernández, A. Sanguino, Z. Peng, A. Crespo, E. Ozturk, X. Zhang, S. Wang, W. Bornmann, G. Lopez-Berestein, Rational drug redesign to overcome drug resistance in cancer therapy: imatinib moving target. Cancer Res. **67**(9), 4028–4033 (2007)

171. A. Fernández, A. Sanguino, Z. Peng, E. Ozturk, J. Chen, A. Crespo, S. Wulf, A. Shavrin, C. Qin, J. Ma, J. Trent, Y. Lin, H.-D. Han, L.S. Mangala, J.A. Bankson, J. Gelovani, A. Samarel, W. Bornmann, A.K. Sood, G. Lopez-Berestein, An anticancer C-Kit kinase inhibitor is reengineered to make it more active and less cardiotoxic. J. Clin. Investig. **117**(12), 4044–4054 (2007)

172. A. Fernández, H.A. Scheraga, Insufficiently dehydrated hydrogen bonds as determinants of protein interactions. Proc. Natl. Acad. Sci. USA **100**(1), 113–118 (2003)

173. A. Fernández, L.R. Scott, Adherence of packing defects in soluble proteins. Phys. Rev. Lett. **91**(4), 18102 (2003)

174. A. Fernández, L.R. Scott, Dehydron: a structurally encoded signal for protein interaction. Biophys. J. **85**, 1914–1928 (2003)

175. A. Fernández, L.R. Scott, Under-wrapped soluble proteins as signals triggering membrane morphology. J. Chem. Phys. **119**(13), 6911–6915 (2003)

176. A. Fernández, L.R. Scott, Modulating drug impact by wrapping target proteins. Expert Opin. Drug Discov. **2**, 249–259 (2007)

177. A. Fernández, L.R. Scott, Drug leads for interactive protein targets with unknown structure. Drug Discov. Today **21**, 531–535 (2016)

178. A. Fernández, L.R. Scott, Three-body interactions in drug design: reconciling "counterintuitive" decisions in lead optimization. Nat. Biotechnol. submitted, 1–4 (2016)

179. A. Fernández, L.R. Scott, R.S. Berry, The nonconserved wrapping of conserved protein folds reveals a trend towards increasing connectivity in proteomic networks. Proc. Natl. Acad. Sci. USA **101**(9), 2823–2827 (2004)

180. A. Fernández, L.R. Scott, R.S. Berry, Packing defects as selectivity switches for drug-based protein inhibitors. Proc. Natl. Acad. Sci. USA **103**, 323–328 (2006)

181. A. Fernández, L.R. Scott, H.A. Scheraga, Amino-acid residues at protein-protein interfaces: why is propensity so different from relative abundance? J. Phys. Chem. B **107**(36), 9929–9932 (2003)

182. A. Fernández, T.R. Sosnick, A. Colubri, Dynamics of hydrogen bond desolvation in protein folding. J. Mol. Biol. **321**(4), 659–675 (2002)

183. J.C. Ferreon, V.J. Hilser, The effect of the polyproline II (PPII) conformation on the denatured state entropy. Protein Sci. **12**(3), 447–457 (2003)

184. R.D. Finn, J. Tate, J. Mistry, P.C. Coggill, S.J. Sammut, H.-R. Hotz, G. Ceric, K. Forslund, S.R. Eddy, E.L. Sonnhammer, A. Bateman, The Pfam protein families database. Nucl. Acids Res. **36**, D281–288 (2008)

185. J.E. Fitzgerald, A.K. Jha, T.R. Sosnick, K.F. Freed, Polypeptide motions are dominated by peptide group oscillations resulting from dihedral angle correlations between nearest neighbors. Biochemistry **46**(3), 669–682 (2007)

186. A.F. Fliri, W.T. Loging, P.F. Thadeio, R.A. Volkmann, Biological spectra analysis: linking biological activity profiles to molecular structure. Proc. Natl. Acad. Sci. USA **102**(2), 261–266 (2005)

187. B. Folch, M. Rooman, Y. Dehouck, Thermostability of salt bridges versus hydrophobic interactions in proteins probed by statistical potentials. J. Chem. Inf. Model. **48**(1), 119–127 (2008)

188. L.R. Forrest, B. Honig, An assessment of the accuracy of methods for predicting hydrogen positions in protein structures. Proteins: Struct. Funct. Bioinform. **62**(2), 296–309 (2005)

189. W.R. Forsyth, J.M. Antosiewicz, A.D. Robertson, Empirical relationships between protein structure and carboxyl pK_a values in proteins. Proteins-Struct. Funct. Genet. **48**(2), 388–403 (2002)

190. R. Franklin, R.G. Gosling, Evidence for 2-chain helix in crystalline structure of sodium deoxyribonucleate. Nature **172**, 156–157 (1953)

191. F. Franks, *Water: a Matrix for Life* (Royal Society of Chemistry, Cambridge, 2000)

192. J.S. Franzen, R.E. Stephens, The effect of a dipolar solvent system on interamide hydrogen bonds. Biochemistry **2**(6), 1321–1327 (1963)

193. C.M. Fraser, A. Fernández, L.R. Scott, Dehydron analysis: quantifying the effect of hydrophobic groups on the strength and stability of hydrogen bonds, in *Advances in Computational Biology* ed. by H.R. Arabnia (Springer, New York, 2010)

194. C.M. Fraser, A. Fernández, L.R. Scott, Wrappa: a screening tool for candidate dehydron identification. Research Report UC/CS TR-2011-5, Department of Computer Science, University, Chicago, 2011

195. E. Freire, The propagation of binding interactions to remote sites in proteins: analysis of the binding of the monoclonal antibody D1.3 to lysozyme. Proc. Natl. Acad. Sci. USA **96**, 10118–10122 (1999)

196. R.H. French, Origins and applications of London dispersion forces and Hamaker constants in ceramics. J. Am. Ceram. Soc. **83**(9), 2117–46 (2000)

197. V. Fridkin, S. Ducharme, *The Ferroelectricity at the Nanoscale* (Springer, Berlin, 2013)

198. E. Fry, R. Acharya, D. Stuart, Methods used in the structure determination of foot-and-mouth disease virus. Acta Crystall. A **49**, 45–55 (1993)

199. R.L. Fulton, The nonlinear dielectric behavior of water: comparisons of various approaches to the nonlinear dielectric increment. J. Chem. Phys. **130**, 204503 (2009)

200. J.P. Gallivan, D.A. Dougherty, Cation-pi interactions in structural biology. Proc. Natl. Acad. Sci. USA **96**(17), 9459–9464 (1999)

201. J. Gao, D.A. Bosco, E.T. Powers, J.W. Kelly, Localized thermodynamic coupling between hydrogen bonding and microenvironment polarity substantially stabilizes proteins. Nat. Struct. Mol. Biol. **16**(7), 684–690 (2009)

202. A.E. García, K.Y. Sanbonmatsu, α-helical stabilization by side chain shielding of backbone hydrogen bonds. Proc. Natl. Acad. Sci. USA **99**(5), 2782–2787 (2002)

203. R.E. Georgescu, E.G. Alexov, M.R. Gunner, Combining conformational flexibility and continuum electrostatics for calculating pK_as in proteins. Biophys. J. **83**(4), 1731–1748 (2002)

204. A.K. Ghosh, S. Gemma, *Structure-based Design of Drugs and Other Bioactive Molecules: Tools and Strategies* (Wiley, New York, 2014)

205. R. Giles, *Mathematical Foundations of Thermodynamics* (Macmillan, New York, 1964)

206. D. Gilis, S. Massar, N.J. Cerf, M. Rooman, Optimality of the genetic code with respect to protein stability and amino-acid frequencies. Genome Biol. **2**(11), 49–1 (2001)

207. I. Gitlin, J.D. Carbeck, G.M. Whitesides, Why are proteins charged? networks of charge-charge interactions in proteins measured by charge ladders and capillary electrophoresis. Angew. Chem. Int. Edition **45**(19), 3022–3060 (2006)

208. F. Glaser, D.M. Steinberg, I.A. Vakser, N. Ben-Tal, Residue frequencies and pairing preferences at protein-protein interfaces. Proteins: Struct. Funct. Genet. **43**, 89–102 (2001)

209. G.A. Grant, C.W. Luetje, R. Summers, X.L. Xu, Differential roles for disulfide bonds in the structural integrity and biological activity of κ-bungarotoxin, a neuronal nicotinic acetylcholine receptor antagonist. Biochemistry **37**(35), 12166–12171 (1998)

210. J.K. Gregory, D.C. Clary, K. Liu, M.G. Brown, R.J. Saykally, The water dipole moment in water clusters. Science **275**(5301), 814–817 (1997)

211. A.V. Grinberg, R. Bernhardt, Effect of replacing a conserved proline residue on the function and stability of bovine adrenodoxin. Protein Eng. **11**(11), 1057–1064 (1998)

212. V. Nick, Grishin, Fold change in evolution of protein structures. J. Struct. Biol. **134**(2–3), 167–185 (2001)

213. J.A. Gruenke, R.T. Armstrong, W.W. Newcomb, J.C. Brown, J.M. White, New insights into the spring-loaded conformational change of influenza virus hemagglutinin. J. Virol. **76**(9), 4456–4466 (2002)

214. J.I. Guijarro, M. Sunde, J.A. Jones, I.D. Campbell, C.M. Dobson, Amyloid fibril formation by an SH3 domain. Proc. Natl. Acad. Sci. USA **95**(8), 4224–4228 (1998)

215. S. Günther, J. Von Eichborn, P. May, R. Preissner, JAIL: a structure-based interface library for macromolecules. Nucleic Acids Res. **37**(suppl 1), D338–D341 (2009)

216. I. Halperin, H. Wolfson, R. Nussinov, Protein-protein interactions; coupling of structurally conserved residues and of hot spots across interfaces. implications for docking. Structure **12**, 1027–1038 (2004)

217. D. Hamelberg, J.A. McCammon, Fast peptidyl cis-trans isomerization within the flexible Gly-rich flaps of HIV-1 protease. J. Am. Chem. Soc. **127**(40), 13778–13779 (2005)

218. Y. Harpaz, M. Gerstein, C. Chothia, Volume changes on protein folding. Structure **2**, 641–649 (1994)

219. M.J. Hartshorn, C.W. Murray, A. Cleasby, M. Frederickson, I.J. Tickle, H. Jhoti, Fragment-based lead discovery using X-ray crystallography. J. Med. Chem. **48**(2), 403–413 (2005)

220. J.B. Hasted, *Aqueous Dielectrics* (Chapman and Hall, London, 1974)

221. B. Hayes, The invention of the genetic code. Am. Sci. **86**(1), 8–14 (1998)

222. J.P. Helfrich, Dynamic laser light scattering technology for the molecular weight and hydrodynamic radius characterization of proteins. Pharm. Lab. **1**, 34–40 (1998)

223. Z.S. Hendsch, B. Tidor, Do salt bridges stabilize proteins? A continuum electrostatic analysis. Protein Sci.: Publ. Protein Soc. **3**(2), 211 (1994)

224. D.H. Herce, L. Perera, T.A. Darden, C. Sagui, Surface solvation for an ion in a water cluster. J. Chem. Phys. **122**(2), 024513 (2005)

225. M. Heuberger, T. Drobek, N.D. Spencer, Interaction forces and morphology of a protein-resistant poly(ethylene glycol) layer. Biophys. J. **88**(1), 495–504 (2005)

226. J. Higo, M. Nakasako, Hydration structure of human lysozyme investigated by molecular dynamics simulation and cryogenic X-ray crystal structure analyses: on the correlation between crystal water sites, solvent density, and solvent dipole. J. Comput. Chem. **23**(14), 1323–1336 (2002)

227. A. Hildebrandt, R. Blossey, S. Rjasanow, O. Kohlbacher, H.-P. Lenhof, Novel formulation of nonlocal electrostatics. Phys. Rev. Lett. **93**(10), 108104 (2004)

228. A. Hildebrandt, R. Blossey, S. Rjasanow, O. Kohlbacher, H.P. Lenhof, Electrostatic potentials of proteins in water: a structured continuum approach. Bioinformatics **23**(2), e99 (2007)

229. M.K. Hill, M. Shehu-Xhilaga, S.M. Crowe, J. Mak, Proline residues within spacer peptide p1 are important for human immunodeficiency virus type 1 infectivity, protein processing, and genomic RNA dimer stability. J. Virol. **76**, 11245–11253 (2002)

230. N.E. Hill, W.E. Vaughan, A.H. Price, M. Davies (eds.), *Dielectric Properties and Molecular Behaviour* (van Nostrand, London, 1969)

231. S. Hirohashi, Y. Kanai, Cell adhesion system and human cancer morphogenesis. Cancer Sci. **94**(7), 575–581 (2003)

232. U. Hobohm, M. Scharf, R. Schneider, C. Sander, Selection of representative protein data sets. Protein Sci. **1**(3), 409–417 (1992)

233. B. Honig, A.S. Yang, Free energy balance in protein folding. Adv. Protein Chem. **46**, 27–58 (1995)

234. A.L. Hopkins, J.S. Mason, J.P. Overington, Can we rationally design promiscuous drugs? Curr. Opin. Struct. Biol. **16**(1), 127–136 (2006)

235. M. Hoshino, H. Katou, Y. Hagihara, K. Hasegawa, H. Naiki, Y. Goto, Mapping the core of the beta(2)-microglobulin amyloid fibril by H/D exchange. Nat. Struct. Biol. **19**, 332–336 (2002)

236. T. Hou, P. Koumoutsakos, Special section on multiscale modeling in materials and life sciences. SIAM J. Multiscale Model. Simul. **4**(1), 213–214 (2005)

237. S. Hovmöller, T. Zhou, T. Ohlson, Conformations of amino acids in proteins. Acta Crystallogr. Sect. D **58**(5), 768–776 (2002)

238. Z. Hu, B. Ma, H. Wolfson, R. Nussinov, Conservation of polar residues as hot spots at protein interfaces. Proteins: Struct. Funct. Genet. **39**, 331–342 (2000)

239. M.A. Huntley, G.B. Golding, Simple sequences are rare in the Protein data bank. Proteins: Struct. Funct. Bioinform. **48**(1), 134–140 (2002)

240. B.M.P. Huyghues-Despointes, T.M. Klingler, R.L. Baldwin, Measuring the strength of side-chain hydrogen bonds in peptide helixes: the Gln·Asp (i, i + 4) interaction. Biochemistry **34**, 13267–13271 (2002). doi:10.1021/bi00041a001

241. L.M. Iakoucheva, K.A. Dunker, Order, disorder, and flexibility: prediction from protein sequence. Structure **11**, 1316–1317 (2003)

242. W. Im, D. Beglov, B. Roux, Continuum solvation model: electrostatic forces from numerical solutions to the Poisson-Boltzmann equation. Comput. Phys. Commun. **111**, 59–75 (1998)

243. R. Improta, L. Vitagliano, L. Esposito, Peptide bond distortions from planarity: new insights from quantum mechanical calculations and peptide/protein crystal structures. PLoS ONE **6**(9), e24533, 09 (2011)

244. J. Israelachvili, *Intermolecular and Surface Forces*, 2nd edn. (Academic Press, London, 1991)

245. A. Jabs, M.S. Weiss, R. Hilgenfeld, Non-proline cis peptide bonds in proteins. J. Mol. Biol. **286**(1), 291–304 (1999)

246. C. Jarzynski, Nonequilibrium equality for free energy differences. Phys. Rev. Lett. **78**(14), 2690 (1997)

247. M. Jaskolski, M. Gilski, Z. Dauter, A. Wlodawer, Stereochemical restraints revisited: how accurate are refinement targets and how much should protein structures be allowed to deviate from them? Acta Crystallogr. Sect. D **63**(5), 611–620 (2007)

248. E.T. Jaynes, R. Smoluchowski, Ferroelectricity. Phys. Today **6**, 17 (1953)

249. G.A. Jeffrey, *An Introduction to Hydrogen Bonds* (Oxford Science Publications, Clarendon Press, Oxford, 1997)

250. G.A. Jeffrey, Hydrogen-bonding: an update. Crystallogr. Rev. **9**, 135–176 (2003). doi:10.1080/08893110310001621754

251. J.H. Jensen, H. Li, A.D. Robertson, P.A. Molina, Prediction and rationalization of protein pK_a values using QM and QM/MM methods. J. Phys. Chem. A **109**(30), 6634–6643 (2005)

252. A.K. Jha, K.F. Freed, Solvation effect on conformations of 1, 2: dimethoxyethane: charge-dependent nonlinear response in implicit solvent models. J. Chem. Phys. **128**, 034501 (2008)

253. C.G. Ji, J.Z.H. Zhang, Quantifying the stabilizing energy of the intraprotein hydrogen bond due to local mutation. J. Phys. Chem. B **115**(42), 12230–12233 (2011)

254. S. Ji, The linguistics of DNA: words, sentences, grammar, phonetics, and semantics. Ann. N. Y. Acad. Sci. **870**, 411–417 (1999)

255. L. Jiang, L. Lai, CH··· O hydrogen bonds at protein-protein interfaces. J. Biol. Chem. **277**(40), 37732 (2002)

256. R.O. Jones, Density functional theory: its origins, rise to prominence, and future. Rev. Mod. Phys. **87**(3), 897 (2015)

257. S. Jones, J.M. Thornton, Principles of protein-protein interactions. Proc. Natl. Acad. Sci. USA **93**, 13–20 (1996)

258. S. Jones, J.M. Thornton, Analysis of protein-protein interaction sites using surface patches. J. Mol. Biol. **272**, 121–132 (1997)

259. S. Jones, J.M. Thornton, Prediction of protein-protein interaction sites using patch analysis. J. Mol. Biol. **272**, 133–143 (1997)

260. F. Jourdan, S. Lazzaroni, B.L. Méndez, P.L. Cantore, M. de Julio, P. Amodeo, N.S. Iacobellis, A. Evidente, A. Motta, A left-handed α-helix containing both L-and D-amino acids: the solution structure of the antimicrobial lipodepsipeptide tolaasin. Proteins: Struct. Funct. Bioinform. **52**(4), 534–543 (2003)

261. U. Kaatze, R. Behrends, R. Pottel, Hydrogen network fluctuations and dielectric spectrometry of liquids. J. Non-Cryst. Solids **305**(1), 19–28 (2002)

262. R. Kaufmann, U. Junker, M. Schilli-Westermann, C. Klötzer, J. Scheele, K. Junker, Meizothrombin, an intermediate of prothrombin cleavage potently activates renal carcinoma cells by interaction with par-type thrombin receptors. Oncol. Rep. **10**, 493–496 (2003)

263. W. Kauzmann, Some factors in the interpretation of protein denaturation, in *Advances in Protein Chemistry*, vol. 14 (Academic Press, New York, 1959), pp. 1–63

264. J. Kemmink, T.E. Creighton, The physical properties of local interactions of tyrosine residues in peptides and unfolded proteins. J. Mol. Biol. **245**(3), 251–260 (1995)

265. F.N. Keutsch, J.D. Cruzan, R.J. Saykally, The water trimer. Chem. Rev. **103**(7), 2533–2578 (2003)

266. A.I. Khinchin, *Mathematical Foundations of Statistical Mechanics* (Dover, New York, 1949)

267. J. Kim, J. Mao, M.R. Gunner, Are acidic and basic groups in buried proteins predicted to be ionized? J. Mol. Biol. **348**(5), 1283–1298 (2005)

268. S.S. Kim, T.J. Smith, M.S. Chapman, M.C. Rossmann, D.C. Pevear, F.J. Dutko, P.J. Felock, G.D. Diana, M.A. McKinlay, Crystal structure of human rhinovirus serotype 1A (HRV1A). J. Mol. Biol. **210**, 91–111 (1989)

269. S. Kim, S.J. Karrila, *Microhydrodynamics: Principles and Selected Applications* (Dover Publications, New York, 2005)

270. C. King, E.N. Garza, R. Mazor, J.L. Linehan, I. Pastan, M. Pepper, D. Baker, Removing T-cell epitopes with computational protein design. Proc. Natl. Acad. Sci. USA **111**(23), 8577–8582 (2014)

271. W. Klopper, J.G.C.M. van Duijneveldt-van, F.B. de Rijdt van Duijneveldt, Computational determination of equilibrium geometry and dissociation energy of the water dimer. Phys. Chem. Chem. Phys. **2**, 2227–2234 (2000)

272. I.M. Klotz, Solvent water and protein behavior: view through a retroscope. Protein Sci. **2**(11), 1992–1999 (1993)

273. I.M. Klotz, J.S. Franzen, Hydrogen bonds between model peptide groups in solution. J. Am. Chem. Soc. **84**(18), 3461–3466 (1962)

274. Z.A. Knight, K.M. Shokat, Features of selective kinase inhibitors. Chem. Biol. **12**(6), 621–637 (2005)

275. O. Koch, M. Bocola, G. Klebe, Cooperative effects in hydrogen-bonding of protein: a systematic analysis of crystal data using secbase. Proteins: Struct. Funct. Bioinform. **61**(2), 310–317 (2005)

276. P. Koehl, H. Orland, M. Delarue, Computing ion solvation free energies using the dipolar Poisson model. J. Phys. Chem. B **113**(17), 5694–5697 (2009)

277. P.A. Kollman, L.C. Allen, The theory of the hydrogen bond. Chem. Rev. **72**(3), 283–303 (1972)

278. J. Kongsted, P. Söderhjelm, U. Ryde, How accurate are continuum solvation models for drug-like molecules? J. Comput.-Aided Mol. Design **23**(7), 395–409 (2009)

279. J. Korlach, P. Schwille, W.W. Webb, G.W. Feigenson, Characterization of lipid bilayer phases by confocal microscopy and fluorescence correlation spectroscopy. Proc. Natl. Acad. Sci. USA **96**, 8461–8466 (1999)

280. A.A. Kornyshev, A. Nitzan, Effect of overscreening on the localization of hydrated electrons. Zeitschrift für Physikalische Cheme **215**(6), 701–715 (2001)

281. A.A. Kornyshev, G. Sutmann, Nonlocal dielectric saturation in liquid water. Phys. Rev. Lett. **79**, 3435–3438 (1997)

282. V. Korobov, A. Yelkhovsky, Ionization potential of the helium atom. Phys. Rev. Lett. **87**(19), 193003 (2001)

283. T. Kortemme, A.V. Morozov, D. Baker, An orientation-dependent hydrogen bonding potential improves prediction of specificity and structure for proteins and protein-protein complexes. J. Mol. Biol. **326**, 1239–1259 (2003)

284. G.V. Kozhukh, Y. Hagihara, T. Kawakami, K. Hasegawa, H. Naiki, Y. Goto, Investigation of a peptide responsible for amyloid fibril formation of β_2-microglobulin by achromobacter protease I. J. Biol. Chem. **277**(2), 1310–1315 (2002)

285. L.M. Krauss, *Fear of Physics* (Basic Books, New York, 1994)

286. S. Krishnaswamy, M.G. Rossmann, Structural refinement and analysis of mengo virus. J. Mol. Biol. **211**, 803–844 (1990)

287. J. Kroon, J.A. Kanters, J.G.C.M. van Duijneveldt-van De, F.B. van Rijdt, J.A.Vliegenthart Duijneveldt, O-H · · O hydrogen bonds in molecular crystals a statistical and quantum-chemical analysis. J. Mol. Struct. **24**, 109–129 (1975)

288. M.F. Kropman, H.J. Bakker, Dynamics of water molecules in aqueous solvation shells. Science **291**(5511), 2118–2120 (2001)

289. S. Kumar, R. Nussinov, Salt bridge stability in monomeric proteins. J. Mol. Biol. **293**(5), 1241–1255 (1999)

290. S. Kumar, R. Nussinov, Relationship between ion pair geometries and electrostatic strengths in proteins. Biophys. J. **83**(3), 1595–1612 (2002)

291. W. Kunz, *Specific Ion Effects* (World Scientific Pub Co Inc., Singapore, 2010)

292. J. Kyte, R.F. Doolittle, A simple method for displaying the hydropathic character of a protein. J. Mol. Biol. **157**, 105–132 (1982)

293. L.D. Landau, E.M. Lifshitz, *Fluid mechanics, in Course of Theoretical Physics*, vol. 6 (Buterworth-Heineman, Oxford, 1987)

294. I. Langmuir, The arrangement of electrons in atoms and molecules. J. Am. Chem. Soc. **41**(6), 868–934 (1919)

295. P.I. Lario, A. Vrielink, Atomic resolution density maps reveal secondary structure dependent differences in electronic distribution. J. Am. Chem. Soc. **125**, 12787–12794 (2003). doi:10.1021/ja0289954

296. R.A. Laskowski, M.W. MacArthur, D.S. Moss, J.M. Thornton, Procheck: a program to check the stereochemical quality of protein structures. J. Appl. Crystallogr. **26**(2), 283–291 (1993)

297. W.M. Latimer, W.H. Rodebush, Polarity and ionization from the standpoint of the Lewis theory of valence. J. Am. Chem. Soc. **42**, 1419–1433 (1920). doi:10.1021/ja01452a015

298. C. Laurence, M. Berthelot, Observations on the strength of hydrogen bonding. Perspect. Drug Discov. Design **18**, 3960 (2000)

299. E. Lax, *The mold in Dr. The Story of the Penicillin Miracle* ((Macmillan, Florey's Coat, 2004)

300. A. Leaver-Fay, Y. Liu, J. Snoeyink, X. Wang, Faster placement of hydrogens in protein structures by dynamic programming. J. Exp. Algorithmics **12**, 2.5:1–2.5:16 (2008)

301. D.N. LeBard, D.V. Matyushov, Ferroelectric hydration shells around proteins: electrostatics of the protein-water interface. J. Phys. Chem. B **114**(28), 9246–9258 (2010)

302. B. Lee, F.M. Richards, The interpretation of protein structures: estimation of static accessibility. J. Mol. Biol. **55**, 379–400 (1971)

303. E.C. Lee, B.H. Hong, J.Y. Lee, J.C. Kim, D. Kim, Y. Kim, P. Tarakeshwar, K.S. Kim, Substituent effects on the edge-to-face aromatic interactions. J. Am. Chem. Soc. **127**(12), 4530–4537 (2005)

304. J. Lehmann, Physico-chemical constraints connected with the coding properties of the genetic system. J. Theor. Biol. **202**(2), 129–144 (2000)

305. S. Leikin, A.A. Kornyshev, Theory of hydration forces. Nonlocal electrostatic interaction of neutral surfaces. J. Chem. Phys. **92**(6), 6890–6898 (1990)

306. P.A. Leland, K.E. Staniszewski, C. Park, B.R. Kelemen, R.T. Raines, The ribonucleolytic activity of angiogenin. Biochemistry **41**(4), 1343–1350 (2002)

307. M. Levitt, A simplified representation of protein conformations for rapid simulation of protein folding. J. Mol. Biol **104**, 59–107 (1976)

308. M. Levitt, M.F. Perutz, Aromatic rings act as hydrogen bond acceptors. J. Mol. Biol. **201**, 751–754 (1988)

309. Y. Levy, J.N. Onuchic, Water and proteins: a love-hate relationship. Proc. Natl. Acad. Sci. USA **101**(10), 3325–3326 (2004)

310. G.N. Lewis, The atom and the molecule. J. Am. Chem. Soc. **38**(4), 762–785 (1916)

311. H. Li, A.D. Robertson, J.H. Jensen, Very fast empirical prediction and rationalization of protein pK_a values. Proteins-Struct. Funct. Bioinform. **61**(4), 704–721 (2005)

312. Y. Li, H. Li, F. Yang, S.J. Smith-Gill, R.A. Mariuzza, X-ray snapshots of the maturation of an antibody response to a protein antigen. Nat. Struct. Mol. Biol. **10**, 482–488 (2003). doi:10.1038/nsb930

313. L. Limozin, E. Sackmann, Polymorphism of cross-linked actin networks in giant vesicles. Phys. Rev. Lett. **89**(16), 168103 (2002)

314. M. Lisal, J. Kolafa, I. Nezbeda, An examination of the five-site potential (TIP5P) for water. J. Chem. Phys. **117**, 8892–8897 (2002)

315. S.J. Littler, S.J. Hubbard, Conservation of orientation and sequence in protein domain-domain interactions. J. Mol. Biol. **345**(5), 1265–1279 (2005)

316. B.A. Liu, K. Jablonowski, M. Raina, M. Arcé, T. Pawson, P.D. Nash, The human and mouse complement of SH2 domain proteins - establishing the boundaries of phosphotyrosine signaling. Mol. Cell **22**(6), 851–868 (2006)

317. J. Liu, L.R. Scott, A. Fernández, Interactions of aligned nearest neighbor protein side chains. J. Bioinform. Comput. Biol. **7** (2006) (submitted)

318. J.-L. Liu, B. Eisenberg, Poisson-fermi model of single ion activities in aqueous solutions. Chem. Phys. Lett. **637**, 1–6 (2015)

319. K. Liu, M.G. Brown, C. Carter, R.J. Saykally, J.K. Gregory, D.C. Clary, Characterization of a cage form of the water hexamer. Nature **381**, 501–503 (1996)

320. L. Lo Conte, C. Chothia, J. Janin, The atomic structure of protein-protein recognition sites. Mol. Biol. **285**, 2177–2198 (1999)

321. V.V. Loladze, G.I. Makhatadze, Energetics of charge-charge interactions between residues adjacent in sequence. Proteins: Struct. Funct. Bioinform. **79**(12), 3494–3499 (2011)

322. F. London, Zur Theorie und Systematik der Molekularkräfte. Zeitschrift für Physik A Hadrons and Nuclei **63**, 245–279 (1930). doi:10.1007/BF01421741

323. S.C. Lovell, J.M. Word, J.S. Richardson, D.C. Richardson, The penultimate rotamer library. Proteins: Struct. Funct. Genet. **40**, 389–408 (2000)

324. H. Lu, L. Lu, J. Skolnick, MULTIPROSPECTOR: an algorithm for the prediction of protein-protein interactions by multimeric threading. Proteins: Struct. Funct. Genet. **49**, 350–364 (2002)

325. P. Luo, R.L. Baldwin, Interaction between water and polar groups of the helix backbone: an important determinant of helix propensities. Proc. Natl. Acad. Sci. USA **96**(9), 4930–4935 (1999)

326. B. Ma, T. Elkayam, H. Wolfson, R. Nussinov, Protein-protein interactions: structurally conserved residues distinguish between binding sites and exposed protein surfaces. Proc. Natl. Acad. Sci. USA **100**(10), 5772–5777 (2003)

327. J.C. Ma, D.A. Dougherty, The cation-π interaction. Chem. Rev. **97**(5), 1303–1324 (1997)

328. M.W. MacArthur, J.M. Thornton, Deviations from planarity of the peptide bond in peptides and proteins. J. Mol. Biol. **264**(5), 1180–1195 (1996)

329. J.D. Madura, J.M. Briggs, R.C. Wade, M.E. Davis, B.A. Luty, A. Ilin, J. Antosciewicz, M.K. Gilson, B. Bagheri, L.R. Scott, J.A. McCammon, Electrostatics and diffusion of molecules in solution: simulations with the University of Houston Brownian dynamics program. Comput. Phys. Commun. **91**, 57–95 (1995)

330. M.W. Mahoney, W.L. Jorgensen, A five-site model for liquid water and the reproduction of the density anomaly by rigid, nonpolarizable potential functions. J. Chem. Phys. **112**, 8910–8922 (2000)

331. M. Maleki, M.M. Aziz, L. Rueda, Analysis of obligate and non-obligate complexes using desolvation energies in domain-domain interactions, in *Proceedings of the Tenth International Workshop on Data Mining in Bioinformatics* (ACM, 2011), p. 2

332. R.J. Mallis, K.N. Brazin, D.B. Fulton, A.H. Andreotti, Structural characterization of a proline-driven conformational switch within the Itk SH2 domain. Nat. Struct. Biol. **9**, 900–905 (2002)

333. N. Mandel, G. Mandel, B.L. Trus, J. Rosenberg, G. Carlson, R.E. Dickerson, Tuna cytochrome c at 2.0å resolution. III. Coordinate optimization and comparison of structures. J. Biol. Chem. **252**(13), 4619–4636 (1977)

334. K. Manikandan, S. Ramakumar, The occurrence of C-H···O hydrogen bonds in α-helices and helix termini in globular proteins. Proteins: Struct. Funct. Bioinform. **56**(4), 768–781 (2004)

335. R. Mannhold, G.I. Poda, C. Ostermann, I.V. Tetko, Calculation of molecular lipophilicity: state-of-the-art and comparison of log P methods on more than 96,000 compounds. J. Pharm. Sci. **98**(3), 861–893 (2009)

336. P.E. Mason, G.W. Neilson, C.E. Dempsey, A.C. Barnes, J.M. Cruickshank, The hydration structure of guanidinium and thiocyanate ions: implications for protein stability in aqueous solution. Proc. Natl. Acad. Sci. USA **100**(8), 4557–4561 (2003)

337. C. Matzler, U. Wegmuller, Dielectric properties of freshwater ice at microwave frequencies. J. Phys. D: Appl. Phys. **20**(12), 1623–1630 (1987)

338. B. McCoy, T.T. Wu, *The Two-Dimensional Ising Model* (Harvard University Press, Harvard, 1973)

339. I.K. McDonald, J.M. Thornton, Satisfying hydrogen bonding potential in proteins. J. Mol. Biol. **238**, 777–793 (1994)

340. L.P. McIntosh, G. Hand, P.E. Johnson, M.D. Joshi, M. Körner, L.A. Plesniak, L. Ziser, W.W. Wakarchuk, S.G. Withers, The pK_a of the general acid/base carboxyl group of a glycosidase cycles during catalysis: a ^{13}C-NMR study of bacillus circulans xylanase. Biochemistry **35**(12), 9958–9966 (1996)

341. V.J. McParland, A.P. Kalverda, S.W. Homans, S.E. Radford, Structural properties of an amyloid precursor of beta(2)-microglobulin. Nat. Struct. Biol. **9**(5), 326–331 (2002)

342. C. Mead, L. Conway, *Introduction to VLSI Systems* (Addison-Wesley, Reading, 1979)

343. S. Mecozzi, A.P. West, D.A. Dougherty, Cation-π interactions in aromatics of biological and medicinal interest: electrostatic potential surfaces as a useful qualitative guide. Proc. Natl. Acad. Sci. USA **93**(20), 10566–10571 (1996)

344. E.L. Mehler, F. Guarnieri, A self-consistent, microenvironment modulated screened Coulomb potential approximation to calculate pH-dependent electrostatic effects in proteins. Biophys. J. **77**, 3–22 (1999)

345. I. Mihalek, I. Reš, O. Lichtarge, On itinerant water molecules and detectability of protein-protein interfaces through comparative analysis of homologues. J. Mol. Biol. **369**(2), 584–595 (2007)

346. C. Millot, A.J. Stone, Towards an accurate intermolecular potential for water. Mol. Phys. **77**(3), 439–462 (1992)

347. D.J. Mitchell, L. Steinman, D.T. Kim, C.G. Fathman, J.B. Rothbard, Polyarginine enters cells more efficiently than other polycationic homopolymers. J. Pept. Res. **56**(5), 318–325 (2000)

348. J.B.O. Mitchell, J. Smith, D-amino acid residues in peptides and proteins. Proteins: Struct. Funct. Genet. **50**(4), 563–571 (2003)

349. S. Miyazawa, R.L. Jernigan, Estimation of effective interresidue contact energies from protein crystal structures: quasi-chemical approximation. Macromolecules **18**(3), 534–552 (1985)

350. D.L. Mobley, A.E. Barber, C.J. Fennell, K.A. Dill, Charge asymmetries in hydration of polar solutes. J. Phys. Chem. B **112**(8), 2405–2414 (2008)

351. A. Möglich, F. Krieger, T. Kiefhaber, Molecular basis for the effect of urea and guanidinium chloride on the dynamics of unfolded polypeptide chains. J. Mol. Biol. **345**(1), 153–162 (2005)

352. C. Momany, L.C. Kovari, A.J. Prongay, W. Keller, R.K. Gitti, B.M. Lee, A.E. Gorbalenya, L. Tong, J. McClure, L.S. Ehrlich, M.F. Carter, M.G. Rossmann, Crystal structure of dimeric HIV-1 capsid protein. Nat. Struct. Biol. **3**, 763–770 (1996)

353. F.A. Momany, R.F. McGuire, A.W. Burgess, H.A. Scheraga, Energy parameters in polypeptides. VII. Geometric parameters, partial atomic charges, nonbonded interactions, hydrogen bond interactions, and intrinsic torsional potentials for the naturally occurring amino acids. J. Phys. Chem. **79**(22), 2361–2381 (1975)

354. J. Montes de Oca, A. Rodriguez Fris, G. Appignanesi, A. Fernández, Productive induced metastability in allosteric modulation of kinase function. FEBS J. **281**(13), 3079–3091 (2014)

355. A.V. Morozov, T. Kortemme, K. Tsemekhman, D. Baker, Close agreement between the orientation dependence of hydrogen bonds observed in protein structures and quantum mechanical calculations. Proc. Natl. Acad. Sci. USA **101**, 6946–6951 (2004)

356. J.A. Morrill, R. MacKinnon, Isolation of a single carboxyl-carboxylate proton binding site in the pore of a cyclic nucleotide-gated channel. J. Gen. Physiol. **114**(1), 71–84 (1999)

357. R. Moss, A. Fernández, Inhibition of MyBP-C binding to myosin as a treatment for heart failure (2015)

358. D.W. Mount, *Bioinformatics* (Cold Spring Harbor Laboratory Press, 2001)

359. A.S. Muresan, H. Diamant, K.-Y. Lee, Effect of temperature and composition on the formation of nanoscale compartments in phospholipid membranes. J. Am. Chem. Soc. **123**, 6951–6952 (2001)

360. R. Nelson, M.R. Sawaya, M. Balbirnie, A.O. Madsen, C. Riekel, R. Grothe, D. Eisenberg, Structure of the cross-β spine of amyloid-like fibrils. Nature **435**, 773–778 (2005). doi:10.1038/nature03680

361. G. Némethy, I.Z. Steinberg, H.A. Scheraga, Influence of water structure and of hydrophobic interactions on the strength of side-chain hydrogen bonds in proteins. Biopolymers **1**, 43–69 (1963)

362. A. Neumeister, N. Praschak-Rieder, B. Hesselmann, O. Vitouch, M. Rauh, A. Barocka, J. Tauscher, S. Kasper, Effects of tryptophan depletion in drug-free depressed patients who responded to total sleep deprivation. Arch. Gen. Psychiatry **55**(2), 167–172 (1998)

363. A. Nicholls, K.A. Sharp, B. Honig, Protein folding and association: insights from the interfacial and thermodynamic properties of hydrocarbons. *Proteins: Struct. Funct. Genet.* **11**(4), 281–296 (1991)

364. K.T. No, O.Y. Kwon, S.Y. Kim, M.S. Jhon, H.A. Scheraga, A simple functional representation of angular-dependent hydrogen-bonded systems. 1. Amide, carboxylic acid, and amide-carboxylic acid pairs. J. Phys. Chem. **99**, 3478–3486 (1995)

365. M.E.M. Noble, J.A. Endicott, L.N. Johnson, Protein kinase inhibitors: insights into drug design from structure. Science **303**(5665), 1800–1805 (2004)

366. I.M.A. Nooren, J.M. Thornton, Diversity of protein-protein interactions. EMBO J. **22**, 3486–3492 (2003)

367. I.M.A. Nooren, J.M. Thornton, Structural characterisation and functional significance of transient protein-protein interactions. J. Mol. Biol. **325**(5), 991–1018 (2003)

368. M. Novotny, G.J. Kleywegt, A survey of left-handed helices in protein structures. J. Mol. Biol. **347**(2), 231–241 (2005)

369. Y. Nozaki, C. Tanford, The solubility of amino acids and two glycine peptides in aqueous ethanol and dioxane solutions. J. Biol. Chem. **246**(7), 2211–2217 (1971)

370. Y. Ofren, B. Rost, Analysing six types of protein-protein interfaces. J. Mol. Biol. **335**, 377–387 (2003)

371. D.O. Omecinsky, K.E. Holub, M.E. Adams, M.D. Reily, Three-dimensional structure analysis of μ-agatoxins: further evidence for common motifs among neurotoxins with diverse ion channel specificities. Biochemistry **35**(9), 2836–2844 (1996)

372. T. Ooi, Thermodynamics of protein folding: effects of hydration and electrostatic interactions. Adv. Biophys. **30**, 105–154 (1994)

373. T. Orlova, L.R. Scott, The role of solvation in hydrogen bond geometry in protein helices. Research Report UC/CS TR-2016-? Department of Computer Science, University Chicago, 2016

374. D.R. Owen, *A First Course in the Mathematical Foundations of Thermodynamics* (Springer, Berlin, 1984)

375. C.N. Pace, G.R. Grimsley, J.M. Scholtz, Protein ionizable groups: pK values and their contribution to protein stability and solubility. J. Biol. Chem. **284**(20), 13285–13289 (2009)

376. C.N. Pace, S. Treviño, E. Prabhakaran, J.M. Scholtz, Protein structure, stability and solubility in water and other solvents. Philos. Trans. R. Soc. B: Biol. Sci. **359**(1448), 1225–1235 (2004)

377. D. Pahlke, D. Leitner, U. Wiedemann, D. Labudde, COPS-cis/trans peptide bond conformation prediction of amino acids on the basis of secondary structure information. Bioinformatics **21**(5), 685–686 (2005)

378. S.K. Pal, J. Peon, A.H. Zewail, Biological water at the protein surface: dynamical solvation probed directly with femtosecond resolution. Proc. Natl. Acad. Sci. USA **99**(4), 1763–1768 (2002)

379. P. Paricaud, M. Předota, A.A. Chialvo, P.T. Cummings, From dimer to condensed phases at extreme conditions: accurate predictions of the properties of water by a Gaussian charge polarizable model. JChemPhys **122**, 244511 (2005)

380. W. Pauli, The connection between spin and statistics. Phys. Rev. **58**(8), 716–722 (1940)

381. W. Pauli, *Thermodynamics and the Kinetic Theory of Gases* (Dover, New York, 2000)

382. L. Pauling, *Nature of the Chemical Bond*, 3rd edn. (Cornell University Press, Ithaca, 1960)

383. L. Pauling, *General Chemistry* (Dover, New York, 1970)

384. L. Pauling, R.B. Corey, H.R. Branson, The structure of proteins: two hydrogen-bonded helical configurations of the polypeptide chain. Proc. Natl. Acad. Sci. USA **37**(4), 205–211 (1951)

385. L. Pauling, E.B. Wilson, *Introduction to Quantum Mechanics with Applications to Chemistry* (Dover, New York, 1985)

386. J. Pavlicek, S.L. Coon, S. Ganguly, J.L. Weller, S.A. Hassan, D.L. Sackett, D.C. Klein, Evidence that proline focuses movement of the floppy loop of Arylalkylamine N-Acetyltransferase (EC 2.3.1.87). J. Biol. Chem. **283**(21), 14552–14558 (2008)

387. M.Y. Pavlov, R.E. Watts, Z. Tan, V.W. Cornish, M. Ehrenberg, A.C. Forster, Slow peptide bond formation by proline and other n-alkylamino acids in translation. Proc. Natl. Acad. Sci. **106**(1), 50–54 (2009)

388. G. Pei, T.M. Laue, A. Aulabaugh, D.M. Fowlkes, B.R. Lentz, Structural comparisons of meizothrombin and its precursor prothrombin in the presence or absence of procoagulant membranes. Biochemistry **31**, 6990–6996 (1992)

389. J.K. Percus, *Kinetic Theory and Statistical Mechanics* (University Courant Institue of Mathemtatical Sciences, New York, 1970)

390. S. Persson, J.A. Killian, G. Lindblom, Molecular ordering of interfacially localized tryptophan analogs in ester- and ether-lipid bilayers studied by ^{2}H-NMR. Biophys. J. **75**(3), 1365–1371 (1998)

391. M.F. Perutz, The role of aromatic rings as hydrogen-bond acceptors in molecular recognition. Philos. Trans. R. Soc. A **345**(1674), 105–112 (1993)

392. D. Petrey, B. Honig, Free energy determinants of tertiary structure and the evaluation of protein models. Protein Sci. **9**(11), 2181–2191 (2000)

393. G.A. Petsko, D. Ringe, *Protein Stucture and Function* (New Science Press, London, 2004)

394. M. Petukhov, D. Cregut, C.M. Soares, L. Serrano, Local water bridges and protein conformational stability. Protein Sci. **8**, 1982–1989 (1999)

395. A. Pierucci-Lagha, R. Feinn, V. Modesto-Lowe, R. Swift, M. Nellissery, J. Covault, H.R. Kranzler, Effects of rapid tryptophan depletion on mood and urge to drink in patients with co-morbid major depression and alcohol dependence. Psychopharmacology **171**(3), 340–348 (2004)

396. N. Pietrosemoli, A. Crespo, A. Fernández, Dehydration propensity of order-disorder intermediate regions in soluble proteins. J. Proteome Res. **6**(9), 3519–3526 (2007)

397. G.C. Pimentel, A.L. McClellan, Hydrogen bonding. Annu. Rev. Phys. Chem. **22**(1), 347–385 (1971)

398. M. Planck, *Treatise on Thermodynamics* (Dover, New York, 2008)

399. L.A. Plesniak, G.P. Connelly, L.P. Mcintosh, W.W. Wakarchuk, Characterization of a buried neutral histidine residue in Bacillus circulans xylanase: NMR assignments, pH titration, and hydrogen exchange. Protein Sci. **5**(11), 2319–2328 (1996)

400. E.V. Pletneva, A.T. Laederach, D.B. Fulton, N.M. Kostic, The role of cation-π interactions in biomolecular association. Design of peptides favoring interactions between cationic and aromatic amino acid side chains. J. Am. Chem. Soc. **123**(26), 6232–6245 (2001)

401. J.W. Ponder, F.M. Richards, Tertiary templates for proteins: use of packing criteria in the enumeration of allowed sequences for different structural classes. J. Mol. Biol. **193**, 775–791 (1987)

402. P.K. Ponnuswamy, M. Prabhakaran, P. Manavalan, Hydrophobic packing and spatial arrangement of amino acid residues in globular proteins. Biochim. Biophys. Acta (BBA) - Protein. Structure **623**(2), 301–316 (1980)

403. P. Popelier, *Atoms in Molecules* (Prentice-Hall, Harlow, 2000)

404. C.J. Preston, *Gibbs States on Countable Sets* (Cambridge University Press, Cambridge, 2008)

405. M.J. Previs, S. Beck Previs, J. Gulick, J. Robbins, D.M. Warshaw, Molecular mechanics of cardiac myosin-binding protein C in native thick filaments. Science **337**(6099), 1215–1218 (2012)

406. S.B. Prusiner, Prions. Proc. Natl. Acad. Sci. USA **95**(23), 13363–13383 (1998)

407. J. Qian, N.M. Luscombe, M. Gerstein, Protein family and fold occurrence in genomes: power-law behaviour and evolutionary model. J. Mol. Biol. **313**(4), 673–681 (2001)

408. S. Radaev, P. Sun, Recognition of immunoglobulins by Fcγ receptors. Mol. Immunol. **38**(14), 1073–1083 (2002)

409. D. Rajamani, S. Thiel, S. Vajda, C.J. Camacho, Anchor residues in protein-protein interactions. Proc. Natl. Acad. Sci. USA **101**, 11287–11292 (2004)

410. C. Ramakrishnan, Ramachandran and his map. Resonance 48–56 (2001)

411. C. Ramakrishnan, G.N. Ramachandran, Stereochemical criteria for polypeptide and protein chain conformations. II Allowed conformations for a pair of peptide units. Biophys. J. **5**, 909–933 (1965)

412. R. Ramesh, *Thin Film Ferroelectric Materials and Devices* (Kluwer Academic Publishers, Boston, 1997)

413. J.J. Ramsden, Review of optical methods to probe protein adsorption at solid-liquid interfaces. J. Stat. Phys. **73**, 853–877 (1993)

414. J.B. Rauch, L.R. Scott, The electrostatics of periodic crystals. to appear (2017)

415. S.W. Rick, R.E. Cachau, The nonplanarity of the peptide group: molecular dynamics simulations with a polarizable two-state model for the peptide bond. J. Chem. Phys. **112**(11), 5230–5241 (2000)

416. R. Riek, G. Wider, M. Billeter, S. Hornemann, R. Glockshuber, K. Wuthrich, Prion protein NMR structure and familial human spongiform encephalopathies. Proc. Natl. Acad. Sci. USA **95**(20), 11667–11672 (1998)

417. D. Ringe, What makes a binding site a binding site? Curr. Opin. Struct. Biol. **5**, 825–829 (1995)

418. G.G. Roberts (ed.), *Langmuir-Blodgett Films* (Plenum Press, New York, 1990)

419. D. Robinson, T. Bertrand, J.-C. Carry, F. Halley, A. Karlsson, M. Mathieu, H. Minoux, M.-A. Perrin, B. Robert, L. Schio, W. Sherman, Differential water thermodynamics determine PI3K-Beta/Delta selectivity for solvent-exposed ligand modifications. J. Chem. Inf. Modeling 0(0):null, 0. PMID: 27144736

420. D.D. Robinson, W. Sherman, R. Farid, Understanding kinase selectivity through energetic analysis of binding site waters. ChemMedChem **5**(4), 618–627 (2010)

421. C. Rodriguez-Aguayo, P. Vivas-Mejia, J.M. Benito, A. Fernandez, F.-X. Claret, A. Chavez-Reyes, A.K. Sood, G. Lopez-Berestein, JNK-1 inhibition leads to antitumor activity in ovarian cancer. Cancer Res. **70**(8 Supplement), 5468–5468 (2010)

422. J. Ròmer, T.H. Bugge, L.R. Lund, M.J. Flick, J.L. Degen, K. Danò, Impaired wound healing in mice with a disrupted plasminogen gene. Nat. Med. **2**(3), 287–292 (1996)

423. M. Rooman, J. Liévin, E. Buisine, R. Wintjens, Cation-π/H-bond stair motifs at protein-DNA interfaces. J. Mol. Biol. **319**, 6776 (2002)

424. P.C. Rosas, Y. Liu, M. Abdalla, C. Thomas, D. Kidwell, R. Kumar, K. Baker, B. Patel, C. Warrens, R. Solaro et al., Phosphorylated cardiac myosin binding protein-c enhances lusitropy. J. Am. Coll. Cardiol. **12**(63), A871 (2014)

425. B. Roux, T. Simonson, Implicit solvent models. Biophys. Chem. **78**(1), 1–20 (1999)

426. D. Ruell, *Thermodynamic Formalism: The Mathematical Structures of Classical Equilibrium Statistical Mechanics* (Addison-Wesley, Reading, 1978)

427. D. Ruell, *Statistical Mechanics: Rigorous Results* (Addison-Wesley, Reading, 1989)

428. S. Samsonov, J. Teyra, M.T. Pisabarro, A molecular dynamics approach to study the importance of solvent in protein interactions. Proteins: Struct. Funct. Bioinform. **73**(2), 515–525 (2008)

429. R. Samudrala, M. Levitt, Decoys RUs: a database of incorrect conformations to improve protein structure prediction. Protein Sci. **9**(07), 1399–1401 (2000)

430. L. Sandberg, R. Casemyr, O. Edholm, Calculated hydration free energies of small organic molecules using a nonlinear dielectric continuum model. J. Phys. Chem. B **106**(32), 7889–7897 (2002)

431. L. Sandberg, O. Edholm, Nonlinear response effects in continuum models of the hydration of ions. J. Chem. Phys. **116**, 2936–2944 (2002)

432. R.T. Sanderson, Electronegativity and bond energy. J. Am. Chem. Soc. **105**, 2259–2261 (1983)

433. J.L.R. Santos, R. Aparicio, I. Joekes, J.L. Silva, J.A.C. Bispo, C.F.S. Bonafe, Different urea stoichiometries between the dissociation and denaturation of tobacco mosaic virus as probed by hydrostatic pressure. Biophys. Chem. **134**(3), 214–224 (2008)

434. B. Santra, J. Klimeš, D. Alfè, A. Tkatchenko, B. Slater, A. Michaelides, R. Car, M. Scheffler, Hydrogen bonds and van der Waals forces in ice at ambient and high pressures. Phys. Rev. Lett. **107**(18), 185701 (2011)

435. C.F. Sanz-Navarro, R. Grima, A. García, E.A. Bea, A. Soba, J.M. Cela, P. Ordejón, An efficient implementation of a QM-MM method in SIESTA. Theor. Chem. Acc. **128**(4–6), 825–833 (2011)

436. V. Sasisekharan, Stereochemical criteria for polypeptide and protein structures, in *Collagen* (Wiley, New York, 1962), pp. 39–78

437. A.J. Saunders, G.B. Young, G.J. Pielak, Polarity of disulfide bonds. Protein Sci. **2**(7), 1183–1184 (1993)

438. L. Sawyer, M.N.G. James, Carboxyl-carboxylate interactions in proteins. Nature **295**, 79–80 (1982). doi:10.1038/295079a0

439. J.A. Schellman, Fifty years of solvent denaturation. Biophys. Chem. **96**(2–3), 91–101 (2002)

440. J.M. Scholtz, D. Barrick, E.J. York, J.M. Stewart, R.L. Baldwin, Urea unfolding of peptide helices as a model for interpreting protein unfolding. Proc. Natl. Acad. Sci. USA **92**(1), 185–189 (1995)

441. I. Schomburg, A. Chang, C. Ebeling, M. Gremse, C. Heldt, G. Huhn, D. Schomburg, BRENDA, the enzyme database: updates and major new developments. Nucl. Acids Res. **32**(database issue), D431–433 (2004)

442. E. Schrodinger, *Statistical Thermodynamics* (Dover, New York, 1989)

443. L. Schwartz, *Théorie des Distributions* (Hermann, Paris, 1966)

444. L.R. Scott, T.W. Clark, B. Bagheri, *Scientific Parallel Computing* (Princeton University Press, Princeton, 2005)

445. L.R. Scott, Nonstandard dielectric response. Research Report UC/CS TR-2010-06, Department of Computer Science, University Chicago, 2010

446. L.R. Scott, The impact of wrapping on sidechain-mainchain hydrogen bonds. Research Report UC/CS TR-2017-3, Department of Computer Science, University Chicago, 2017

447. L.R. Scott, M. Boland, K. Rogale, A. Fernández, Continuum equations for dielectric response to macromolecular assemblies at the nano scale. J. Phys. A: Math. Gen. **37**, 9791–9803 (2004)

448. L.R. Scott, Y. Jiang, X. Dexuan, Comparison of a nonlocal dielectric model with a discontinuous model. Research Report UC/CS TR-2015-06, Department of Computer Science, University Chicago, 2015

449. L.R. Scott, T. Orlova, M. Golubitsky, Parameterizations of hydrogen bond quality. Research Report UC/CS to appear, Department of Computer Science, University Chicago, 2017

450. L.R. Scott, A. Fernández Stigliano, A disruptive dipole-dipole alignment promotes a stable molecular association. Research Report UC/CS TR-2013-10, Department of Computer Science, University Chicago, 2013

451. R. Scott, Finite element convergence for singular data. Numer. Math. **21**, 317–327 (1973)

452. R. Scott, Optimal L^∞ estimates for the finite element method on irregular meshes. Math. Comput. **30**, 681–697 (1976)

453. J. Seelig, A. Seelig, Lipid conformation in model membranes and biological membranes. Q. Rev. Biophys. **13**(1), 19–61 (1980)

454. Z. Shi, C.A. Olson, N.R. Kallenbach, Cation-π interaction in model α-helical peptides. J. Am. Chem. Soc. **124**(13), 3284–3291 (2002)

455. Z. Shi, K. Chen, Z. Liu, T.R. Sosnick, N.R. Kallenbach, PII structure in the model peptides for unfolded proteins - studies on ubiquitin fragments and several alanine-rich peptides containing QQQ, SSS, FFF, and VVV. Proteins - Struct. Funct. Bioinform. **63**(2), 312–321 (2006)

456. K. Shimomura, T. Fukushima, T. Danno, K. Matsumoto, M. Miyoshi, Y. Kowa, Inhibition of intestinal absorption of phenylalanine by phenylalaninol. J. Biochem. **78**(2), 269–275 (1975)

457. H. Shull, G.G. Hall, Atomic units. Nature **184**, 1559–1560 (1959). doi:10.1038/1841559a0

458. M.B. Sierra, S.R. Accordino, J.A. Rodriguez-Fris, M.A. Morini, G.A. Appignanesi, A. Fernández, Stigliano, Protein packing defects "heat up" interfacial water. Eur. Phys. J. E **36**(6), 1–8 (2013)

459. A. Singer, Z. Schuss, R.S. Eisenberg, Attenuation of the electric potential and field in disordered systems. J. Stat. Phys. **119**(5–6), 1397–1418 (2005)

460. U.C. Singh, P.A. Kollman, A water dimer potential based on ab initio calculations using Morokuma component analyses. J. Chem. Phys. **83**(8), 4033–4040 (1985)

461. N. Sinha, S. Mohan, C.A. Lipschultz, S.J. Smith-Gill, Differences in electrostatic properties at antibody-antigen binding sites: implications for specificity and cross-reactivity. Biophys. J. **83**(6), 2946–2968 (2002)

462. O. Sipahioglu, S.A. Barringer, I. Taub, A.P.P. Yang, Characterization and modeling of dielectric properties of turkey meat. J. Food Sci. **68**(2), 521–527 (2003)

463. G.R. Smith, M.J. Sternberg, Prediction of protein-protein interactions by docking methods. Curr. Opin. Struct. Biol. **12**, 28–35 (2002)

464. J.D. Smith, C.D. Cappa, K.R. Wilson, B.M. Messer, R.C. Cohen, R.J. Saykally, Energetics of hydrogen bond network rearrangements in liquid water. Science **306**, 851–853 (2004)

465. P. Sondermann, R. Huber, V. Oosthuizen, U. Jacob, The 3.2-Å crystal structure of the human IgG1 Fc fragment-FcγRIII complex. Nature **406**, 267–273 (2000)

466. A.K. Soper, K. Weckström, Ion solvation and water structure in potassium halide aqueous solutions. Biophys. Chem. **124**(3), 180–191 (2006)

467. B.J. Stapley, T.P. Creamer, A survey of left-handed polyproline II helices. Protein Sci. **8**, 587–595 (1999)

468. T. Steiner, Competition of hydrogen-bond acceptors for the strong carboxyl donor. Acta Crystallogr. Sect. B **57**(1), 103–106 (2001)

469. T. Steiner, G. Koellner, Hydrogen bonds with π-acceptors in proteins: frequencies and role in stabilizing local 3D structures. J. Mol. Biol. **305**(3), 535–557 (2001)

470. D.F. Stickle, L.G. Presta, K.A. Dill, G.D. Rose, Hydrogen bonding in globular proteins. J. Mol. Biol. **226**(4), 1143–1159 (1992)

471. A. Fernández Stigliano, *Biomolecular Interfaces: Interactions, Functions and Drug Design* (Springer, Berlin, 2015)

472. A.J. Stone, Distributed multipole analysis, or how to describe a molecular charge distribution. Chem. Phys. Lett. **83**(2), 233–239 (1981)

473. A.J. Stone, *The Theory of Intermolecular Forces* (Oxford University Press, USA, 1997)

474. A.J. Stone, M. Alderton, Distributed multipole analysis: methods and applications. Mol. Phys. **56**(5), 1047–1064 (1985)

475. K. Subramanian, S. Lakshmi, K. Rajagopalan, G. Koellner, T. Steiner, Cooperative hydrogen bond cycles involving O-H$\cdots$π and C-H$\cdots$O hydrogen bonds as found in a hydrated dialkyne. J. Mol. Struct. **384**(2–3), 121–126 (1996)

476. P.A. Suci, G.G. Geesey, Comparison of adsorption behavior of two mytilus edulis foot proteins on three surfaces. Coll. Surf. B Biointerfaces **22**, 159–168 (2001)

477. S. Sudarsanam, S. Srinivasan, Sequence-dependent conformational sampling using a database of ϕ_{i+1} and ψ_i angles for predicting polypeptide backbone conformations. Protein Eng. **10**(10), 1155–1162 (1997)

478. M. Sundd, N. Iverson, B. Ibarra-Molero, J.M. Sanchez-Ruiz, A.D. Robertson, Electrostatic interactions in ubiquitin: stabilization of carboxylates by lysine amino groups. Biochemistry, 7586–7596 (2002)

479. A. Szabo, N.S. Ostlund (eds.), *Modern Quantum Chemistry* (Dover, New York, 1996)

480. I. Szalai, S. Nagy, S. Dietrich, Nonlinear dielectric effect of dipolar fluids. J. Chem. Phys. **131**, 154905 (2009)

481. C. Tanford, *Hydrophobic Effect* (Wiley, New York, 1973)

482. C. Tanford, J. Reynolds, *Nature's Robots, a History of Proteins* (Oxford University Press, Oxford, 2001)

483. R. Taylor, O. Kennard, Hydrogen-bond geometry in organic crystals. Acc. Chem. Res. **17**(9), 320–326 (1984)

484. C.J. Thompson *Mathematical Statistical Mechanics* (Macmillan, New York, 1972)

485. C.J. Thompson, *Classical Equilibrium Statistical Mechanics* (Oxfrd University Press, Oxfrd, 1988)

486. R.L. Thurlkill, G.R. Grimsley, J.M. Scholtz, C.N. Pace, Hydrogen bonding markedly reduces the pK of buried carboxyl groups in proteins. J. Mol. Biol. **362**(3), 594–604 (2006)

487. R.L. Thurlkill, G.R. Grimsley, J.M. Scholtz, C.N. Pace, pK_a values of the ionizable groups of proteins. Protein Sci. **15**(5), 1214–1218 (2006)

488. I.J. Tickle, Experimental determination of optimal root-mean-square deviations of macromolecular bond lengths and angles from their restrained ideal values. Acta Crystallogr. Sect. D **63**(12), 1274–1281 (2007)

489. C.E. Tognon, P.H.B. Sorensen, Targeting the insulin-like growth factor 1 receptor (IGF1R) signaling pathway for cancer therapy. Expert Opin. Ther. Targets **16**(1), 33–48 (2012)

490. T. Tokushima, Y. Harada, O. Takahashi, Y. Senba, H. Ohashi, L.G.M. Pettersson, A. Nilsson, S. Shin, High resolution X-ray emission spectroscopy of liquid water: the observation of two structural motifs. Chem. Phys. Lett. **460**(4–6), 387–400 (2008)

491. J. Tomasi, B. Mennucci, R. Cammi, Quantum mechanical continuum solvation models. Chem. Rev. **105**(8), 2999–3094 (2005)

492. C. Truesdell, S. Bharatha, *The Concepts and Logic of Classical Thermodynamics as a Theory of Heat Engines* (Rigorously Constructed Upon the Foundation Laid by S. Carnot and F, Reech (Springer, Berlin, 1977)

493. C.-J. Tsai, S.L. Lin, H.J. Wolfson, R. Nussinov, Studies of protein-protein interfaces: a statistical analysis of the hydrophobic effect. Protein Sci. **6**, 53–64 (1997)

494. J. Tsai, R. Taylor, C. Chothia, M. Gerstein, The packing density in proteins: standard radii and volumes. J. Mol. Biol. **290**, 253–266 (1999)

495. G.E. Uhlenbeck, L.S. Ornstein, On the theory of the Brownian motion. Phys. Rev. **36**(5), 823–841 (1930)

496. H. Umeda, M. Takeuchi, K. Suyama, Two new elastin cross-links having pyridine skeleton. J. Biol. Chem. **276**(16), 12579–12587 (2001)

497. A. Unzue, M. Xu, J. Dong, L. Wiedmer, D. Spiliotopoulos, A. Caflisch, C. Nevado, *Fragment-based design of selective nanomolar ligands of the CREBBP bromodomain* J. Med. Chem. (2015)

498. D.W. Urry, The gramicidin a transmembrane channel: a proposed π (L, D) helix. Proc. Natl. Acad. Sci. USA **68**(3), 672–676 (1971)

499. J.D. Van der Waals, *Over de Continuiteit van den Gas-en Vloeistoftoestand*, vol. 1 (Sijthoff, 1873)

500. P.T. Van Duijnen, M. Swart, Molecular and atomic polarizabilities: thole's model revisited. J. Phys. Chem. A **102**(14), 2399–2407 (1998)

501. E. Vanden-Eijnden, J. Weare, Rare event simulation of small noise diffusions. Commun. Pure Appl. Math. **65**(12), 1770–1803 (2012)

502. M. Vieth, R.E. Higgs, D.H. Robertson, M. Shapiro, E.A. Gragg, H. Hemmerle, Kinomics-structural biology and chemogenomics of kinase inhibitors and targets. Biochim. Biophys. Acta **1697**, 243–257 (2004)

503. J.A. Vila, D.R. Ripoll, M.E. Villegas, Y.N. Vorobjev, H.A. Scheraga, Role of hydrophobicity and solvent-mediated charge-charge interactions in stabilizing alpha-helices. Biophys. J. **75**, 2637–2646 (1998)

504. D. Vitkup, E. Melamud, J. Moult, C. Sander, Completeness in structural genomics. Nat. Struct. Mol. Biol. **8**, 559–566 (2001). doi:10.1038/88640

505. D. Voet, J.G. Voet, *Biochemistry* (Wiley, New York, 1990)

506. C. Vogel, C. Berzuini, M. Bashton, J. Gough, S.A. Teichmann, Supra-domains: evolutionary units larger than single protein domains. J. Mol. Biol. **336**(3), 809–823 (2004)

507. J. Von Eichborn, S. Günther, R. Preissner, Structural features and evolution of protein-protein interactions. Genome Inf. **22**, 1–10 (2010)

508. W.L. Wang, Y. Deng Yujie, B. Kim, L. Pierce, G. Krilov, D. Lupyan, S. Robinson, M.K. Dahlgren, J. Greenwood et al., Accurate and reliable prediction of relative ligand binding potency in prospective drug discovery by way of a modern free-energy calculation protocol and force field. J. Am. Chem. Soc. **137**(7), 2695–2703 (2015)

509. X. Wang, M. Bogdanov, W. Dowhan, Topology of polytopic membrane protein subdomains is dictated by membrane phospholipid composition. EMBO J. **21**(21), 5673–5681 (2002)

510. A. Warshel, A. Papazyan, Electrostatic effects in macromolecules: fundamental concepts and practical modeling. Curr. Opin. Struct. Biol. **8**, 211–217 (1998)

511. A. Warshel, Electrostatic origin of the catalytic power of enzymes and the role of preorganized active sites. J. Biol. Chem. **273**, 27035–27038 (1998)

512. J.D. Watson, F.H.C. Crick, Molecular structure of nucleic acids: a structure for deoxyribose nucleic acid. Nature **171**, 737–738 (1953)

513. G.J. Wedemayer, P.A. Patten, L.H. Wang, P.G. Schultz, R.C. Stevens, Structural insights into the evolution of an antibody combining site. Science **276**(5319), 1665–1669 (1997)

514. S. Weggler, V. Rutka, A. Hildebrandt, A new numerical method for nonlocal electrostatics in biomolecular simulation. J. Comput. Phys. **229**(11), 4059–4074 (2010)

515. A.K.H. Weiss, T.S. Hofer, B.R. Randolf, B.M. Rode, Guanidinium in aqueous solution studied by quantum mechanical charge field-molecular dynamics (QMCF-MD). Phys. Chem. Chem. Phys. **14**(19), 7012–7027 (2012)

516. Ph Wernet, D. Nordlund, U. Bergmann, M. Cavalleri, M. Odelius, H. Ogasawara, L.Å. Näslund, T.K. Hirsch, L. Ojamäe, P. Glatzel, L.G.M. Pettersson, A. Nilsson, The structure of the first coordination shell in liquid water. Science **304**(5673), 995–999 (2004)

517. M. Wilkins, A.R. Stokes, H.R. Wilson, Molecular structure of deoxypentose nucleic acids. Nature **171**, 738–740 (1953)

518. R.W. Williams, A. Chang, D. Juretic, S. Loughran, Secondary structure predictions and medium range interactions. Biochim. Biophys. Acta (BBA) - Protein Struct. Mol. Enzymol. **916**(2), 200–204 (1987)

519. R. Wintjens, J. Liévin, M. Rooman, E. Buisine, Contribution of cation-π interactions to the stability of protein-DNA complexes. J. Mol. Biol. **302**, 394–410 (2000)

520. S.T. Wlodek, T.W. Clark, L.R. Scott, J.A. McCammon, Molecular dynamics of acetylcholinesterase dimer complexed with tacrine. J. Am. Chem. Soc. **119**, 9513–9522 (1997)

521. S.T. Wlodek, T. Shen, J.A. McCammon, Electrostatic steering of substrate to acetylcholinesterase: analysis of field fluctuations. Biopolymers **53**(3), 265–271 (2000)

522. G. Wohlfahrt, Analysis of pH-dependent elements in proteins: geometry and properties of pairs of hydrogen-bonded carboxylic acid side-chains. Proteins: Struct. Funct. Bioinform. **58**, 396–406 (2005)

523. Y.I. Wolf, S.E. Brenner, P.A. Bash, E.V. Koonin, Distribution of protein folds in the three superkingdoms of life. Genome Res. **9**(1), 17–26 (1999)

524. Y.I. Wolf, N.V. Grishin, E.V. Koonin, Estimating the number of protein folds and families from complete genome data. J. Mol. Biol. **299**(4), 897–905 (2000)

525. R. Wolfenden, L. Andersson, P.M. Cullis, C.C.B. Southgate, Affinities of amino acid side chains for solvent water. Biochemistry **20**(4), 849–855 (1981)

526. J. Wong, Coevolution theory of the genetic code at age thirty. BioEssays **27**(4), 416–425 (2005)

527. J.M. Word, S.C. Lovell, J.S. Richardson, D.C. Richardson, Asparagine and glutamine: using hydrogen atom contacts in the choice of side-chain amide orientation. J. Mol. Biol. **285**(4), 1735–1747 (1999)

528. C.K. Wu, B. Hu, J.P. Rose, Z.-J. Liu, T.L. Nguyen, C. Zheng, E. Breslow, B.-C. Wang, Structures of an unliganded neurophysin and its vasopressin complex: implications for binding and allosteric mechanisms. Protein Sci. **10**(9), 1869–1880 (2001)

529. S. Wuchty, Scale-free behavior in protein domain networks. Mol. Biol. Evolut. **18**(9), 1694–1702 (2001)

530. D. Xie, Y. Jiang, A nonlocal modified Poisson-Boltzmann equation and finite element solver for computing electrostatics of biomolecules. J. Comput. Phys. **322**, 1–20 (2016)

531. D. Xie, Y. Jiang, P. Brune, L.R. Scott, A fast solver for a nonlocal dielectric continuum model. SISC **34**(2), B107–B126 (2012)

532. D. Xie, Y. Jiang, L.R. Scott, Efficient algorithms for solving a nonlocal dielectric model for protein in ionic solvent. SISC **35**(6), B1267B1284 (2014)

533. D. Xie, J.-L. Liu, B. Eisenberg, Nonlocal poisson-fermi model for ionic solvent. Phys. Rev. E **94**, 012114 (2016)

534. D. Xie, J.-L. Liu, B. Eisenberg, L.R. Scott, A nonlocal Poisson-Fermi model for ionic solvent. arXiv:1603.05597 [physics.chem-ph], 2016

535. X. Dong, C.-J. Tsai, R. Nussinov, Hydrogen bonds and salt bridges across protein-protein interfaces. Protein Eng. **10**(9), 999–1012 (1997)

536. M. Xu, A. Unzue, J. Dong, D. Spiliotopoulos, C. Nevado, A. Caflisch, *Discovery of CREBBP bromodomain inhibitors by high-throughput docking and hit optimization guided by molecular dynamics* (J. Med, Chem, 2015)

537. B. Xue, R.L. Dunbrack, R.W. Williams, A.K. Dunker, V.N. Uversky, PONDR-FIT: a meta-predictor of intrinsically disordered amino acids. Biochim. Biophys. Acta (BBA)-Proteins. Proteomics **1804**(4), 996–1010 (2010)

538. W.-M. Yau, W.C. Wimley, K. Gawrisc, S.H. White, The preference of tryptophan for membrane interfaces. Biochemistry **37**, 14713–14718 (1998)

539. W. Yu, S.K. Lakkaraju, E.P. Raman, L. Fang, A.D. MacKerell Jr., Pharmacophore modeling using site-identification by ligand competitive saturation (SILCS) with multiple probe molecules. J. Chem. Inf. Model. **55**(2), 407–420 (2015)

540. H. Yuki, Y. Tanaka, M. Hata, H. Ishikawa, S. Neya, T. Hoshino, Implementation of π-π interactions in molecular dynamics simulation. J. Comput. Chem. **28**(6), 1091–1099 (2007)

541. R. Zahn, A. Liu, T. Luhrs, R. Riek, C. von Schroetter, F. Lopez Garcia, M. Billeter, L. Calzolai, G. Wider, K. Wüthrich, NMR solution structure of the human prion protein. Proc. Natl. Acad. Sci. USA **97**(1), 145–150 (2000)

542. D.T. Zallen, Despite franklin's work, wilkins earned his nobel. Nature **425**(6953), 15–15 (2003)

543. A. Zarrine-Afsar, A. Mittermaier, L.E. Kay, A.R. Davidson, Protein stabilization by specific binding of guanidinium to a functional arginine-binding surface on an SH3 domain. Protein Sci. **15**, 162–170 (2006)

544. Z. Zeng, H. Shi, W. Yun, Z. Hong, *Survey of natural language processing techniques in bioinformatics* (Comput. Math, Methods Med, 2015)

545. Q. Zhai, M.B. Landesman, H. Robinson, W.I. Sundquist, C.P. Hill, Identification and structural characterization of the ALIX-binding late domains of simian immunodeficiency virus SIVmac239 and SIVagmTan-1. J. Virol. **85**(1), 632–637 (2011)

546. H.-X. Zhao, X.-J. Kong, H. Li, Y.-C. Jin, L.-S. Long, X.C. Zeng, R.-B. Huang, L.-S. Zheng, Transition from one-dimensional water to ferroelectric ice within a supramolecular architecture. Proc. Natl. Acad. Sci. **108**(9), 3481–3486 (2011)

547. W. Zhong, J.P. Gallivan, Y. Zhang, L. Li, H.A. Lester, D.A. Dougherty, From ab initio quantum mechanics to molecular neurobiology: a cation-π binding site in the nicotinic receptor. Proc. Natl. Acad. Sci. USA **95**(21), 12088–12093 (1998)

548. R. Zhou, X. Huang, C.J. Margulis, B.J. Berne, Hydrophobic collapse in multidomain protein folding. Science **305**(5690), 1605–1609 (2004)

549. Z.H. Zhou, M.L. Baker, W. Jiang, M. Dougherty, J. Jakana, G. Dong, G. Lu, W. Chiu, Electron cryomicroscopy and bioinformatics suggest protein fold models for rice dwarf virus. Nat. Struct. Biol. **8**, 868–873 (2001)

550. Z.H. Zhou, W. Chiu, K. Haskell, H. Spears, J. Jakana, F.J. Rixon, L.R. Scott, Parallel refinement of herpesvirus B-capsid structure. Biophys. J. **73**, 576–588 (1997)

551. Z.H. Zhou, M. Dougherty, J. Jakana, J. He, F.J. Rixon, W. Chiu, Seeing the herpesvirus capsid at 8.5 Å. Science **288**, 877–880 (2000)

552. Z.H. Zhou, S.J. Macnab, J. Jakana, L.R. Scott, W. Chiu, F.J. Rixon, Identification of the sites of interaction between the scaffold and outer shell in herpes simplex virus-1 capsids by difference electron imaging. Proc. Natl. Acad. Sci. USA **95**, 2778–2783 (1998)

553. M. Zorko, Ü. Langel, Cell-penetrating peptides: mechanism and kinetics of cargo delivery. Adv. Drug Deliv. Rev. **57**(4), 529–545 (2005)

Index

© Springer International Publishing AG 2017
L.R. Scott and A. Fernández, *A Mathematical Approach to Protein Biophysics*,
Biological and Medical Physics, Biomedical Engineering,
DOI 10.1007/978-3-319-66032-5